Development of Sensory and Neurosecretory Cell Types

Series Preface (Evolutionary Cell Biology)

In recent decades, evolutionary principles have been integrated into biological disciplines such as developmental biology, ecology and genetics. As a result, major new fields emerged, chief among which are Evolutionary Developmental Biology (or Evo-Devo) and Ecological Developmental Biology (or Eco-Devo). Inspired by the integration of knowledge of change over single life spans (ontogenetic history) and change over evolutionary time (phylogenetic history), Evo-Devo produced a unification of developmental and evolutionary biology that generated unanticipated synergies: Molecular biologists employ computational and conceptual tools generated by developmental biologists and by systematists, while evolutionary biologists use detailed analysis of molecules in their studies. These integrations have shifted paradigms and enabled us to answer questions once thought intractable.

Major highlights in the development of modern Evo-Devo are a comparison of the evolutionary behavior of cells, evidenced in Stephen J. Gould's 1979 proposal of changes in the timing of the activity of cells during development — heterochrony — as a major force in evolutionary change, and numerous studies demonstrating how conserved gene families across numerous cell types 'explain" development and evolution. Advances in technology and in instrumentation now allow cell biologists to make ever more detailed observations of the structure of cells and the processes by which cells arise, divide, differentiate and die. In recent years, cell biologists have increasingly asked questions whose answers require insights from evolutionary history. As just one example: How many cell types are there and how are they related? Given this conceptual basis, cell biology — a rich field in biology with history going back centuries — is poised to be reintegrated with evolution to provide a means of organizing and explaining diverse empirical observations and testing fundamental hypotheses about the cellular basis of life. Integrating evolutionary and cellular biology has the potential to generate new theories of cellular function and to create a new field, "*Evolutionary Cell Biology*."

Mechanistically, cells provide the link between the genotype and the phenotype, both during development and in evolution. Hence the proposal for a series of books under the general theme of "*Evolutionary Cell Biology: Translating Genotypes into Phenotypes*", to document, demonstrate and establish the central role played by cellular mechanisms in in the evolution of all forms of life.

Brian K. Hall and Sally A. Moody

Development of Sensory and Neurosecretory Cell Types

Vertebrate Cranial Placodes, Vol. 1

Gerhard Schlosser

CRC Press
Taylor & Francis Group
Boca Raton London New York

CRC Press is an imprint of the
Taylor & Francis Group, an **informa** business

First edition published 2021
by CRC Press
6000 Broken Sound Parkway NW, Suite 300, Boca Raton, FL 33487-2742
and by CRC Press

2 Park Square, Milton Park, Abingdon, Oxon, OX14 4RN

ISBN: 978-1-138-06146-0 (hbk)
ISBN: 978-0-367-74778-7 (pbk)
ISBN: 978-1-315-16231-7 (ebk)

Typeset in Times
by KnowledgeWorks Global Ltd.

Contents

Preface

Our senses provide us with a richness of experiences that we all too often take for granted. Without our sense organs, our world would be a silent and dark place (not a place at all, really). Yet, we share our sophisticated ears, eyes and noses only with our fellow vertebrates. The senses of their closest living relatives, the tunicates (including sea squirts) and lancelets (or amphioxus), are much simpler. Accordingly, these animals live in a much simpler sensory world. Many other animals get around with an even more basic outfit of sensory cells - isolated cells scattered through their outer tissues - which do not form complex sense organs at all. So how did the vertebrate head become equipped with these complicated new sensory organs? How do they develop in the vertebrate embryo? And how did they evolve from the simpler sensory systems of our ancestors? These are the core questions that this book tries to answer.

Most of the cranial sense organs of vertebrates arise from embryonic structures known as cranial placodes, which in turn arise from a common embryonic precursor in development. In addition to sense organs, cranial placodes also give rise to sensory neurons that transmit the sensory information to the brain as well as to many neurosecretory cells – neuron-like cells that produce hormones (including the cells of the anterior pituitary, the major hormonal control organ of the vertebrate body). Due to the close developmental and evolutionary links between sensory, neuronal and neurosecretory cells derived from cranial placodes, they will be considered jointly in this book. While the photoreceptors of the vertebrate retina and pineal organ are not derived from cranial placodes, they will nevertheless be briefly discussed here. The main reason is the close evolutionary relationship between photoreceptors and other sensory cell types (e.g. mechano- and chemoreceptors) that develop from placodes.

The first volume of this book will focus on the development of sensory and neurosecretory cell types from the cranial placodes in vertebrates. Chapter 1 introduces the vertebrate head with its sense organs and neurosecretory organs. Chapter 2 then provides an overview of the various cranial placodes and their derivatives. Chapter 3 presents evidence that all cranial placodes develop from a common embryonic primordium and discusses how the latter is established in the early embryo. How individual placodes originate from the common primordium is addressed in Chapter 4. Chapters 5 to 8 then review, how the various placodally derived sensory and neurosecretory cell types as well as photoreceptors develop.

The second volume of the book will then consider the evolutionary origin of the sensory and neurosecretory cell types developing from cranial placodes and of photoreceptors. Chapter 1 summarizes our current understanding of vertebrate evolution and presents a brief survey of body plans and embryonic development of their closest relatives (tunicates, amphioxus, hemichordates and echinoderms). Chapter 2 clarifies conceptual issues relating to homology and evolutionary innovation and introduces an evolutionary concept of cell types. Chapters 3 to 5 then compare the sensory and neurosecretory cell types of the vertebrate head with similar cell types in other animals to get insights into their evolutionary origins. While Chapter 3 reviews the evolution of mechano- and chemosensory cells, Chapter 4 considers photosensory cells and Chapter 5 neurosecretory cells. These comparative chapters will show that many sensory cell types have a long evolutionary history. The final chapter, Chapter 6, then addresses the question of how cranial placodes evolved as novel structures in vertebrates by redeploying pre-existing and sometimes evolutionarily ancient cell types.

The book is aimed at a wide audience ranging from graduate students to fellow scientists in the fields of developmental and evolutionary biology. To make it accessible also to newcomers to these fields, I attempted to provide relatively broad background information in the introductory chapters of each volume (Chapter 1 of Volume 1; Chapters 1 and 2 of Volume 2). Because these introductory chapters cover a lot of ground, it was impossible to always provide original references and I have mostly cited reviews there. In the other chapters, I tried to cite relevant original papers in addition to reviews, wherever possible. Nevertheless, I will almost certainly have missed some relevant sources or may have had to limit the number of papers cited due to space constraints. Therefore, I ask all my colleagues, whose relevant papers I left out, to please accept my apologies. Some of the chapters on development in the first volume may be too detailed for those readers with more evolutionary interests. However, I provide summaries of the main points at the end of each major section of each chapter, and it should be possible for those readers to get the main message by reading the summaries and skipping the rest.

I wish to express my gratitude to all my colleagues who have inspired me over the years and without whose ideas and research advances this book could never have been written. I am also very grateful to the following colleagues for providing valuable comments on parts of the book: Eric Bellefroid, Bernd Fritzsch, Benjamin Grothe, Volker Hartenstein, Nick Holland, Anne-Helene Monsoro-Burq, Sally Moody, Seb Shimeld, Andrea Streit, and Günter Wagner. I specifically want to thank my wife, Elke Rink, who read and commented on the entire manuscript, apart from having to bear recurrent spells of absentmindedness over the greater part of two years. Furthermore, I acknowledge my editor, Chuck Crumly, and his team at CRC Press for their support during the publication process. Finally, it must be noted that parts of Chapter 6 in Volume 1 on hair cells and of Chapter 3 in Volume 2 on the evolution of mechanosensory cells have been previously published in a chapter entitled "Evolution of hair cells" in: Fritzsch, B. (Ed.) and Grothe, B. (Volume Editor), *The Senses: A Comprehensive Reference, Vol. 2*. Elsevier, Academic Press 2020, pp. 302–336.

1 The Vertebrates' New Head

One of the most striking features of vertebrates is their elaborate head, complete with a variety of sense organs, a large brain, and a skull. Vertebrates are, therefore, also known as craniates – "skull animals". At first glance, having a head may not seem like a very remarkable distinction. Almost all bilaterally symmetrically animals – animals of the group Bilateria – have a front end with a mouth and a back end with the anus. Sensory organs tend to be concentrated at the front end, because this is the direction, into which the animal moves. Muscles and appendages associated with feeding or other environmental interactions are also concentrated there. The higher density of sensory and motor organs at the front end also requires larger neural control centers, such as enlarged cranial ganglia (concentration of nerve cells) or a brain. There may be special structures serving to protect the sensory organs and cranial ganglia from environmental impacts as well. We honor these specializations of the front end by calling it a "head". However, not all bilaterian heads are equally complex. Truly sophisticated heads with an assortment of complex specialized sense organs evolved only a few times, most importantly in the arthropods and vertebrates.

The vertebrates, together with the tunicates (sea squirts and allies) and amphioxus (lancelets) belong to a larger group, the phylum chordates. These are defined by the presence of a dorsal skeletal rod, the notochord, in some of their life stages as well as a number of other shared characters. The chordates, in turn, together with the hemichordates and the highly specialized echinoderms belong to one of the two major branches of bilaterian animals, the deuterostomes. Deuterostomes got their name ("second mouth") from the fact that the first opening of the embryonic gut will form the anus, while the mouth breaks through secondarily, in distinction to the other major branch of bilaterians, the protostomes ("first mouth"). This brief sketch shall suffice here. I will consider these phylogenetic relationships in more detail in the second volume (Schlosser 2021).

The vertebrate head is equipped with complex paired sense organs – eyes, ears lateral line system (in fishes and amphibians), and nose – a series of cranial nerves, and a cartilaginous or bony skull that are notably absent from other chordates or deuterostomes. This suggests that these structures are evolutionary novelties that only originated in the vertebrate lineage. In a highly influential series of papers, Northcutt and Gans suggested in the 1980s that these novelties, referred to as the "New Head" of vertebrates, originated when early vertebrates shifted from filter-feeding to a more active and predatory life-style (Gans and Northcutt 1983; Northcutt and Gans 1983). Northcutt and Gans (1983) also proposed that many of these novelties develop from only two embryonic tissues, the neural crest and the cranial placodes, which also originated first in the vertebrate lineage. The neural crest contributes cartilage and bone cells to the skull and forms glial cells and some sensory neurons of the cranial nerves in addition to other cell types. The cranial placodes, in turn, generate the remaining sensory neurons to the cranial nerves and in addition give rise to many of the sense organs as well as the pituitary gland, the major hormonal control organ of the vertebrate body.

The developmental and evolutionary origin of placodes and the deep evolutionary history of the various sensory, neuronal, and neurosecretory (hormone producing) cell types derived from them will be at the core of this book. The origin of the neural crest is covered extensively in other reviews and books (Meulemans and Bronner-Fraser 2004; Sauka-Spengler and Bronner-Fraser 2008; Medeiros 2013; Green, Simoes-Costa, and Bronner 2015; Cheung et al. 2019; York and McCauley 2020; Eames, Medeiros, and Adameyko 2020) and I will only occasionally touch on this topic here, whenever necessary for a better understanding of placode evolution. Before dealing with the development of placodes in the remainder of this book and with their evolutionary history in Volume 2, I will provide some essential general background information in this first chapter. First I will give a brief overview of vertebrate head development followed by a short survey of the placode derived sensory and neurosecretory organs of the vertebrate head and an introduction to cranial nerves.

1.1 AN OVERVIEW OF VERTEBRATE HEAD DEVELOPMENT

The early development of different vertebrates differs significantly between different vertebrate groups mostly depending on the size and yolk content of the egg. However, the basic phases and principles of development are largely conserved. For the purpose of this general introduction, I will here gloss over these differences and focus on the similarities using the frog embryo for illustration because it lacks many of the specializations found in particularly yolk-poor (as in mammals) or yolk-rich (as in reptiles and birds) eggs. Moreover, I will focus on the morphological changes here, postponing any discussions of the molecular pathways regulating the acquisition of different cell fates to later chapters (reviewed in Barresi and Gilbert 2019; Wolpert, Tickle, and Martinez-Arias 2019).

1.1.1 Early Embryonic Development

After fertilization, the frog egg undergoes cleavage, a series of rapid synchronous cell division resulting in a ball of cells, which hollows out to form the blastula. The cavity inside is called the blastocoel. During subsequent gastrulation, cells move into the embryo to form the three germ layers – the ectoderm (outer layer), the endoderm (inner layer) lining the embryonic gut (or archenteron), and the mesoderm sandwiched in between (Fig. 1.1A). These movements create an opening, the blastopore, which connects the archenteron to the outside and later develops into the anus. The blastopore, thus, marks the posterior (or caudal) end of the embryo. The mouth breaks through at the opposite anterior (or rostral) side of the embryo much later in development.

On the dorsal side of the blastopore is the dorsal blastopore lip, which will form the notochord, the dorsal-most part of the mesoderm, after gastrulation. It is also an important source of diffusible signals that impinge on the overlying ectoderm on

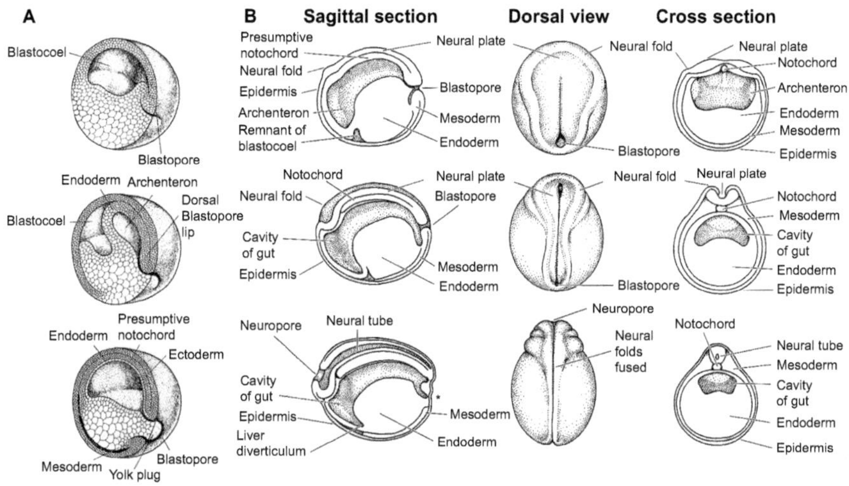

FIGURE 1.1 An overview of early vertebrate development. Early developmental processes are here illustrated for frog embryos. (**A**) Three germ layers and the embryonic gut (archenteron) are established during gastrulation. Upper, middle and lower panel show sagittal sections of embryos in early, mid and late stages of gastrulation (anterior to the left, dorsal to the top). (**B**) After gastrulation, the neural plate is induced on the dorsal side (upper row) and folds up to form the neural tube during subsequent neurulation (middle and lower row). After Balinsky, 1970. ([A] Reprinted with permission from Rugh 1951; [B] Reprinted with permission from Elinson 1997.)

the dorsal side thereby inducing it to form the neural plate. Ectoderm that does not receive these inductive signals will develop into epidermis. This was first discovered by Hans Spemann and his student Hilde Mangold in 1924 (Spemann and Mangold 1924). Grafting the dorsal blastopore lip from one newt species to the ventral side of another newt species, they discovered that a second neural plate (and some other dorsal tissues derived from the mesoderm) was formed on the ventral side of the host. Since tissues of donor and host species could be distinguished by their degree of pigmentation, they were further able to conclude that the second neural plate was derived from the tissue of the host species. The dorsal blastopore lip must, therefore, be a source of signals acting on the adjacent tissues and changing their fate. To reflect this, Spemann christened the dorsal blastopore lip the "organizer" and in his honor, it is known as "Spemann's organizer" today. Subsequent studies have shown that the organizer is composed of different parts, which induce different structures along the anterior-posterior axis (Stern 2001; Niehrs 2004; Carron and Shi 2016). The part of the dorsal blastopore lip that moves into the embryo first and ends up in the most anterior position, induces head structures, while the parts that move into the embryo later induce trunk structures.

At the end of gastrulation, the neural plate ectoderm thickens and thereby becomes distinguishable from the epidermis (Fig. 1.1B). Differences between the head and trunk region of the embryo become first apparent at this stage with the anterior part of the neural plate, which will give rise to the brain, being wider than its

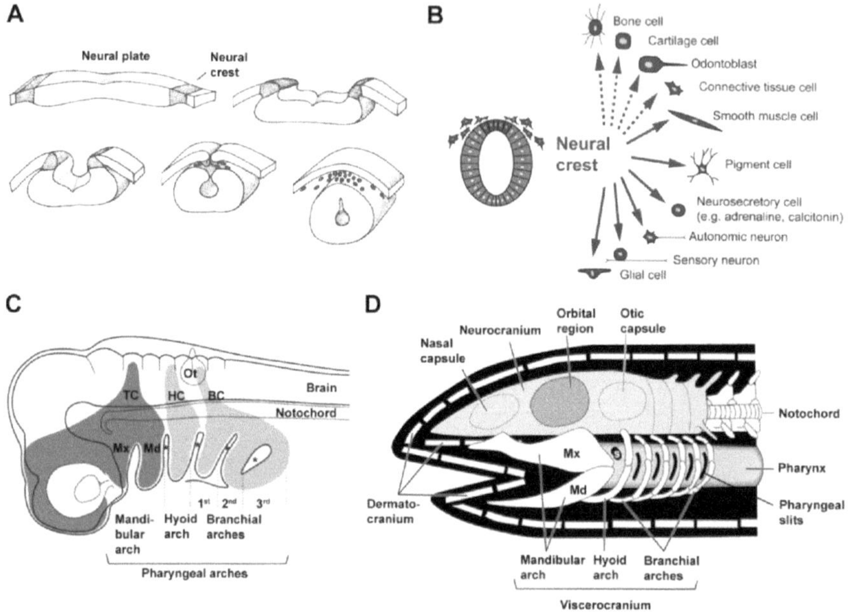

FIGURE 1.2 Contribution of neural crest to vertebrate head development. (**A**) Origin of neural crest cells (grey) during neurulation. (**B**) Derivatives of neural crest cells. Broken arrows indicate derivatives only formed by the cranial neural crest. (**C**) Neural crest streams populating the pharyngeal arches in a generalized vertebrate embryo. Asterisks mark ectodermal pharyngeal grooves, which will fuse with the underlying pharyngeal pouches to form the pharyngeal slits. (**D**) Pharyngeal arch derivatives in a generalized vertebrate. ([A] Reprinted with permission from Elinson 1997. After Balinsky 1970; [B] Modified from Schlosser 2008; [C] Reprinted with permission from Kuratani et al. 2012; [D] Reprinted with permission from Kuratani, Matsuo, and Aizawa 1997. BC, Branchial crest; HC, Hyoid crest; Md, Mandible (lower jaw); Mx, maxilla (upper jaw); Ot, otic vesicle; s, Spiracle; TC, Trigeminal crest.)

posterior part, which forms the spinal cord. In the next stage of development, known as neurulation, elevation of the rim of the neural plate produces neural folds. These ultimately fuse to form the neural tube, which will give rise to the entire central nervous system (CNS), i.e. brain and spinal cord. During neurulation, a special cell population of cells originating at the border of the neural plate, will lose their epithelial character and migrate along several routes into the embryo. These are the neural crest cells (Fig. 1.2A). The precise onset of neural crest migration differs for different vertebrates. In frogs this happens already before neural tube closure, whereas in salamanders and birds, for example, it coincides with the fusion of the neural folds (Kulesa, Ellies, and Trainor 2004).

1.1.2 Neural Crest Development

Since its discovery by Wilhelm His in 1868 (His 1868; Glover et al. 2018), the neural crest has attracted the special attention of vertebrate embryologists, mostly due to two peculiar aspects of its development. First, the migratory neural crest cells spread extensively throughout the body populating almost every region, contrary to most other embryonic tissues, which develop by folding or short range migration and, thus, stay relatively local. Second, the neural crest gives rise to a large variety of different cell types, including cell types that normally arise from the mesoderm rather than the ectoderm, such as cartilage, bone, and muscle cells (Fig. 1.2 B) (reviewed in Le Douarin and Kalcheim 1999; Simoes-Costa and Bronner 2015; Dupin et al. 2018; Rothstein, Bhattacharya, and Simoes-Costa 2018). This versatility has earned the neural crest the recognition as a "fourth germ layer" (Hall 2000), although it never forms a separate layer in the embryo.

At trunk levels, the neural crest gives rise to pigment cells, the somatosensory neurons forming the dorsal root ganglia (DRG) of the spinal nerves, which supply the skin, and the visceromotor neurons of the sympathetic and parasympathetic ganglia, which innervate the inner organs. The Schwann cells, a type of glial cells forming insulating myelin sheaths around the axons in the peripheral nervous system (PNS), and the cells of the adrenal medulla, which produce the stress hormones epinephrine and norepinephrine, are also neural crest derived (Fig. 1.2B). At cranial levels, the list of neural crest derivatives is even longer (Fig. 1.2B). Apart from the somatosensory neurons of the cranial nerves, the enteric neurons of the digestive tract and the cranial Schwann cells and pigment cells, the cranial neural crest also gives rise to many cartilages and bones of the skull as well as to some smooth muscles and the heart outflow tract.

The neural crest originates from the border of the neural plate at all levels along the anterior-posterior axis with exception of the transverse neural fold, which girds the neural plate anteriorly. However, at trunk levels, the neural crest comprises a relatively small number of cells compared to the massive population of cells in the cranial region. In the trunk, neural crest cells migrate ventrally as individual cells along two different pathways – a superficial pathway underneath the skin and a deep pathway along the surface of the neural tube. Cells migrating along the superficial pathway will predominantly differentiate into pigment cells, while cells migrating along the deep pathway will form neurons and glial cells.

In contrast, in the head neural crest cells migrate as a collective into the wall of the pharynx where they form several neural crest streams (Fig. 1.2C). Cranial neural crest cells are initially partitioned into distinct streams when they emerge from the neural tube by local foci of increased cell death (Graham, Heyman, and Lumsden 1993) and are kept separate by the developing pharyngeal pouches. These outpocketings of the endoderm play the decisive role in subdividing the wall of the pharynx into regularly spaced pharyngeal arches and can do so even in the absence of neural crest cells (Veitch et al. 1999; Piotrowski and Nüsslein-Volhard 2000; Graham 2008). In fishes and amphibians, the pharyngeal pouches will fuse with the adjacent ectoderm to form the pharyngeal slits, whereas in amniotes they usually never perforate (Fig. 1.2C). The first pharyngeal slit (separating the first and the second pharyngeal arch) is typically smaller than the following ones or completely absent. It forms the so called spiracle in cartilaginous fishes (Fig. 1.2D).

The neural crest streams migrate between the endodermal and ectodermal epithelia lining the pharyngeal wall and a core of mesodermal cells, thereby populating the pharyngeal arches (reviewed in Noden 1991; Helms and Schneider 2003; Santagati and Rijli 2003; Noden and Trainor 2005; Kuratani 2005; Minoux and Rijli 2010; Dash and Trainor 2020). While the mesodermal cells at the core of the pharyngeal arches will develop into a subset of cranial muscles, the neural crest will form the cartilaginous or bony skeletal elements of the viscerocranium, the ventral part of the skull associated with the pharynx (Fig. 1.2D). Further dorsally, neural crest cells also contribute together with the mesoderm to the remaining parts of the skull, the neurocranium (or braincase) enclosing the brain and the dermatocranium comprising the dermal bones (bones that are not prefigured by cartilage), which provide much of the outside cover of the skull (Fig. 1.2D). These dorsal neural crest cells will also contribute somatosensory neurons to those cranial nerves that innervate the pharyngeal arches (so-called branchiomeric nerves) as described in more detail in section 1.3 below.

The anterior-most neural crest stream in all vertebrates, also known as trigeminal crest, will populate the first pharyngeal arch (mandibular arch), from which the upper and lower jaws will develop in jawed vertebrates (gnathostomes) (Fig. 1.2C, D). In addition, the anterior portion of the trigeminal crest migrates anteriorly around the developing eyes and nose, where it contributes to other parts of the skull. The trigeminal crest also contributes to the ganglia of the profundal and trigeminal nerves, which innervate the first pharyngeal arch and more anterior parts of the head (see section 1.3 below). The following neural crest stream, known as hyoid crest, will populate the second pharyngeal arch (hyoid arch), which helps to attach the jaw to the skull in many fishes (Fig. 1.2C, D). The hyoid crest also contributes a small ganglion to the facial nerve, which innervates the second pharyngeal arch. Posterior to the hyoid crest stream, there are a varying number of branchial crest streams, which populate the posterior pharyngeal arches. These are also referred to as "branchial arches" because they are equipped with gills in fishes and amphibians[1] (Fig. 1.2C, D). The number of branchial arches varies between vertebrate groups. While sharks and other cartilaginous fishes have five to seven branchial arches, most other vertebrates have a smaller number. The branchial neural crest contributes to the ganglia of the nerves innervating the third (glossopharyngeal nerve) and more posterior branchial arches (vagal nerve).

1.1.3 Cranial Placode Development

After completion of neurulation and migration of the neural crest, specialized regions of the head ectoderm known as cranial placodes appear in a number of different locations. Placodes were first described by Jan Willem van Wijhe as thickenings of the ectoderm in shark embryos that contribute to the developing cranial nerves (van Wijhe 1883). Karl Wilhelm von Kupffer later introduced the term "placodes" for these thickenings and provided detailed descriptions of their development in lamprey embryos

[1] Note that the term "branchial arches" is sometimes used to denote the entire series of pharyngeal arches, while here I use "branchial arches" more narrowly to specifically refer to the third and more posterior pharyngeal arches.

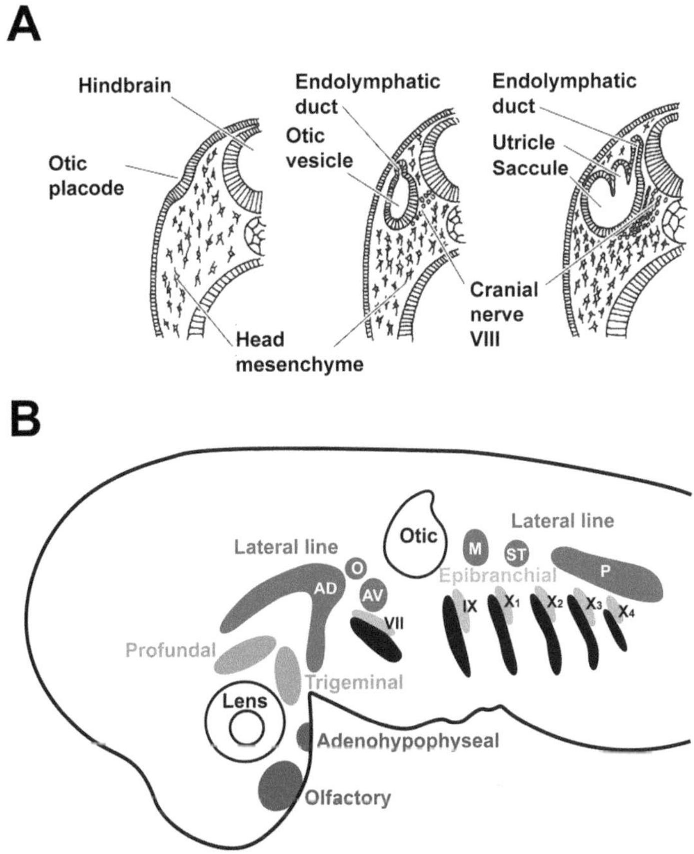

FIGURE 1.3 Contribution of cranial placodes to vertebrate head development. (**A**) Invagination of otic placode to form the otic vesicle. (**B**) Cranial placodes in a generalized vertebrate embryo include the adenohypophyseal, olfactory, lens and otic placode, the profundal and trigeminal placodes, and several epibranchial and lateral line placodes. Epibranchial placodes are closely associated with the pharyngeal pouches (shown in black). AD, anterodorsal lateral line placode; AV, anteroventral lateral line placode; O, otic lateral line placode; M, middle lateral line placode; P, posterior lateral line placode; ST, supratemporal lateral line placode; VII, epibranchial placode of facial nerve; IX, epibranchial placode of glossopharyngeal nerve; X_{1-4}, epibranchial placodes of vagal nerve. ([A] Reprinted with permission from Elinson 1997. After Rugh 1951; [B] Redrawn and modified from Northcutt 1997.)

(von Kupffer 1894). Whereas some placodes such as the otic placode invaginate to form sense organs (Fig. 1.3A), others such as the lateral line placodes form sensory organs on the surface. In addition, most placodes give rise to sensory neurons, which migrate away from the ectoderm to form ganglia of the cranial nerves in the adjacent mesenchyme. I will only briefly sketch the development of cranial placodes here, because it will be described in great detail in the remaining chapters of this book. The cranial nerves will be introduced more thoroughly in section 1.3 below.

According to their location in the embryo and fate, the following cranial placodes can be recognized in most vertebrates (Fig. 1.3B) (reviewed in Webb and Noden 1993; Baker and Bronner-Fraser 2001; Schlosser 2006; Schlosser 2010; Park and Saint-Jeannet 2010):

- The adenohypophyseal placode, which gives rise to the hormone-producing cells of the anterior pituitary (adenohypophysis);
- The olfactory placode, which forms the olfactory epithelia of the nose and the olfactory nerve;
- The lens placode, which invaginates to form the lens of the eye;
- The profundal and trigeminal placodes, which contribute somatosensory neurons to the profundal and trigeminal nerves;
- A series of epibranchial placodes, located just above the pharyngeal pouches, which contribute viscerosensory neurons to the facial, glossopharyngeal and vagal nerves;
- The otic placode, which invaginates to form the inner ear and the sensory neurons of the vestibulocochlear nerve;
- A series of lateral line placodes, which form the lateral line sensory system and the sensory neurons of the lateral line nerves.

All of these cranial placodes are now known to have a common origin from a region of ectoderm surrounding the anterior neural plate and neural crest termed the "pre-placodal ectoderm" and express a common set of transcription factors providing them with a generic placodal identity (or regulatory state) shared between the different placodes (see Chapter 3 for details).

Although cranial placodes were initially described as ectodermal thickenings and are often composed of columnar (elongated) epithelial cells, some placodes such as the profundal placode in the chick (Stark et al. 1997) are in fact never thickened. More generally, therefore, cranial placodes have been defined as "specialized areas of the cranial non-neural ectoderm (i.e. ectoderm outside of neural plate and neural crest), where cells undergo pronounced cell shape changes (which may result in thickening, invagination, and/or cell delamination) and which give rise to various non-epidermal cell types" (Schlosser 2006).

In addition all proper cranial placodes are united by their common origin and shared identity, which distinguishes them from other thickenings or specializations of the cranial ectoderm. The latter include, for example the developing amphibian cement and hatching glands (Drysdale and Elinson 1992; Sive and Bradley 1996). In the bilayered ectoderm of amphibians, the cement and hatching glands both arise from the superficial layer, whereas all placodes arise from the deep layer (Northcutt, Catania, and Criley 1994; Northcutt and Brändle 1995; Schlosser and Northcutt 2000). Nevertheless there may be an evolutionary link between cement gland and placodes, as will be briefly discussed in Chapters 5 and 6 of the second volume (Schlosser 2021). Other ectodermal thickenings such as the anlagen of teeth, hairs, and feathers are also often described as "placodes" (Pispa and Thesleff 2003), but these do not belong to the cranial placodes due to their different regions of origin and gene expression patterns.

1.1.4 Head Segmentation

The pharyngeal arches with their associated neural crest streams, mesodermal cores, and epibranchial placodes subdivide the vertebrate head into a number of repeated units (or metameres). Similarly, the vertebrate trunk is formed from repeated segments. These become first evident in the paraxial mesoderm – the dorsal part of the mesoderm, immediately lateral to the notochord – which becomes subdivided into a series of block-like somites after gastrulation (Fig. 1.4A). Later the somites develop into the segmental vertebrae and ribs, trunk muscles, and the mesodermal part of the skin (dermis). Because the migrating neural crest cells and motor neurons growing out of the spinal cord avoid the posterior part of each somite (Keynes and Stern 1984; Krull et al. 1997; Bonanomi and Pfaff 2010), somites also impose their segmental arrangements onto the developing sensory DRG and motor roots of the spinal nerves.

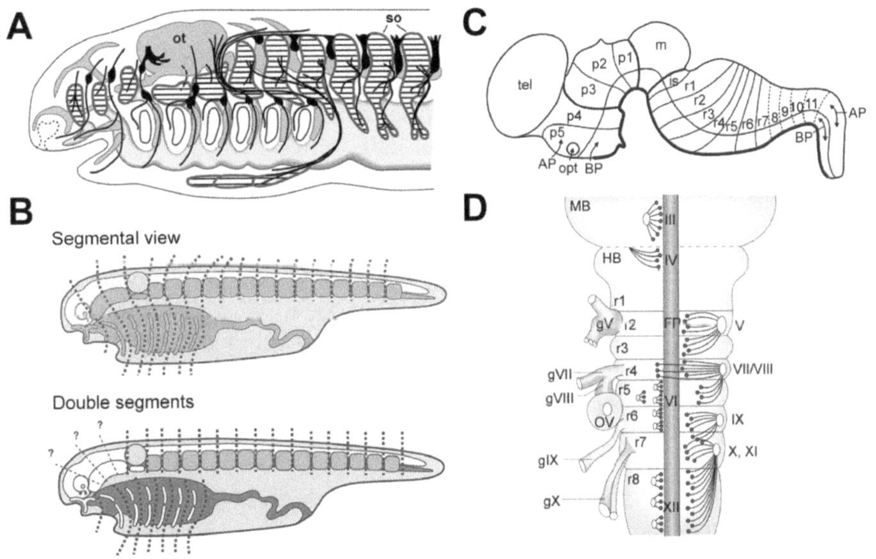

FIGURE 1.4 Vertebrate head segmentation. (**A**) Goodrich's (1930) model of head segmentation proposing that head segments correspond to somites (so; hatched). Ot, otic capsule. (**B**) The segmental view of Goodrich (upper drawing) is compared with the alternative view of Romer and others that pharyngeal segmentation of the head is independent of somitic segmentation of the trunk (lower drawing). Both series overlap but are not in register in the posterior head region. (**C**) Segmentation of the vertebrate forebrain into prosomeres (pl-p5) and of the hindbrain into rhombomeres (rl-rl1). AP, alar plate; BP, basal plate; is, isthmus; m, mesencephalon (midbrain); opt, optic nerve; tel, telencephalon. (**D**) Origin of cranial nerve ganglia (left) and motor neurons (right) from rhombomeres. MB, midbrain; HB, hindbrain; OV, otic vesicle; rl-8, rhombomeres 1-8; roman numerals, cranial nerves. ([A,B] Reprinted with permission from Kuratani 2008; [C] Reprinted with permission from Nieuwenhuys 2011; [D] Reprinted with permission from Guthrie 2007.)

The metameric organization of both head and trunk in the vertebrate body has long fueled speculations that the segmentation of the vertebrate trunk continues into the head (reviewed in Kuratani 2003; Kuratani 2008; Onai, Irie, and Kuratani 2014). This was first proposed by Johann Wolfgang von Goethe, who suggested in 1790 that the vertebrate skull can be understood as a series of enlarged vertebrae. Similar vertebral theories of the skull were later put forward by Lorenz Oken (1807) and Richard Owen (1848). Several comparative embryologists followed in the footsteps of these morphologists and proposed identical (or serially homologous) developmental building blocks underlying head and trunk development. Edwin S. Goodrich (1930) summarized this view in a famous drawing that shows precise registration between the series of somites, which extends into the head and the pharyngeal arches (Fig. 1.4A). However, this idealized scheme could not be confirmed in empirical studies (see Kuratani 2003; Kuratani 2008; Onai et al. 2014). An alternative theory, first proposed by Alfred Romer (1972), suggests that head and trunk of vertebrates are organized by two independent series of segments, which are not in registration: a "visceral" series of pharyngeal arches in the head and a "somatic" series of somites in the trunk (Fig. 1.4B) and this view is now widely accepted.

Metameric units have also been identified in the developing brain, both in the forebrain (prosencephalon) where they are called prosomeres and in the hindbrain (rhombencephalon) where they are known as rhombomeres (Fig. 1.4C). There is a long history of debate on the relation of these units to each other and to other tissues in the head or trunk region. This has been reviewed extensively elsewhere (Kiecker and Lumsden 2005; Nieuwenhuys 2011; Puelles et al. 2013). I only want to point out here that prosomeres are organized very differently than rhombomeres and do not appear to be related to them. The segmentation of the rhombomeres, however, while independently established from the segmental arrangement of the pharyngeal pouches (Graham et al. 2014), appears to be linked to the pharyngeal segmentation of the head with two rhombomeres corresponding to one pharyngeal arch.

This link is established because many neural crest cells emanating from odd numbered rhombomeres 3 and 5 are selectively eliminated by programmed cell death (Graham et al. 1993). As a consequence, the cranial neural crest is separated into distinct trigeminal, hyoid, and branchial streams, which contribute to the formation of the mandibular, hyoid, and branchial arches, respectively (Fig. 1.2C). Proximally, the neural crest streams remain attached to the even numbered rhombomeres where they prefigure the ganglia and entry points (roots) for the axons of sensory neurons of the profundal/trigeminal, facial, and glossopharyngeal cranial nerves (Fig. 1.4 D). The cranial nerves will be introduced in more detail in section 1.3 below. The axons of motor neurons of these nerves also exit at even numbered rhombomeres thereby forming the motor roots of these cranial nerves (Fig. 1.4D). Posterior to rhombomere 6, the registration of rhombomeres with pharyngeal arches becomes less clear. The borders between rhombomeres 7 and 11 are often not well defined and the number of pharyngeal arches supplied by the neural crest cells and motor axons that emerge from this domain and contribute to the vagal nerve, is variable between vertebrates.

1.2 SENSORY AND NEUROSECRETORY ORGANS OF THE VERTEBRATE HEAD

Some of the most remarkable features of the vertebrate head are the complex sense organs, which allow vertebrates to survey their environment. Less familiar, but of pivotal importance for the control and endocrine coordination of many vertebrate behaviors and physiological processes such as growth, reproduction, metabolism, and stress, is the anterior pituitary. Most of the cranial sense organs as well as the anterior pituitary are entirely derived from cranial placodes. A notable exception is the eye. The retina of the eye, which contains the actual photoreceptive cells, forms as an outpocketing from the forebrain and only the lens of the eye is derived from a cranial placode. Nevertheless, I will include the eye in this survey of placode derived cranial organs for two reasons. First, the lens, while not involved in photoreception is crucial for the excellent image forming capacities of the vertebrate eye, which are of central importance for the vertebrates' evolutionary success. And second, the photoreceptors of the eye have some interesting evolutionary relationships to other, placode-derived sensory cells, as I will explore further in Volume 2.

1.2.1 Eye

Apart from cave dwellers, which have secondarily lost eyes, all vertebrates are equipped with a pair of large, image forming eyes with a multilayered retina and a lens that can change its position or shape to form detailed projections of nearby objects or distant scenes on the retina (Fig. 1.5A). During embryonic development these eyes appear soon after the completion of neurulation, when two optic vesicles, one on each side, bulge out on the ventral side of the posterior embryonic forebrain (diencephalon) (Graw 2010) (Fig. 1.5B). After folding back on itself, each optic vesicle forms an optic cup, which remains connected to the forebrain by the optic stalk. The outer layer of the cup gives rise to the pigment layer and the inner layer to the sensory layer of the retina. At the so-called ciliary margin, where both layers meet, the ciliary body, which secretes aqueous humor and extracellular matrix, and the iris develop. The lens placode forms as a thickening of the ectoderm opposite the optic cup and invaginates to form the lens vesicle and ultimately the lens of the eye (Fig. 1.5B). The ectoderm next to the lens differentiates into the cornea. The gel-like vitreous body forms between the neural retina and the lens as an accumulation of extracellular matrix molecules (including collagen and hyaluronic acid) secreted by the ciliary body and lens (Halfter et al. 2005) (Fig. 1.5A).

The neural retina is initially a monolayered epithelium like the rest of the neural tube, but will soon develop into a multilayered structure with several sensory and neuronal cell types (Fig. 1.5C) (Livesey and Cepko 2001; Hatakeyama et al. 2004; Reh 2018). Only the Müller glia cells of the retina continue to span the entire width of the epithelium even at later stages. The photoreceptors (rods and cones) of the retina will come to occupy the outermost layer adjacent to the pigment epithelium. These photoreceptors are packed with discs of membranes containing photopigments (e.g. rhodopsin), which respond to light by undergoing conformational changes (for more detail see Chapter 7). There is only one type of rods, highly sensitive to light of

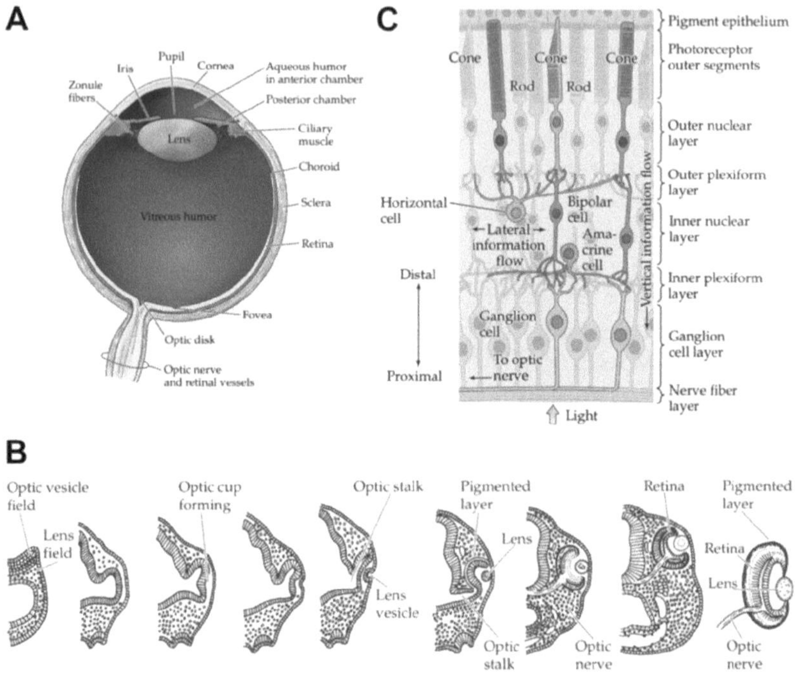

FIGURE 1.5 The paired eyes. (**A**) Cross-section through a human eye. (**B**) Development of the eye in a frog embryo. After Rugh 1951. (**C**) Cell types in the vertebrate retina. ([A] Reprinted with permission from Purves et al. 2008; [B] Reprinted with permission from Elinson 1997; [C] Reprinted with permission from Purves et al. 2008. See text for details.)

an intermediate wavelength, which is used mainly for night time vision. In contrast, there are several types of cones (with numbers varying between different vertebrates), which reach their highest density in the fovea of the retina. Cones are less sensitive than rods but each type responds to a different range of wavelengths, thereby allowing color vision when light is sufficiently bright. The photoreceptors transmit signals to the bipolar cells and these to the retinal ganglion cells in the innermost layer of the retina (Fig. 1.5C). The horizontal and amacrine cells form connections with multiple photoreceptors and retinal ganglion cells in the plane of the retina and play important roles for the first steps of visual information processing (Fig. 1.5C). The retinal ganglion cells send out their axons towards the center of the eye where these coalesce and leave the retina by growing through the optic stalk into the diencephalon, thereby forming the optic nerve. Because there are no photoreceptors in this area, we have a blind spot in our visual field (Fig. 1.5A). We normally don't realize this because our brain fills in information extrapolated from adjacent areas.

A second photoreceptive organ found in many vertebrates develops from the pineal complex, which forms by the fusion of several small unpaired outpocketings from the dorsal diencephalon – including epiphysis (pineal organ) and parietal organ (Vigh et al. 2002). Either pineal or parietal organ or both contain photoreceptors involved in regulating daily or seasonal behavioral rhythms in many vertebrates. However, in mammals and birds the pineal complex has lost its photoreceptivity and serves exclusively as an endocrine organ.

1.2.2 Inner Ear and Lateral Line

Vertebrates possess two complex mechanosensory organs derived from cranial placodes, the inner ear and the lateral line system. The inner ear develops from the otic placode and is present in all vertebrates. It is involved in the perception of balance and movement and in most groups also mediates hearing. The lateral line system gets its name from the arrangement of its receptor organs in lines along the body surface (Fig. 1.6). It develops from a series of lateral line placodes

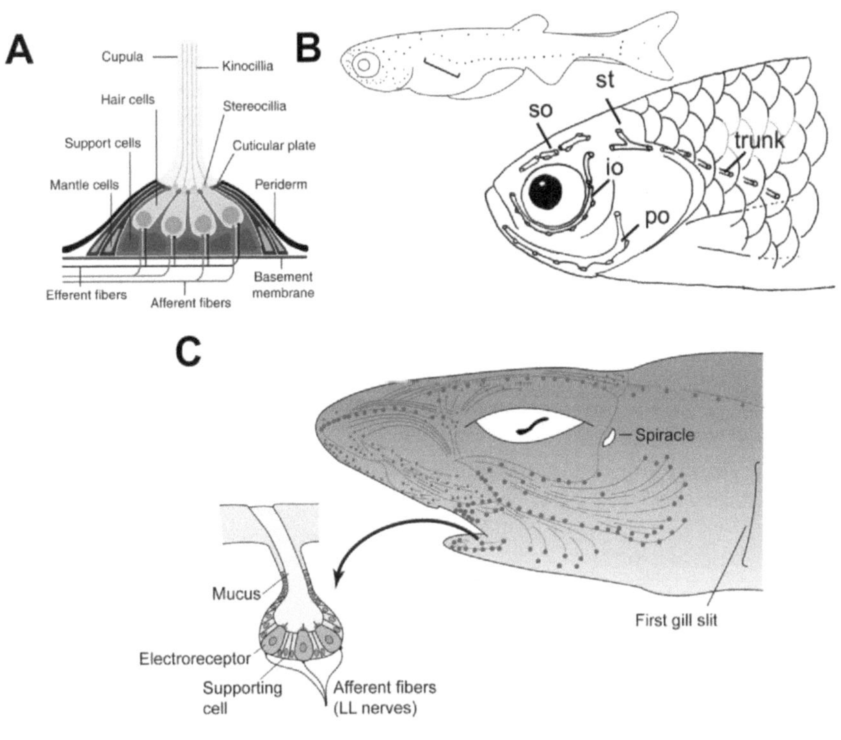

FIGURE 1.6 The lateral line system. (**A**) Mechanoreceptive lateral line organ (neuromast) of zebrafish with hair cells and support cells. (**B**) Distribution of neuromasts in larval (upper panel) and adult (lower panel) zebrafish. In larvae all neuromasts are superficial but some of them (e.g. those indicated by a bracket) will become integrated into canals (e.g. trunk canal) in the adult. io: infraorbital canal; po: preopercular canal; so: supraorbital canal; st: supratemporal canal. (**C**) Distribution of electroreceptive ampullary organs (ampullae of Lorenzini) in a shark with detail of one ampullary organ shown below. LL, lateral line. Drawing of shark by Chris Huh. ([A] Reprinted with permission from Ma and Raible 2009; [B] Reprinted with modification with permission from Wada, Iwasaki, and Kawakami 2014; [C] Reprinted with permission from Wikimedia Commons. Drawing of ampullary organ adapted from Jørgensen 2005 and Baker et al. 2013.)

and is present in all vertebrates except a few groups that have lost it secondarily such as amniotes. It allows perception of water movements on the body surface and in some groups is also involved in electroreception. Both the inner ear and the lateral line use a peculiar type of mechanosensory cell, the so-called hair cell for sensory perception (Fig. 1.6A), which I will characterize in greater detail in Chapter 6 (Coffin et al. 2005; Goodyear, Kros, and Richardson 2006; Fritzsch et al. 2007; McPherson 2018). Hair cells are so-called secondary sensory cells because they lack an axon. Instead they release neurotransmitters from their basal side thereby activating the special somatosensory neurons of the vestibulocochlear or lateral line nerves (see section 1.3) which originate from the same placodes. On their apical side, they carry a "hair bundle" comprising a non-motile cilium placed next to a staircase of microvilli (known as stereovilli or by the misleading term "stereocilia"). Bending of the hair bundle either mechanically opens or closes ion channels thereby activating or inhibiting the hair cell (Fritzsch et al. 2007; Peng et al. 2011; Fettiplace and Kim 2014). Due to the asymmetric arrangement of the hair bundle, hair cells show directional sensitivity, meaning that they are maximally activated by mechanical forces from one particular direction and maximally inhibited by forces from the opposite direction.

The deployment of a similar receptor cell type in both inner ear and lateral line together with similarities in development and innervation has led to the suggestion that the inner ear has evolved from an evolutionarily older lateral line system by invagination (Ayers 1892). This so-called "acousticolateralis" hypothesis has been rejected because there is currently no evidence that the lateral line system preceded the inner ear in evolution (Popper, Platt, and Edds 1992; Duncan and Fritzsch 2012). The similarities in cell type composition and development between inner ear and lateral line system nevertheless suggest that they may have originated from a common developmental progenitor in evolution. This will be discussed in more detail in Volume 2. Here I will confine myself to briefly introducing the development, structure and function of the lateral line and inner ear.

1.2.2.1 Lateral Line

Comparative studies indicate that six lateral line placodes were present in the ancestors of gnathostomes (jawed vertebrates), the anterodorsal, anteroventral, and otic lateral line placodes (not to be confused with the otic placode that gives rise to the inner ear) located before the ear (preotically), and the middle, supratemporal and posterior lateral line placode located behind the ear (postotically) (Fig. 1.3B) (Northcutt 1992; Schlosser 2002b). However, some of these placodes have been lost in some vertebrate groups, for example the otic lateral line placode in most amphibians.

During development (reviewed in Northcutt 1997; Schlosser 2002a; Ghysen and Dambly-Chaudière 2007; Chitnis, Nogare, and Matsuda 2012), lateral line placodes first give rise to precursors of sensory neurons, which migrate away from the placode into the adjacent mesenchyme, where they form the ganglion of the respective lateral line nerve (Fig. 1.7). The remaining placode either elongates or migrates along the basement membrane to establish a sensory ridge (in the head) or migratory primordium (in the trunk). The sensory ridges and migratory primordia then break up into rosette-like clusters of cells, which will differentiate into the mechanosensory organs

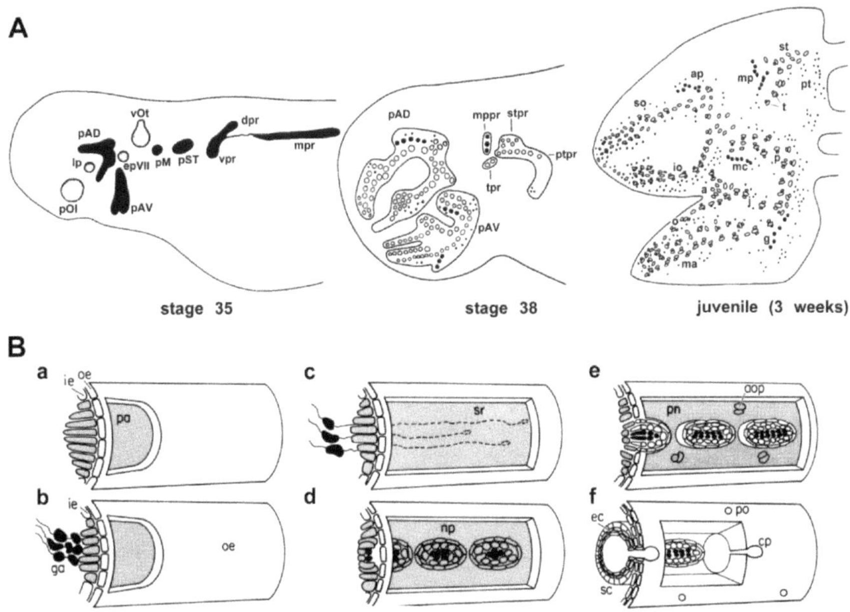

FIGURE 1.7 Development of the lateral line system. (**A**) Three stages of lateral line development in the axolotl. Neuromasts are indicated by open circles, ampullary organs by black dots. (**B**) Schematic illustration of lateral line development in a generalized vertebrate. After the thickening of the inner ectodermal layer (**a**), ganglion cells emerge (**b**), and a lateral line primordium forms by elongation and is tracked by neurites growing out from lateral line ganglia (**c**). Following the formation of primary neuromast primordia in medial zones (**d**), ampullary organ primordia develop in lateral zones (**e**). In amphibians, development ends at this stage, but in many other vertebrates ampullary organs invaginate and neuromasts become enclosed in canals (**f**). aop, ampullary organ primordium; ap, anterior pit line; cp, canal pore; dpr, migratory primordium of dorsal trunk lateral line; ec, epithelial canal; epVII, facial epibranchial placode; ga, ganglion, ie, inner epithelial layer; io, infraorbital lateral line; j, jugal lateral ine; lp, lens placode; ma, mandibular lateral line; mc, middle cheek pit line; mp, middle pit line; mppr, primordium of middle pit line; mpr, migratory primordium of middle trunk lateral line; np, primary neuromast primordium; oe, outer epithelial layer; pa, placode; pAD, anterodorsal lateral line placode; pAV, anteroventral lateral line placode; pn, primary neuromast; po, ampullary pore; pM, middle lateral line placode; pOl, olfactory placode; pST, supratemporal lateral line placode; pt, posttemporal lateral line; sc, secondary connective tissue canal; so, suprorbital lateral line; sr, sensory ridge; st, supratemporal line; stpr, primordium of supratemporal lateral line; t, temporal lateral line; tpr, primordium of temporal lateral line; vpr, migratory primordium of ventral trunk lateral line. ([A,B] Reprinted with modification with permission from Northcutt et al. 1994.)

of the lateral line, so-called neuromasts (Figs. 1.6A and 1.7). Each neuromast is composed of several hair cells and supporting cells and covered by a gelatinous cupula. Because sensory ridges and migratory primordia are tracked by the outgrowing neurites of lateral line ganglion cells, the neuromasts of each lateral line are innervated by sensory neurons of a lateral line nerve derived from the same placode. In fishes some of the neuromasts remain on the surface whereas others become enclosed in canals, which later ossify (Figs. 1.6B and 1.7B).

In many vertebrates, lateral line placodes also give rise to electroreceptive sensory organs (Fig. 1.6C). These are composed of electoreceptive cells and supporting cells recessed below the skin earning them the name ampullary receptors (reviewed in Jørgensen 2005; Baker, Modrell, and Gillis 2013). They develop from the lateral parts of the lateral line primordia, in contrast to the neuromasts, which develop from the central part (Fig. 1.7) (Northcutt, Brändle, and Fritzsch 1995; Modrell et al. 2011). In many cartilaginous fishes and some bony fishes (e.g. the paddlefish, a relative of sturgeons) electroreceptors are widely distributed over the body surface and play an important role in locating prey via their electric fields. In other groups, notably frogs and teleost fishes, ampullary receptors have been completely lost. However, at least two groups of teleost fishes have independently evolved a different type of electroreceptors responding to anodal stimuli (positive on the outside of the animal) rather than cathodal stimuli as in other vertebrates (Bullock, Bodznick, and Northcutt 1983; Bodznick and Montgomery 2005; Baker et al. 2013). These are found either in ampullary organs resembling those of other vertebrates (Fig. 1.6C) or in tuberous organs, in which a plug of epidermal cells separates the receptor cells from the surface. Teleost electroreceptors develop in close neighborhood to lateral line neuromasts suggesting that they are also derived from lateral line placodes but this has not yet been confirmed experimentally.

1.2.2.2 Inner ear

Similar to the lateral line system, the otic placode gives rise to the entire inner ear including the sensory hair cells and the sensory neurons that innervate them (reviewed in Alsina, Giráldez, and Pujades 2009; Wu and Kelley 2012; Maier et al. 2014; Whitfield 2015). In a first step, the otic placode invaginates to form the otic vesicle (Fig. 1.3A). The sensorineural progenitors located in the ventromedial part of the vesicle divide and some of their progeny then migrate away into the adjacent mesenchyme where they differentiate as sensory neurons and form the ganglion of the vestibulocochlear nerve. The otic vesicle will then undergo a series of complicated morphogenetic movements forming the aptly named membraneous labyrinth (Fig. 1.8). During that process, progenitors remaining in the vesicle will become separated into multiple sensory areas, which come to occupy the various pockets and canals of the labyrinth (Fig. 1.8A). The number, arrangement, and function of these sensory areas are quite variable between different vertebrate groups and it is still somewhat contentious, which of these areas are homologous between groups.

Generally, the dorsal part of the inner ear, which is dedicated mostly to detecting position in space and movement (vestibular functions) is more conserved than the ventral part, which has evolved additional roles in hearing (auditory functions) several times independently. This suggests that the ear evolved initially as a mechanoreceptive organ allowing to monitor movements and the position of the body in space providing opportunities for the repeated redeployment of mechanoreceptors for new functions in hearing. The dorsal part of the inner ear comprises a pouch, the utricle, and the semicircular canals emanating from it, whereas the ventral part comprises another pouch, the saccule, from which in

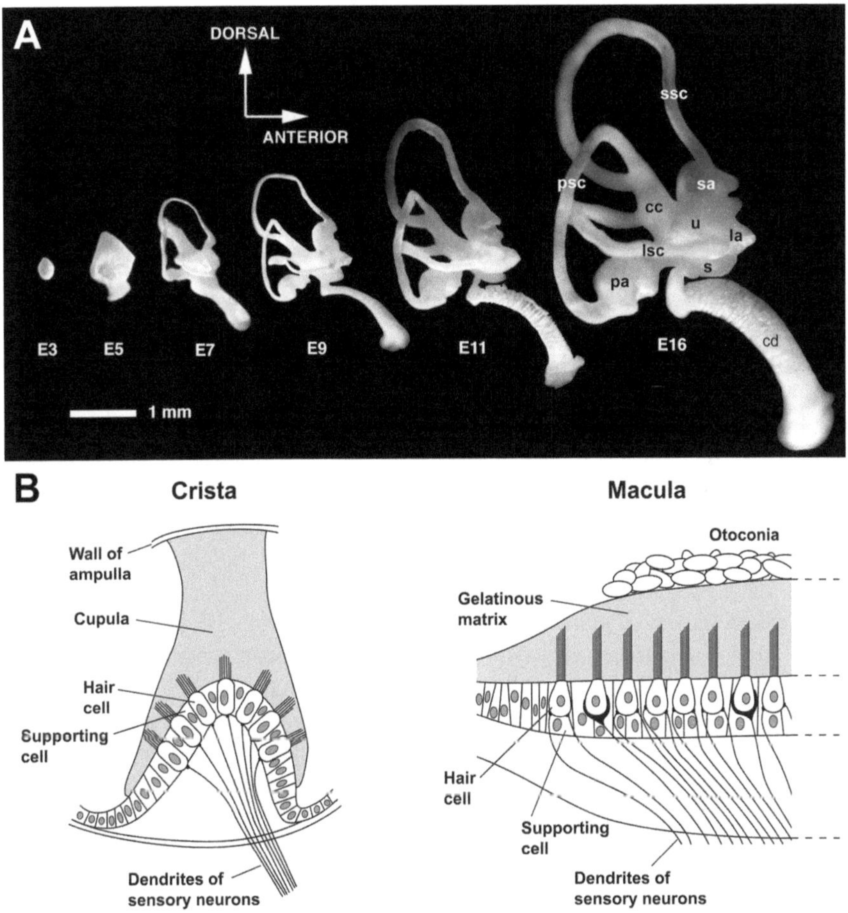

FIGURE 1.8 Development of the vertebrate inner ear. (**A**) Paint filled chicken inner ears from otic vesicle stage to a fully formed membraneous labyrinth. Cc, common crus; cd, cochlear duct; la, lateral ampulla; lsc, lateral semicircular canal; pa, posterior ampulla; psc, posterior semicircular canal; s, saccule; sa, superior ampulla; ssc, superior semicircular canal; u, utricle. (**B**) Different types of sensory areas, each composed of hair cells and supporting cells, are found in the inner ear. These include the cristae of the semicircular canal (left) and the maculae of utricle and saccule (right). ([A] Reprinted with modification with permission from Bissonnette and Fekete 1996; [B] Redrawn and modified from Parker 1980.)

tetrapods the lagena (or lagenar recess) develops an outpocketing. In mammals this recess becomes the cochlear duct. The canals and pouches harbour three types of sensory organs (reviewed in Fritzsch et al. 2002; Fritzsch et al. 2013; Fritzsch and Straka 2014).

First, there are the cristae found in a widened area (ampulla) of each semicircular canal. They resemble neuromasts with hair cells and supporting cells attached to a cupula (Fig. 1.8B). The hair cells in cristae are stimulated when the cupula is bent by movement of the endolymph in the semicircular canal. They are, thus, able to detect angular acceleration due to rotation of the head or body in the plane of the semicircular canal. The three semicircular canals of gnathostomes are positioned orthogonal to each other thereby allowing detection of rotations in all three spatial dimension. In contrast, jawless vertebrates have only two cristae located in one (hagfishes) or two (lampreys) semicircular canals and always lack the horizontal canal.

Second, there are up to three maculae in gnathostomes, while the ears of jawless vertebrates have only one macula. Maculae comprise hair cells and supporting cells embedded in a gelatinous matrix, which is weighed down by many small concretions of calcium carbonate called otoconia or a single otolith (Fig. 1.8B). These cause displacement of hair bundles when the head is tilted or accelerated linearly allowing the detection of body position and movement. The utricular macula is located in the utricle and the saccular macula in the saccule. A third macula, the lagenar macula, can be either located in the saccule (e.g. many ray finned fishes) or in a separate lagenar recess (e.g. elasmobranchs, neopterygian fishes including teleosts, tetrapods); it has been completely lost in therian (marsupial and placental) mammals (Fritzsch et al. 2013). In many fishes and amphibians, some maculae and in particular the saccular macula also respond to sound and play a role in hearing (Ladich and Popper 2004).

Third, in the ear of gnathostomes there are up to three different papillae, which are located outside the semicircular canals, lack otoconia and have been implicated in hearing. The neglected papilla (also called macula neglecta) is located in the utricle, whereas the amphibian papilla and the basilar papilla are located in the lagena. The neglected papilla, which probably contributes to hearing in sharks and rays is very small in many bony fishes and has been lost in frogs and salamanders. As the name suggests, the amphibian papilla is only found in amphibians and probably evolved by budding of the neglected papilla (Fritzsch et al. 2013). Together with the basilar papilla it serves as auditory receptor organ in amphibians. In several lineages of amniotes the basilar papilla (with contentious homology to the equally named papilla of amphibians) expanded and evolved into a specialized auditory organ.

The most complex of these basilar papillae is the organ of Corti in mammals (reviewed in Fettiplace and Hackney 2006; Hudspeth 2014) (Fig. 1.9). It is located in the cochlear duct (or scala media), a coiled version of the lagenar recess filled with endolymph like the rest of the inner ear. The accompanying ducts, scala vestibuli, and scala tympani, which develop from an adjacent fluid filled sac, are instead filled with perilymph, a fluid with a different composition. In placental mammals, the organ of Corti consists of one row of inner hair cells and three rows of outer hair cells, which sit on the basilar membrane and are covered by the tectorial membrane, which makes contact with the hair bundles of the outer hair cells. These membranes are stiff acellular structures composed of extracellular matrix molecules. Only the inner hair cells are involved in hearing and receive the majority of sensory (afferent) innervation. The outer hair cells, which receive mostly efferent innervation, can change their length in response to deflection of their hair bundle. As a consequence, the position of the tectorial membrane is altered and the current of endolymph

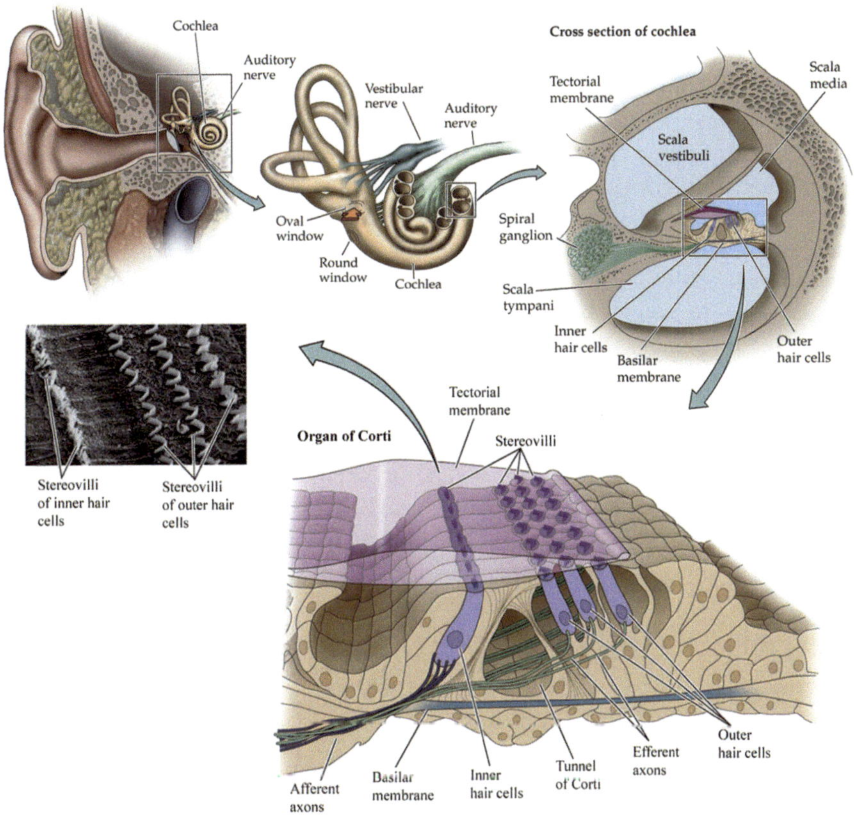

FIGURE 1.9 The mammalian inner ear. The organ of Corti in the mammalian cochlea contains the inner hair cells involved in hearing. The inner ear (yellow) with the cochlea is shown face-on (upper left) and in cross-sections (other panels). The ossicles of the middle ear transfer force from the tympanic membrane to the oval window. The cross section of the cochlea (upper left) shows the scala intermedia with the organ of Corti between the scalae vestibuli and tympani. The tectorial membrane is shown transparent in the detailed view of the organ of Corti (lower right) and is removed in the scanning electron micrograph (lower left). Note the inner and outer hair cells located between the basilar membrane and the tectorial membrane. (Reprinted with permission from Purves et al. 2008.)

flowing over the inner hair cells and bending their hair bundles is increased leading to stimulus amplification (see Chapter 6.3.1).

Hair cells in the mammalian cochlea are stimulated by sound-induced vibrations of the middle ear bones, which exert pressure on the oval window of the scala vestibuli (the round window of the scala tympani on the other end of the perilymphatic space acts as a pressure relief window) (review in Purves et al. 2008; Moller 2012). The resulting vibrations of the perilymph cause displacements of the basilar membrane. This leads to shearing forces between basement and tectorial membrane, which stimulate the hair cells. The louder the sound the more the basilar membrane is displaced and hair cells activated. Because the properties of the basilar membrane change between the oval window and the apex of the cochlea, sounds of different frequencies (pitch) cause maximal displacements in different parts of the basilar membrane. Near the oval window the basilar membrane is relatively narrow and stiff and responds to high frequencies, whereas closer to the apex it is wider and more flexible and responds to low frequencies.

In many vertebrates, accessory structures have evolved to amplify sounds before they reach the sensory areas of the inner ear. In mammals, as just mentioned, there are three middle ear bones (malleus, incus, and stapes), which connect the tympanum and the oval window and serve to amplify the sound pressure waves impinging on the tympanum (Fig. 1.9). The external ear, with the protruding pinna (what we normally refer to as "ear"), further aids in sound amplification and localization. Other tetrapods have evolved tympanic ears (including an air filled middle ear cavity and a tympanum) with only a single bone (stapes, also known as columella), several times independently (Clack and Allin 2004; Manley 2017). Several different accessory structures evolved to amplify sound in fishes, for example the Weberian ossicles, which are derived from the vertebral column and connect the swim bladder to the inner ear (Ladich and Popper 2004).

1.2.3 Olfactory Organs

Apart from the taste buds, which are not placode-derived, the olfactory system is the most important chemoreceptive organ in vertebrates (Fig. 1.10). Whereas taste allows to test food and other objects for their quality in a contact-dependent way, olfaction has multiple functions and responds to general odorants or to molecules involved in intra- or inter-specific communication. For this purpose, the nose of vertebrates contains two specialized chemosensory epithelia. The main olfactory epithelium is dedicated to detect general odorants, often represented by small and volatile or water soluble molecules. The vomeronasal epithelium instead is specialized to detect larger molecules, which may be picked up by direct contact with a substrate and often trigger species-specific and innate behaviors. They typically serve as pheromones – molecules that affect the behavior of members of the same species – or play a role in inter-specific communication although there are many exceptions to this rule (Fig. 1.10A) (reviewed in Mombaerts 2004; Breer, Fleischer, and Strotmann 2006; Baxi, Dorries, and Eisthen 2006; Touhara and Vosshall 2009; Munger, Leinders-Zufall, and Zufall 2009; Silva and Antunes 2017).

The vomeronasal epithelium only originated in the lineage giving rise to lungfishes and tetrapods. However, receptor families related to the tetrapod vomeronasal receptor families V1R and V2R are present in the olfactory epithelium in lampreys, sharks, and teleosts. This suggests that some functional separation between olfactory and vomeronasal receptor cells may have originated in early gnathostomes or even in ancestral vertebrates (Grus and Zhang 2009; Silva and Antunes 2017). Both epithelia are derived from the olfactory placode, which partly

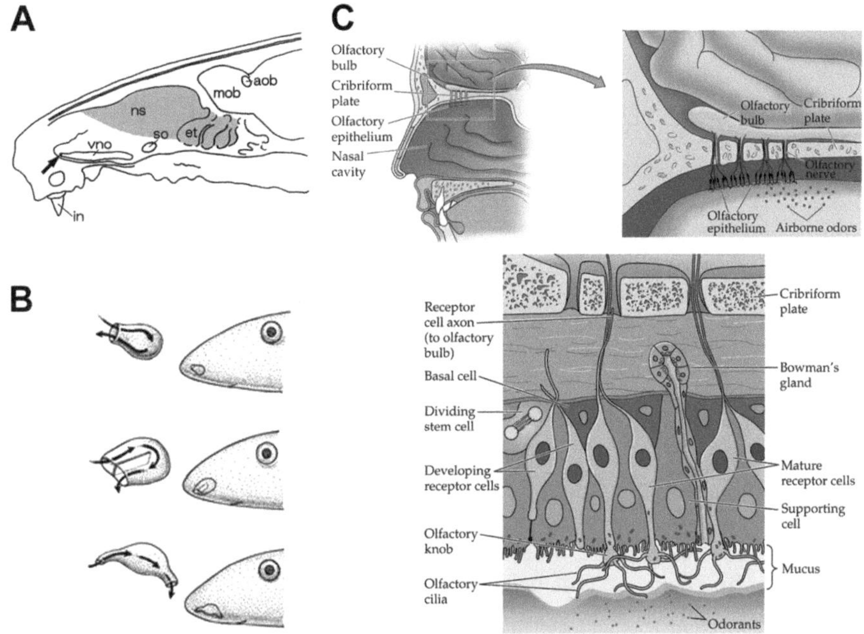

FIGURE 1.10 The olfactory sense organs. (**A**) Separate olfactory and vomeronasal epithelia in the nose of a rat. Olfactory epithelium indicated by shaded area. Arrow indicates opening of the vomeronasal organ. aob, accessory olfactory bulb; et, endoturbinates; in, incisor; mob, main olfactory bulb; ns, nasal septum; so, septal olfactory organ of Masera; vno, vomeronasal organ. (**B**) Nostrils in fishes. Single external nostril (upper panel), external nostrils with incurrent and excurrent opening (middle panel) and combination of external with internal nostrils (choanae; lower panel) are shown. (**C**) Olfactory epithelium in the human nose. ([A] Reprinted with permission from Taniguchi and Taniguchi 2014; [B] Reprinted with permission from Kardong 2009; [C] Reprinted with permission from Purves et al. 2008.)

invaginates to form the nasal cavity (reviewed in Balmer and LaMantia 2005; Maier et al. 2014; Sokpor et al. 2018). Whether the neural crest also contributes cells to these epithelia is still controversial (Suzuki and Osumi 2015) and will be discussed later.

In jawless fishes where the nasal epithelium originates together with the adenohypophysis from a single unpaired placode, there is only a single unpaired medial nostril (Uchida et al. 2003; Oisi et al. 2013), while in gnathostomes paired nasal cavities develop from a pair of olfactory placodes. In many fishes each nasal cavity opens to the outside with either a single nostril or one incurrent and one excurrent opening. In tetrapods (and, independently, in several lineages of fishes) the latter has been displaced into the mouth cavity forming internal nostrils or choanae (Zhu and Ahlberg 2004) (Fig. 1.10B).

The olfactory and vomeronasal receptor cells located in the respective epithelia are primary sensory cells (which have an axon). Bundled together into the olfactory and vomeronasal nerves, their axons grow into the adjacent forebrain, where they terminate in the main and accessory olfactory bulb, respectively (Fig. 1.10A, C). Olfactory and vomeronasal receptor cell reach into the mucus covering the epithelia with cilia, microvilli, or both protruding from their apical side (Eisthen 1992, 1997; Elsaesser and Paysan 2007). The membrane of these extensions contains receptor proteins, which bind in a lock-key mechanism to specific odorants or pheromones. As will be discussed in more detail later in the book (Chapter 6), olfactory and vomeronasal receptor cells employ different types of receptors and signal transduction mechanisms (reviewed in Buck 2004; Touhara and Vosshall 2009; Kaupp 2010; Silva and Antunes 2017). Several hundred odorant receptor proteins are encoded in the vertebrate genome and each receptor cell of the main olfactory epithelium expresses only one of those. This versatility of receptor cells, together with the combinatorial possibilities opened up by the simultaneous activation of multiple receptor cells, accounts for the ability of many vertebrates to discriminate between a large number of odors.

In addition to the olfactory and vomeronasal receptor cells, the olfactory placode also gives rise to other cell types (reviewed in Balmer and LaMantia 2005; Sokpor et al. 2018). These include the sustentacular cells and the mucus producing Bowman's glands of the olfactory and vomeronasal epithelia as well as migratory cells that migrate into the so-called pre-optic and hypothalamic regions of the forebrain. There they differentiate into neurosecretory neurons that release neuropeptides such as gonadotropin releasing hormone (GnRH), which controls the secretion of gonadotropins from the adjacent anterior pituitary, or neuropeptide Y. In some vertebrates, these migratory cells form ganglia of a distinct cranial nerve called the terminal nerve with neuromodulatory effects on olfactory receptor cells and other neurons (Von Bartheld 2004; Kawai, Oka, and Eisthen 2009).

1.2.4 Pituitary

Different from the sensory organs of the vertebrate head, the pituitary gland is an endocrine organ that secretes hormones into the blood stream to coordinate a diverse array of body functions, including growth, metabolism, reproduction, and response to stress. The pituitary gland is attached to the ventral part of the hypothalamus in the forebrain and has two components that are clearly distinct functionally and have an independent developmental origin (Fig. 1.11). The posterior pituitary or neurohypophysis develops from an outpocketing of the hypothalamus (Fig. 1.11A) and the hormones released there (adiuretin and oxytocin) are actually produced in hypothalamic neurons that send their axons into the posterior pituitary. In contrast, the anterior pituitary or adenohypophysis develops from the unpaired adenohypophyseal placode originating in the anterior midline just in front of the neural plate (reviewed in Rizzoti and Lovell-Badge 2005; Zhu, Gleiberman, and Rosenfeld 2007; Kelberman et al. 2009; Davis et al. 2013). In most gnathostomes (except teleosts) the placode becomes first incorporated into the ectodermal anlage of the mouth cavity (stomodeum) and then invaginates to form Rathke's pouch (Fig. 1.11A). This

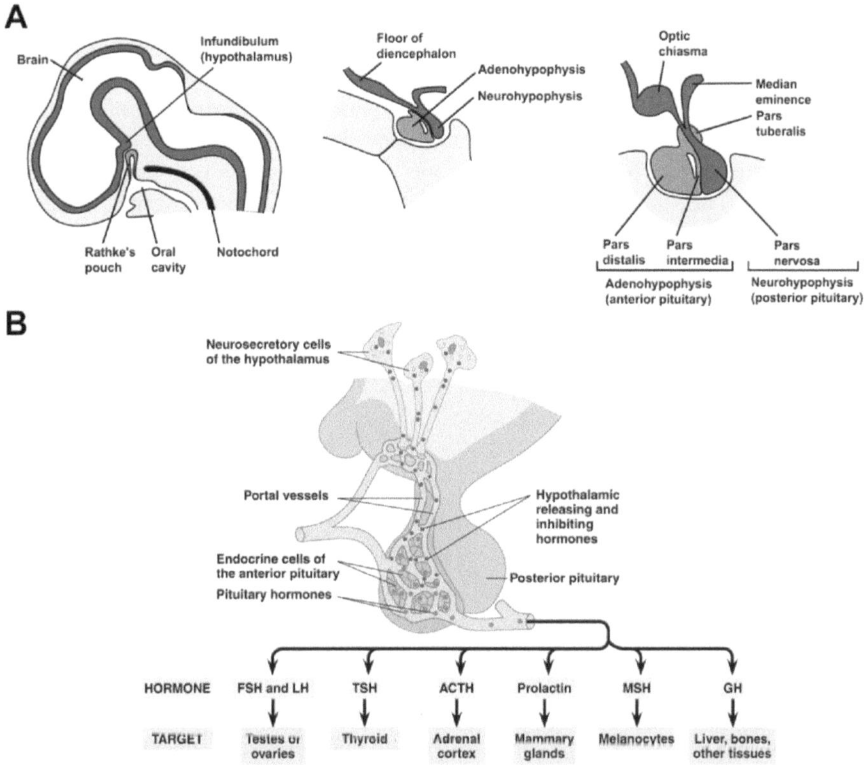

FIGURE 1.11 The pituitary. (**A**) Development of the pituitary. While the anterior pituitary (adenohypophysis) develops from an outpocketing of oral ectoderm (Rathke's pouch), the posterior pituitary (neurohypophysis) develops from the ventral hypothalamus (infundibulum). (**B**) Six different neurosecretory cell types are found in the adenohypophysis of gnathostomes. ([B] Reprinted with permission from Reece, et al. 2011).

pouch becomes closely apposed to the developing neurohypophysis, detaches from the oral ectoderm and differentiates into different types of neurosecretory cells. While jawless vertebrates have only four types of hormone producing cells in the adenohypophysis (Kawauchi and Sower 2006), there are six types in gnathostomes (Fig. 1.11B):

- Corticotropes producing adrenocorticotropic hormone (ACTH);
- Melanotropes producing melanocyte stimulating hormone (MSH);
- Gonadotropes producing the gonadotropic hormones, luteinizing hormone (LH) and follicle stimulating hormone (FSH);
- Thyrotropes producing thyroid stimulating hormone (TSH);
- Lactotropes producing prolactin;
- Somatotropes producing somatotropin also known as growth hormone (GH).

These hormones are released into the blood stream and distributed to the entire body. Many of the hormones produced in the adenohypophysis act on other endocrine glands, in particular the cortex of the adrenal gland (ACTH), which mediates the long term stress response; the gonads (LH/FSH), which produce the sex hormones such as estrogen, progesteron and testosterone; and the thyroid gland (TSH), which produces the important metabolic hormone thyroxine (Norris and Carr 2020). The adenohypophyseal hormones, in turn, are positively or negatively regulated by so-called releasing or inhibiting hormones, respectively. TSH production, for example, is activated by thyroid releasing hormone (TRH), while LH/FSH production is promoted by GnRH. These releasing hormones are produced in the hypothalamus and released at the so-called median eminence into the capillaries of portal vessels, which then transport them a short stretch to the anterior pituitary (Fig. 1.11 B).

1.3 THE CRANIAL NERVES

Concluding our overview of vertebrate head anatomy, we have to turn our attention to the cranial nerves. These form the PNS of the head, connecting its sensory and motor organs to the brain (Fig. 1.12) (reviewed in Northcutt 1993; Butler and Hodos 2005; Porras-Gallo et al. 2019; Striedter and Northcutt 2020). In contrast to the spinal nerves, which are mixed sensory and motor nerves that supply the trunk and are similarly organized for each trunk segment, each cranial nerve is different. Cranial nerves were first described and numbered with roman numerals I–XII based on human anatomy and these names and numbers are now also used for other vertebrates. In addition, there are nerves innervating the lateral line organs in fishes and amphibians, which are not found in humans or other amniotes. These have been considered as branches of other cranial nerves by neuroanatomists of the 19th and early-mid 20th century, but are now considered as a separate series of lateral line nerves (Northcutt 1989).

I will briefly sketch the organization and development of cranial nerves here because many of them receive a contribution from cranial placodes and/or are involved in transmitting sensory information to the brain and, therefore, will be frequently encountered in this book. I need to summarize a lot of information here and apologize in advance for the dense writing to follow. The overwhelmed reader may skip this section but may find it useful to come back to it later for some quick "fact checking" about cranial nerves.

Based on their function, cranial nerves can be classified into three different groups. First, there are those sensory nerves that directly connect the large cranial sense organs to the brain (Fig. 1.12A). These include the olfactory nerve (N. I) from the nose, which enters into the anterior forebrain; and is derived from the olfactory placode. Closely associated with the olfactory nerve are the vomeronasal nerve and the terminal nerve (with both sensory and neuromodulatory functions). These two nerves are absent in humans, and therefore are not part of the numbered series. The sensory nerves also include the vestibulocochlear nerve (N. VIII) from the inner ear and several lateral line nerves from the different lateral lines. The sensory neurons of these nerves are derived from the otic and lateral line placodes,

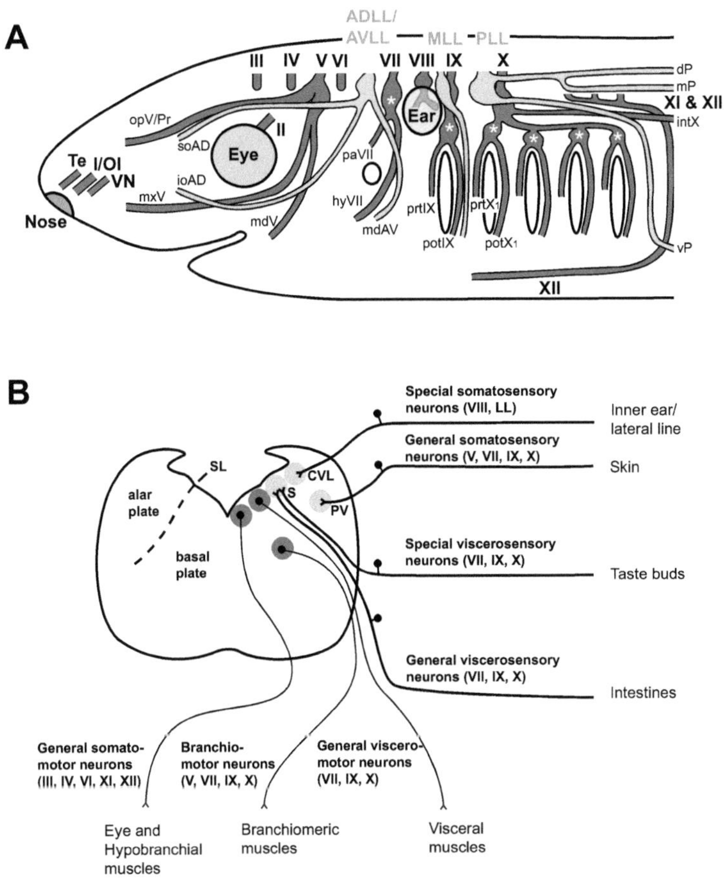

FIGURE 1.12 The cranial nerves. (**A**) Overview of cranial nerves in a generalized vertebrate. I/Ol, olfactory (and vomeronasal) nerve; II, optic nerve; III, oculomotor nerve; IV, trochlear nerve; V, profundal/trigeminal nerve; VI, abducens nerve; VII, facial nerve; VIII, vestibulocochlear nerve; VN, vomeronasal nerve IX, glossopharyngeal nerve; X,vagal nerves, XI, accessory nerve; XII, hypoglossal nerve; ADLL, anterodorsal lateral line nerve; AVLL, anteroventral lateral line nerve; MLL, middle lateral line nerve; PLL, posterior lateral line nerve; Te, terminal nerve. Otic and supratemporal lateral line nerves are often small or reduced and are not shown here. Asterisks indicate the geniculate ganglion of VII and the distal ganglia of IX and X, which contain viscerosensory neurons; all other ganglia depicted contain somatosensory neurons. Major nerve branches are depicted: dP, dorsal ramus of PLL; hyVII, hyoid ramus of VII; intX, intestinal ramus of X; ioAD, infraorbital ramus of AD; mdAV, mandibular ramus of AV; mdV, mandibular ramus of V; mxV, maxillary ramus of V; mP, middle ramus of P; pavi, palatine ramus of VII; opV/Pr, ophthalmic ramus of V/profundal nerve potIX, posttrematic ramus of IX; potX$_1$, posttrematic ramus of X$_1$; prtIX, pretrematic ramus of IX; prtX$_1$, pretrematic ramus of X$_1$; soAD, supraorbital ramus of AD; vP, ventral ramus of P. (**B**) Sensory and motor nuclei of cranial nerves in a cross section through the vertebrate hindbrain. Sensory nuclei are located in the alar plate (lateral), while motor nuclei are located in the basal plate (medial). Cranial nerves indicated as in A. CVL, cochlear, vestibular, and lateral line nuclei; PV, principal nucleus of V; S, solitary nucleus; SL, sulcus limitans. ([B] Redrawn and modified from Nieuwenhuys 2011.)

respectively. They are located in distinct ganglia and send their axons into the hindbrain (Fig. 1.12B). The optic nerve (N. II), which connects the retina to the brain, presents a special case because it is not derived from a placode (only the lens of the eye develops from a placode) and is, strictly speaking, not even part of the peripheral nervous system. The retina develops in the embryo by an outpocketing from the forebrain. It is, therefore, part of the central nervous system and the axons of the retinal ganglion cells, which form the optic "nerve" by growing through the optic stalk connecting the retina to the forebrain, are, strictly speaking, an axon tract of the CNS.

Second, there is a group of nerves providing both sensory and motor innervation to the pharyngeal arches, the so called "branchiomeric" nerves. These comprise the profundal/trigeminal (N. V), facial (N. VII), glossopharyngeal (N. IX), and vagal nerves (N. X), which innervate the first, second, third, and more posterior pharyngeal arches, respectively (Fig. 1.12A). Note that the terminology used to describe the complex of profundal and trigeminal nerves in different vertebrates can be confusing (for review see Schlosser and Northcutt 2000; Schlosser 2006). Separate profundal and trigeminal nerves, each with a separate ganglion are recognized in some anamniotes. The ganglia of the profundal and trigeminal nerves are fused in other lineages including amniotes, where the entire ganglionic complex is known as "trigeminal" (or Gasserian) ganglion and the composite nerve as the "trigeminal nerve". However, the so-called ophthalmic and maxillomandibular parts of the latter are most likely homologous to the profundal and trigeminal ganglia of anamniotes, respectively. For consistency, I will here refer to the anterior (ophthalmic) portion of this nerve and ganglia complex, which innervates the anterior part of the head, as the profundal ganglion/nerve and to the posterior (maxillomandibular) portion, which innervates the first pharyngeal arch, as the trigeminal ganglion/nerve for all vertebrates.

The third and last group of cranial nerves contains exclusively motor axons. These innervate the eye muscles (N. III, oculomotor nerve; N. IV, trochlear nerve; N. VI, abducens nerve) and muscles in the ventral and posterior part of the head (N. XI, accessory nerve; N. XII, hypoglossal nerve). These motor nerves contain the axons of general somatomotor neurons, located in the ventralmost part of the hindbrain (Fig. 1.12B).

The branchiomeric nerves have a very complex organization and development and require some further discussion here. Both neural crest and placodes contribute sensory neurons to these nerves. The cell bodies of these sensory neurons reside in ganglia outside the CNS and send their axons into the dorsal half (the so-called alar plate) of the hindbrain, where they form synapses with CNS neurons (Fig. 1.12A, B). Sensory neurons of the branchiomeric nerves, which mediate temperature, touch, and pain sensation from the skin as well as information on muscle position from the muscle spindles and joints are classified as general somatosensory neurons (GSS) (or general somatic afferents). They send their axons into the dorsal part of the alar plate, where they synapse onto a group of neurons forming a longitudinal column along the hindbrain, the so-called principal nucleus of the trigeminal nerve.

In the periphery, general somatosensory neurons either form free nerve endings (especially for pain and temperature sensation) or innervate special sensory cells (e.g. the epidermally derived Merkel cell) or sense organs (e.g. Meissner, Pacinian,

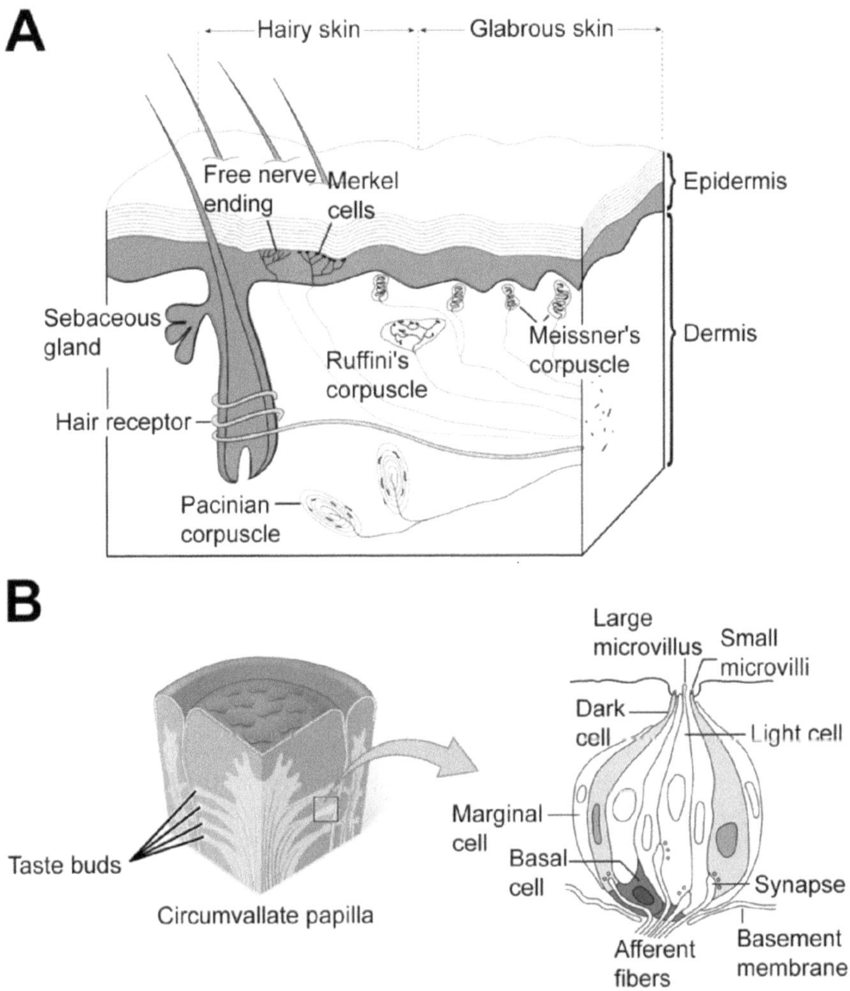

FIGURE 1.13 Cutaneous mechanoreceptors and taste buds. (**A**) Cutaneous mechanoreceptors and free nerve endings of special somatosensory neurons. (**B**) Taste buds innervated by general viscerosensory neurons. Left panel shows taste papilla on tongue with multiple taste buds. Right panel shows detail of one taste bud. ([A] Modified from a drawing by Thomas Haslwanter licensed under the Creative Commons Attribution-Share Alike 3.0 Unported (https://creativecommons.org/licenses/by-sa/3.0/deed.en) license; [B] Left panel reprinted from the Textbook OpenStax Anatomy and Physiology licensed under the Creative Commons Attribution 4.0 International (https://creativeco mmons.org/licenses/by/4.0/deed.en) license. Right panel reprinted with modification with permission from Northcutt 2004; based on Mistretta 1989.)

and Ruffini corpuscles) that respond to different mechanical stimuli (Fig. 1.13A) (Zimmerman, Bai, and Ginty 2014). Other general somatosensory neurons have nerve endings that wrap around special muscle fibers and collagen fibers to form the muscle spindles or Golgi tendon organs (Proske and Gandevia 2012). These provide information about muscle contraction and joint positions thereby allowing to monitor posture and movements of the own body (propioception). In contrast to the general somatosensory neurons of the branchiomeric nerves, the sensory neurons from the inner ear and the lateral line organs, which reach the hindbrain via the vestibulocochlear and lateral line nerves, are classified as special somatosensory neurons (SSS) (or special somatic afferents). They project to distinct columns in the alar plate (the cochlear, vestibular, and lateral line nuclei) and innervate the hair cells of the inner ear and lateral line that are derived from the same placodes Fig. 1.12B).

Another group of sensory neurons of the branchiomeric nerves mediate pain and mechanical sensation from the intestines and inner organs as well as oxygen and pressure sensation, for example from the carotid and aortic bodies associated with arteries in amniotes and the gill epithelia in fish (Fig. 1.12B) (Robinson and Gebhart 2008; Hockman et al. 2017). These sensory neurons are categorized as general viscerosensory (GVS) neurons (or general visceral afferents) and send their axons into the ventral part of the alar plate, where they synapse onto another column of neurons, the so-called nucleus of the solitary tract. In peripheral tissues they either terminate as free nerve endings or are associated with chemo- or mechano-sensory cells such as the neural crest derived glomus cells of the carotid body or the endoderm derived "neuroepithelial cells" in fish gills (Hockman et al. 2017). A final group of sensory neurons innervates the taste buds found in mouth cavity and pharynx (and sometimes on the skin in some fishes). These gustatory (taste) neurons are classified as special viscerosensory (SVS) neurons (or special visceral afferents) and their axons also terminate in the nucleus of the solitary tract. They receive their peripheral input from taste buds, small clusters of chemosensory cells, which develop from local, endo- or ecto dermally derived epithelium (Fig. 1.13B) and respond to five categories of chemosensory stimuli: sweet, bitter, sour, salty, and umami (the savory taste of glutamate) (Finger 1997; Northcutt 2004; Barlow 2015).

The cell bodies of the motor neurons of the branchiomeric nerves are located in the ventral part of the hindbrain (the basal plate), where they also form longitudinal columns, with the motor neurons innervating the smooth muscles of the intestines (visceromotor neurons or visceral efferents) being more dorsally located than the motor neurons innervating the striated muscles of the head (branchiomotor neurons) (Fig. 1.12B). The organization of the hindbrain into these functional columns was first proposed more than a century ago by Herrick (1899) and Johnston (1902) but with some modification retains its value today (reviewed in Nieuwenhuys 2011; Puelles 2019).

During the formation of branchiomeric nerves, placodes interact closely with the adjacent neural crest, which guides proper separation and positioning of different placodes and cooperates with placodal cells during ganglion formation (Begbie and Graham 2001; Shiau et al. 2008; Coppola et al. 2010; Shiau and Bronner-Fraser 2010; Freter et al. 2013; Theveneau et al. 2013). The neural crest also gives rise to all glial cells of the branchiomeric nerves (D'Amico-Martel and Noden 1983). The origin of the various categories of sensory neurons in the branchiomeric nerves could be determined experimentally by grafting small pieces of tissue between chick and quail embryos, whose cells can be distinguished by their different nuclei (Narayanan and Narayanan 1980; Ayer-LeLièvre and Le Douarin 1982; D'Amico-Martel and Noden 1983) (Fig. 1.12B, C). Only general somatosensory neurons are found in the profundal and trigeminal ganglia and these were shown to be partly derived from the neural crest and partly from the profundal and trigeminal placodes. In contrast, the general somatosensory neurons of the glossopharyngeal and vagal nerves (the facial nerve has almost none of those), which occupy the proximal ganglia (near the hindbrain) of these nerves – the superior and jugular ganglion, respectively – were exclusively derived from the neural crest. The general and special viscerosensory neurons of the facial, glossopharyngeal, and vagal nerves, which occupy the distal ganglia of these nerves – the geniculate, petrosal, and nodose ganglion, respectively – are instead derived from the epibranchial placodes, whereas the special somatosensory neurons residing in the ganglia of the vestibulocochlear and lateral line nerves are exclusively derived from the otic and lateral line placodes (Stone 1922; Northcutt and Brändle 1995; D'Amico-Martel and Noden 1983).

2 The Cranial Placodes of Vertebrates – An Overview

After providing some general background on the structure of the vertebrate head with its specialized cranial sense and neurosecretory organs (Chapter 1), I will now present a more detailed review of the mechanisms underlying the embryonic development of cranial sensory and endocrine (neurosecretory) organs and the differentiation of sensory and neurosecretory cells in vertebrates with a focus on those cell types and organs arising from cranial placodes. This survey will be a prerequisite for further exploring the evolutionary history of these cell types and the evolutionary origin of novel vertebrate sense and neurosecretory organs in the second volume (Schlosser 2021).

As I have described briefly in Chapter 1, all vertebrate cranial sense organs with exception of the retina, the taste buds and cutaneous mechanoreceptors develop from a series of embryonic ectodermal thickenings, the cranial placodes. The anterior pituitary – the major endocrine control organ of the vertebrate body – is equally placode derived. I will focus in this first volume on summarizing the development of sensory and endocrine organs derived from cranial placodes (for further review see Baker and Bronner-Fraser 2001; Schlosser 2006; Streit 2007; Park and Saint-Jeannet 2010; Schlosser 2010; Grocott, Tambalo, and Streit 2012; Schlosser 2014; Saint-Jeannet and Moody 2014; Moody and LaMantia 2015; Singh and Groves 2016; Aguillon, Blader, and Batut 2016; Streit 2018). However, I will also briefly consider the differentiation of photoreceptors because of their evolutionary relatedness to some of the placodal cell types.

In this chapter, I will give an overview of the various cranial placodes and their derivatives (Figs. 2.1, 2.2). In Chapter 3, I will then present evidence that all cranial placodes develop from a common embryonic primordium. I will also review recent studies elucidating the gene regulatory networks (GRNs) that control how this common primordium is established in the early embryo. In Chapter 4, I will then address how individual placodes segregate from the common primordium. Finally, I will discuss how the various placodally derived sensory and neurosecretory cell types as well as photoreceptors differentiate (Chapters 5–8).

Before embarking on this long journey, however, I need to briefly explain what I consider to be a "cell type". This will then be discussed more thoroughly in Chapter 2 of the second volume (Schlosser 2021).

2.1 A BRIEF PRIMER ON CELL TYPES

Traditional definitions of cell types, which are based on similarities in morphology, physiology, or gene expression, are of limited use for comparing cell types between species, because they do not allow to distinguish cell types that are related

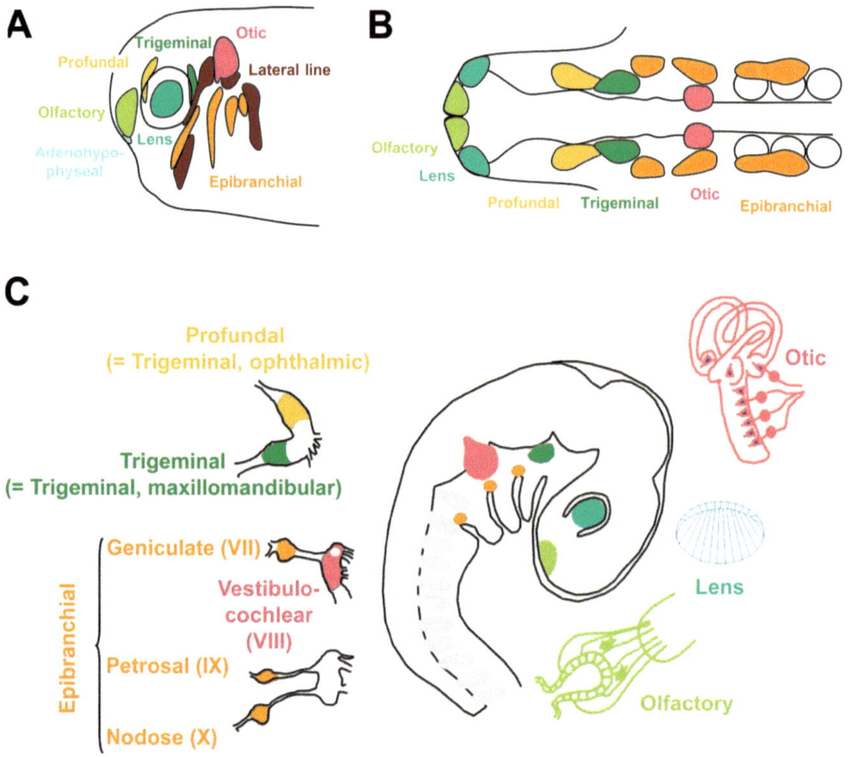

FIGURE 2.1 Cranial placodes in vertebrates. (**A**) Tailbud stage *Xenopus* embryo in lateral view. (**B**) 10–13 Somite stage chick embryo in dorsal view. In amniotes, profundal and trigeminal placode are often referred to as ophthalmic and maxillomandibular placode of the trigeminal nerve, respectively. (**C**) Placodal derivatives in a 3-day-old chick embryo. The color coded placodes of a chick embryo (lateral view) give rise to the ganglia of cranial nerves (left side) and sense organs (right side) shown in the same color. ([A] Modified from Schlosser and Northcutt 2000; [B] Modified from Streit 2004; based on D'Amico-Martel and Noden 1983; Bhattacharyya et al. 2004; [C] Modified from Streit 2008, adapted from Webb and Noden 1993. See text for details.)

by common ancestry (homologous cell types) from those that independently acquired similar phenotypes in evolution. Since one of the main aim of these two volumes is to trace the evolutionary lineages of cells, I will therefore adopt a different, evolutionary concept of cell types according to which a cell type is "a set of cells in an organism that change in evolution together, partially independent of other cells, and are evolutionarily more closely related to each other than to other cells" (Arendt et al. 2016).

The identity of such cell types is typically defined by a core regulatory network (CoRN) of genes, often genes encoding transcription factors. When a cell type changes and acquires different adaptations during evolution, its CoRN typically remains the same, while downstream genes change, allowing to recognize the same (homologous) cell type in different species.

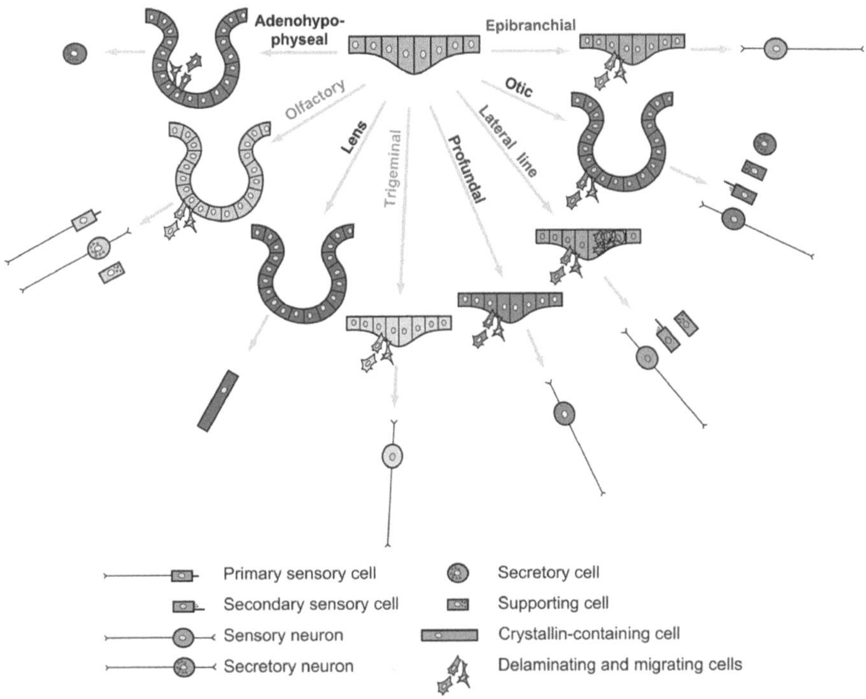

FIGURE 2.2 Schematic overview of cranial placode development in vertebrates. Cellular derivatives and morphogenetic movements (invagination, delamination, and migration) of various cranial placodes are depicted. (Modified from Schlosser 2005.)

CoRNs usually include so-called "terminal selector genes" (Hobert 2008, 2011), which control the expression of large batteries of genes encoding the proteins responsible for the differentiation of the cell, i.e. its adoption of a specific phenotype. However, terminal selectors may specify different cell types when activated in different lineages of cells. This is due to the fact that different lineages of cells (e.g. mesodermal versus ectodermal cells) differ from each other with respect to their so-called competence to activate one or the other set of differentiation genes. This competence is determined by some of the transcription factors they already express (which may be required as co-factors of terminal selectors for the activation of differentiation genes) and by the regions of their DNA that are located in non-condensed chromatin and are, thus, accessible for transcriptional activation. The distribution of accessible chromatin regions, in turn, depends on the prior activity of pioneer factors. The latter are transcription factors capable of attaching to their DNA binding sites even when methylated (repressed) and located in condensed chromatin, which then make these regions accessible for other transcription factors (Cirillo et al. 2002; Zaret and Carroll 2011; Iwafuchi-Doi and Zaret 2016; Iwafuchi-Doi 2019). Thus, cell type identity is specified by terminal selectors in combination with pioneer and other competence factors (for more detail see Chapters 2 and 5 in Schlosser 2021).

2.2 CRANIAL PLACODES: DEVELOPMENTAL ORIGIN AND DERIVATIVE CELL TYPES

2.2.1 Adenohypophyseal Placode

The adenohypophyseal placode forms as an unpaired structure in the rostral-most ectoderm of the embryo just anterior to the neural plate (Figs. 2.1 and 2.2) (Couly and Le Douarin 1985; Eagleson, Jenks, and Van Overbeeke 1986; Kawamura and Kikuyama 1992). In the embryos of many gnathostomes including amniotes it then moves ventrally and becomes part of the roof of the stomodeum, which later will form the mouth cavity (Fig. 1.11). It then invaginates from the roof of the stomodeum to form Rathke's pouch, which becomes closely attached to the infundibulum, a ventral evagination of the hypothalamus (Kouki et al. 2001; Sasaki et al. 2003; Sánchez-Arrones et al. 2015). Whereas the infundibulum will subsequently develop into the neurohypophysis or posterior pituitary, the pouch forms the primordium of the adenohypophysis or anterior pituitary. A recent genetic lineage tracing study in mice has suggested that neural crest cells invade the primordium of the anterior pituitary and contribute to all of its neurosecretory cell types (Ueharu et al. 2017). However, this finding requires confirmation by proper lineage tracing studies using tissue grafting or dye labeling, since genetic lineage tracing studies are prone to errors due to leaky expression of reporter genes (see discussion below).

In many stem and some extant gnathostomes (some teleosts, larval amphibians), the adenohypophysis remains connected with the stomodeum by the buccohypophyseal duct (Khonsari et al. 2013). However, in most gnathostomes, the adenohypophysis detaches completely from the mouth cavity and the duct degenerates. The residual lumen of the pouch separates the so-called intermediary lobe of the pituitary, which is closely apposed to the neurohypophysis, from the anterior lobe (Fig. 1.11A). The proliferative progenitor cells of the adenohypophysis surround this residual lumen and produce the six neurosecretory cell types of the adenohypophysis (hormones produced in parentheses), the corticotropes (ACTH), melanotropes (MSH), gonadotropes (LH, FSH), thyrotropes (TSH), lactotropes (prolactin), and somatotropes (growth hormone) (Fig. 1.11 B) (Dasen and Rosenfeld 2001; Zhu et al. 2007; Kelberman et al. 2009; Pogoda and Hammerschmidt 2009; Davis et al. 2013, Rizzoti 2015). The lineage decisions between these different cell types are controlled by multiple transcription factors as will be discussed in Chapter 8. There is no evidence for regular cell turnover in the adult adenohypophysis. However, recent evidence suggests that a few quiescent stem cells are maintained into adult stages conferring limited regenerative potential to the mammalian pituitary (Fauquier et al. 2008; Andoniadou et al. 2013; Rizzoti, Akiyama, and Lovell-Badge 2013; Ho et al. 2020).

In some groups of gnathostomes, including amphibians and teleosts, the adenohypophyseal placode develops differently without forming a pouch but ultimately gives rise to the same cell types. In amphibians, the placode forms a solid epithelial rod growing inwards just dorsal to the stomodeum (Eagleson et al. 1986; Kawamura and Kikuyama 1992; Dickinson and Sive 2007), while in teleosts prospective adenohypophyseal and stomodeal cells are initially intermingled and become separated from each other by delamination of adenohypophyseal cells prior to and during invagination of the stomodeum (Herzog et al. 2003; Chapman et al. 2005).

In the extant jawless vertebrates, no distinct adenohypophyseal placode forms. Instead, the anterior pituitary arises together with the olfactory epithelia from a common nasohypophyseal placode. Although, a previous study suggested that the adenohypophysis of hagfishes is endodermally derived (Gorbman 1983), a recent study has demonstrated that it develops from an ectodermal nasohypophyseal placode, similar to lampreys (Oisi et al. 2013). The nasohypophyseal placode forms rostral to the stomodeum and becomes separated from it by the developing upper lips in lampreys or the oronasohypophyseal septum in hagfishes (Uchida et al. 2003; Oisi et al. 2013). The nasohypophyseal placode invaginates and remains connected to the outside world by a single opening (monorhiny). The placode subsequently gives rise to a distinct adenohypophysis from its posterior part with four types of neurosecretory cells (Kawauchi and Sower 2006). Its anterior part develops into an olfactory epithelium, which is partly separated into a left and a right domain and gives rise to paired olfactory nerves (Zaidi et al. 1998).

2.2.2 Olfactory Placode

In most gnathostomes, the olfactory placode invaginates incompletely to form the nasal cavity with respiratory and olfactory epithelia (Figs. 1.10, 2.1, 2.2) (reviewed in Brunjes and Frazier 1986; Balmer and LaMantia 2005; Maier et al. 2014; Moody and LaMantia 2015; Sokpor et al. 2018). In many gnathostomes with a bilayered ectoderm (many fishes and amphibians), the superficial and deep ectodermal layers interdigitate during this process, and the superficial layer contributes to supporting cells (Klein and Graziadei 1983; Zeiske et al. 2003). However, in teleosts, all cell types of the respiratory and olfactory epithelia develop from the deep ectodermal layer and the olfactory pit develops by cell rearrangements rather than by invagination (Hansen and Zeiske 1993). In tetrapods and their closest relatives, the lungfishes, the olfactory placode also gives rise to the vomeronasal epithelium (Figs. 1.10, 3.4B) (Gonzalez et al. 2010). The latter is distinct and spatially segregated from the olfactory epithelium and appears to be specialized for contact-dependent detection of large molecules including many pheromones (reviewed in Breer, Fleischer, and Strotmann 2006; Grus and Zhang 2009; Silva and Antunes 2017).

These epithelia contain many different cell types including the mucus producing cells of the Bowman glands, the sustentacular or supporting cells, and the primary sensory cells (possessing an axon) (e.g. Klein and Graziadei 1983; Couly and Le Douarin 1985; Hansen and Zeiske 1993; Zeiske et al. 2003). The olfactory and vomeronasal sensory cells carry transmembrane receptors of the odorant and vomeronasal receptor families, respectively, on their surface (see Chapter 6) (Buck 2004; Touhara and Vosshall 2009; Kaupp 2010). Their axons project directly into the olfactory bulb of the forebrain, forming the olfactory and vomeronasal nerves.

The olfactory placode also gives rise to several types of cells releasing neuropeptides, such as gonadotropin releasing hormone (GnRH), neuropeptide Y and FMRFamide (Schwanzel-Fukuda and Pfaff 1989; Wray, Grant, and Gainer 1989; el Amraoui and Dubois 1993; Murakami and Arai 1994; Northcutt and Muske 1994,

Yamamoto et al. 1996; reviewed in Somoza et al. 2002; Whitlock et al. 2006; Schwarting, Wierman, and Tobet 2007; Wray 2010; Forni and Wray 2015). These cells migrate into the forebrain along the olfactory and vomeronasal nerves. One population of migratory cells continues to migrate caudally in the forebrain before settling in the septo-preoptic nuclei of the hypothalamus. Their axons subsequently secrete GnRH into the portal blood vessels, which supply the adenohypophysis, thereby triggering release of the adenohypopyseal hormones luteinizing hormone (LH) and follicle-stimulating hormone (FSH) (Fig. 1.11B). The latter, in turn stimulate the gonads and are therefore also known as gonadotropic hormones. Another population of migratory cells forms the ganglia and axons of a distinct nerve, the so-called terminal nerve (see Chapter 6; reviewed in Muske and Moore 1988; Demski 1993; Von Bartheld 2004). The association of olfactory deficits (anosmia) with reduced gonad size and infertility (hypogonadism) in human patients suffering from Kallmann syndrome can be explained by perturbations of olfactory placode development affecting both GnRH expressing and olfactory receptor cells (Whitlock et al. 2006; Forni and Wray 2015).

Based on lineage tracing experiments in zebrafish, the origin of GnRH cells from the olfactory placode has recently been questioned. Instead, it was proposed that septo-preoptic and terminal GnRH cells arise from adenohypophyseal placode and neural crest, respectively (Whitlock, Wolf, and Boyce 2003, Whitlock et al. 2005). However, all of these cell populations are closely apposed in early zebrafish embryos and it cannot be ruled out that the olfactory placode was inadvertently labeled in these experiments. Indeed, a recent video-microscopical study, which backtracked the terminal nerve population of GnRH cells to their region of origin concluded that all of these GnRH cells are derived from the anterior placodal area of the pre-placodal ectoderm, from which the olfactory placode develops (Aguillon et al. 2018). This suggests that the neural crest does not contribute to GnRH cells in teleosts.

Nevertheless, the question, whether neural crest cells contribute to the population of GnRH cells is still contentious due to conflicting studies in tetrapods. Several recent studies in mice, which used genetic lineage tracing of neural crest cells (analyzing reporter genes under the control of putatively neural crest-specific enhancers), reported that a subpopulation of GnRH cells is neural crest derived (Forni et al. 2011). In contrast, a study in chick embryos using tissue grafting concluded that all GnRH cells originate in the olfactory placode (Sabado et al. 2012). The discrepancy between these studies may reflect species-specific differences. More likely, however, it may result from leaky placodal expression of the neural crest reporter genes used in mice, a notorious problem of genetic lineage tracing studies. To rule out such artefacts, findings of genetic lineage tracing studies should always be corroborated by true lineage tracing using tissue grafting or dye labeling. Genetic lineage tracing studies in mice and zebrafish have also concluded that a subset of progenitor cells and olfactory receptor cells in the olfactory epithelium are neural crest derived (Katoh et al. 2011; Forni et al. 2011; Saxena, Peng, and Bronner 2013; Suzuki et al. 2013). However, careful backtracking of olfactory receptor cells, expressing the putatively neural crest-specific reporter construct *Tg(-4.9sox10:eGFP)* in zebrafish demonstrated that all of these cells were derived from the anterior placodal area (Aguillon et al. 2018).

The olfactory placode was long thought to give rise to another population of migratory cells, which will form the glial cell of the olfactory nerve, the so-called olfactory ensheathing cells (Couly and Le Douarin 1985; Chuah and Au 1991; Norgren, Ratner, and Brackenbury 1992). However, lineage tracing experiments in chick and mouse have now clearly shown that olfactory ensheathing cells are neural crest derived like all other glial cells in the peripheral nervous system (Barraud et al. 2010; Forni et al. 2011).

In contrast to most other placodes, the olfactory placode retains stem and progenitor cells (horizontal and globose basal cells) throughout life, which allow tissue turnover and regeneration throughout vertebrates including mammals (Beites et al. 2005; Leung, Coulombe, and Reed 2007; Sokpor et al. 2018).

2.2.3 Lens Placode

In most vertebrates the lens placode invaginates to form a lens vesicle, which then forms the lens of the eye (Figs. 1.5, 2.1, 2.2) (reviewed in Chow and Lang 2001; Lovicu and McAvoy 2005; Cvekl and Duncan 2007; Medina-Martinez and Jamrich 2007; Kondoh 2008; Cvekl and Zhang 2017). However, in teleosts the lens forms differently by delamination of cells (Greiling and Clark 2012). Whereas the outer cells of the lens vesicle form the lens epithelium, the inner cells, which face the retina, differentiate into the primary lens fiber cells. This process involves elimination of cellular organelles and nuclei, accumulation of large amounts of crystallin proteins, which makes the lens transparent, and elongation of the lens fiber cells, which thereby fill the vesicle (see Chapter 8). The lens epithelium remains proliferative and cells of the epithelium that are displaced to the equator will differentiate into secondary lens fiber cells allowing the growth of the lens throughout life.

2.2.4 Profundal and Trigeminal Placode

In gnathostomes, cells delaminate from the profundal and trigeminal placodes and together with neural crest cells form the ganglia of the profundal and trigeminal nerves, which innervate the anterior part of the head and the first pharyngeal arch respectively (Figs. 1.12, 2.1, 2.2). Whereas the neural crest contributes both glial cells and a subset of general somatosensory neurons to these ganglia, the placodes only contribute another subset of general somatosensory neurons (Knouff 1927; Hamburger 1961; Noden 1980a, 1980b; Ayer-LeLièvre and Le Douarin 1982; D'Amico-Martel and Noden 1983; Schlosser and Northcutt 2000; Begbie, Ballivet, and Graham 2002).

The ganglia of the profundal and trigeminal nerves, which remain separate in many fishes (e.g. Allis 1897; Norris and Hughes 1920; Norris 1925; Song and Northcutt 1991; Northcutt and Bemis 1993; Northcutt 1993b; Piotrowski and Northcutt 1996) fuse in amphibians and amniotes (Hamburger 1961; Ayer-LeLièvre and Le Douarin 1982; D'Amico-Martel and Noden 1983; Northcutt and Brändle 1995; Schlosser and Roth 1997; Begbie et al. 2002). In amniotes the fused ganglion is known as the trigeminal (or Gasserian) ganglion. Based on gene expression and innervation pattern, the ophthalmic and maxillomandibular subdivisions of this ganglion, which develop by the fusion of neural crest cells with cells derived from ophthalmic and maxillomandibular placodes, are most likely homologous to the profundal and trigeminal placodes of other gnathostomes (see discussion in Schlosser and Northcutt 2000; Schlosser and Ahrens 2004). In lampreys, which also have a distinct profundal and trigeminal nerve, a single placode contributes to the profundal ganglion, but two placodes contribute to the trigeminal ganglion – these placodes contribute neurons to those parts of the ganglion innervating the upper and lower lip, respectively (Kuratani et al. 1997; Modrell et al. 2014).

The profundal and trigeminal placodes are the only placodes giving rise to general somatosensory neurons, which are exclusively neural crest derived in other cranial ganglia and the dorsal root ganglia (see below; D'Amico-Martel and Noden 1983; Teillet, Kalcheim, and Le Douarin 1987). Both the neural crest and the placode derived general somatosensory neurons of the profundal and trigeminal nerves mediate touch, temperature, and pain sensation from the mouth cavity and the rostral part of the face (Darian-Smith 1973; Butler and Hodos 2005). Placodally derived trigeminal neurons appear to be larger, develop earlier, and occupy the more distal part of the profundal and trigeminal ganglia compared to the neural crest derived neurons (Hamburger 1961; Noden 1980a, 1980b; D'Amico-Martel and Noden 1980, 1983). Since mechanoreceptive neurons tend to be larger than nociceptors (Marmigere and Ernfors 2007), this suggests that there may be some bias for placodes and neural crest to contribute preferentially to mechanoreceptors and nociceptors, respectively. However, so far no clear-cut difference in function or gene expression between placode- and neural crest-derived populations could be detected. Instead, several studies suggest that gene expression profiles of neural crest derived neurons in the dorsal root ganglia are very similar to the neurons in the mixed profundal/trigeminal ganglion, suggesting that the same type of general somatosensory neuron can be generated from either profundal/trigeminal placodes or neural crest (Eng et al. 2007; Dykes et al. 2010; Lopes, Denk, and McMahon 2017).

2.2.5 Otic Placode

In most vertebrates the otic placode invaginates to form the otic vesicle (Figs. 1.3 A, 2.1, 2.2) (reviewed in Fritzsch, Barald, and Lomax 1998; Barald and Kelley 2004; Bok, Chang, and Wu 2007; Whitfield 2015; Sai and Ladher 2015). In teleosts, the otic vesicle forms by a slightly different process involving cavitation (Haddon and Lewis 1996). The otic vesicle gives rise to the entire inner ear and the sensory neurons transmitting information from the ear to the brain (Li, Van De Water, and Ruben 1978; Kil and Collazo 2001; Bell et al. 2008; Sánchez-Guardado, Puelles, and Hidalgo-Sánchez 2014; reviewed in Fekete and Wu 2002; Bok et al. 2007; Alsina, Giráldez, and Pujades 2009; Magarinos et al. 2012; Wu and Kelley 2012; Maier et al. 2014; Whitfield 2015).

Sensory neurons are generated very early soon after the otic vesicle has formed, migrate out of the vesicle and coalesce with neural crest cells to form the vestibulocochlear ganglion (see Chapter 6). The neural crest cells subsequently will differentiate into glial cells, while the placodally derived cells will differentiate into the special somatosensory neurons (D'Amico-Martel and Noden 1983; Breuskin et al. 2010;

Sandell et al. 2014). The latter will innervate the hair cells, which develop from the otic vesicle after the emergence of the sensory neurons (see Chapter 6). As discussed in Chapter 1, these hair cells are concentrated in various sensory areas committed to the detection of gravity, angular acceleration, and sound, which vary between different groups of vertebrates (Figs. 1.8, 3.4A) (reviewed in Müller and Littlewood-Evans 2001; Gao 2003; Frolenkov et al. 2004; Groves, Zhang, and Fekete 2013). Apart from the hair cells of the various sensory areas, the otic vesicle also gives rise to all other cell types of the inner ear including various types of supporting cells and the basal and marginal cells of the endolymph-producing stria vascularis (reviewed in Kelly and Chen 2009).

The neural crest has long been known to contribute melanocyte-like intermediate cells to the stria vascularis, which migrate in between the placode-derived basal and marginal cells (Hilding and Ginzberg 1977; Steel and Barkway 1989). A recent genetic lineage tracing study in mice reported that neural crest cells also contribute to hair cells, supporting cells, and sensory neurons derived from the otic placode (Freyer, Aggarwal, and Morrow 2011). However, the possibility of artefactual labeling of placodally derived otic cells by the transgenic lines used in this study must be seriously considered (see discussion above), since extensive fate mapping studies of the neural crest in chick and other vertebrate embryos conducted over several decades fail to support any major neural crest contribution to these cell types (D'Amico-Martel and Noden 1983; Sadaghiani and Thiebaud 1987; Le Douarin and Kalcheim 1999; Sandell et al. 2014).

2.2.6 Lateral Line Placodes

Comparative analyses suggest that six lateral line placodes were probably primitively present in gnathostomes (Fig. 1.3B) (Northcutt 1992, 1993a, 1993b, 1997). However, the number of lateral line placodes or some of their derivative receptor organs has been reduced in several gnathostome lineages and they have been completely lost in amniotes (Fritzsch 1989; Northcutt 1992, 1997; Schlosser 2002b). A recent study showed, however, that neuromast-like structures can form in mammalian embryos when apoptosis is blocked, suggesting latent retention of the developmental program for lateral line placode development in mammals (Washausen and Knabe 2018). Lateral line placodes in lampreys have not been described in detail but based on the distribution of lateral line organs and nerves, there may also be six lateral line placodes, like in some gnathostomes (Northcutt 1989). However, in hagfishes the lateral line system is either partly or completely reduced (Braun 1996; Braun and Northcutt 1997) probably in association with their burrowing life style.

Similar to the otic placode, the lateral line placodes present in most fishes and amphibians give rise to both the axonless sensory receptors of the lateral line system and to the sensory neurons innervating them (Figs. 2.1, 2.2) (see Chapter 6) (Stone 1922; Northcutt, Catania, and Criley 1994; Northcutt, Brändle, and Fritzsch 1995; Modrell et al. 2011; reviewed in Winklbauer 1989; Northcutt 1992; Smith 1996; Northcutt 1997, Schlosser 2002a; Ghysen and Dambly-Chaudière 2004, 2007; Gibbs 2004; Ma and Raible 2009; Piotrowski and Baker 2014). Sensory neurons are generated early, delaminate from the lateral line placodes and intermingle with gliogenic (glia-forming) neural crest cells to form the ganglia of the lateral line nerves (Fig. 1.7). The remaining placodes become lateral line primordia. These elongate or migrate along defined pathways in the head and trunk, regularly depositing clusters of cells (Fig. 1.7). These subsequently will develop into the hair cells and supporting cells of mechanosensory receptor organs (neuromasts) of the lateral line (Flock 1967; Russell 1976; Blaxter 1987; Gibbs 2004; Coffin et al. 2005). In many vertebrates, the primordia will also give rise to electrosensitive receptor organs (ampullary organs) composed of electroreceptor cells (possibly modified hair cells) and supporting cells (Fig. 1.7) (reviewed in Baker, Modrell, and Gillis 2013). Whether the secondarily re-evolved electroreceptors of some teleosts (tuberous or ampullary organs) are also derived from lateral line placodes has still to be confirmed (Baker et al. 2013). The lateral line placodes of lampreys also give rise to neuromasts as well as to special electrosensory "end buds" (Ronan and Bodznick 1986) and to solitary microvillous cells, which are light sensitive and probably serve as extraocular photoreceptors (Whitear and Lane 1983; Ronan and Bodznick 1991).

Although the neural crest has also been claimed to contribute to neuromasts and ampullary organs (Collazo, Fraser, and Mabee 1994; Freitas et al. 2006), there are no conclusive experimental data to support this. The proposed contribution of neural crest to neuromasts is based on a fate-mapping study in zebrafish (Collazo et al. 1994), which could not rule out that the neural crest contribution was limited to glial cells that have independently been shown to be neural crest derived (Gilmour, Maischein, and Nüsslein-Volhard 2002). The proposed contribution of neural crest to ampullary organs (Freitas et al. 2006) is merely based on their expression of putative neural crest markers (Sox8, HNK1), which are however not neural crest-specific (see Schlosser 2002a). A neural crest contribution to lateral line organs could also not be verified in studies using different neural crest reporter lines in zebrafish embryos (e.g. Gilmour et al. 2002), suggesting that lateral line organs are exclusively placode derived.

2.2.7 Epibranchial Placodes

Epibranchial placodes get their name from their location immediately dorsal to the pharyngeal pouches (Figs. 2.1, 2.2). The number of epibranchial placodes in different vertebrates, thus, varies with the number of pharyngeal pouches with seven or more pouches developing in cartilaginous fishes and cyclostomes but only four to five in amniotes (Graham 2008; O'Neill et al. 2012; Frisdal and Trainor 2014; Modrell et al. 2014). Cells delaminating from these placodes intermingle with neural crest cells to form the distal ganglia of the facial (geniculate ganglion), glossopharyngeal (petrosal ganglion), and vagal nerves (nodose ganglia) (Fig. 1.12) (Yntema 1937, 1943, 1944; Narayanan and Narayanan 1980; Ayer-LeLièvre and Le Douarin 1982; D'Amico-Martel and Noden 1983; Couly and Le Douarin 1990; reviewed in Northcutt 2004; Krimm 2007; Harlow and Barlow 2007; Baker, O'Neill, and McCole 2008; Ladher, O'Neill, and Begbie 2010). The geniculate and petrosal ganglia arise from a single placode each, associated with the first and second pharyngeal pouch, respectively. The vagal ganglia, which often fuse into a single ganglion (nodose ganglion), arise from the remaining epibranchial placodes, associated with the posterior pharyngeal pouches, which vary in number between vertebrates.

The epibranchial placode derived cells in these distal ganglia will differentiate into gustatory (=special viscerosensory) neurons innervating taste buds and into general viscerosensory neurons (see Chapter 6), whereas neural crest cells will develop into glial cells. The taste buds themselves are not derived from any placodes but instead develop from ectodermal and/or endodermal epithelia of the mouth cavity (Barlow and Northcutt 1995; Okubo, Clark, and Hogan 2009; Thirumangalathu et al. 2009). In contrast, both the general somatosensory neurons and the glial cells of the proximal ganglia of the glossopharyngeal and vagal nerves (superior and jugular ganglia, respectively) are exclusively neural crest derived with no placodal contribution.

2.2.8 Other Placodes

In addition to the placodes described so far, which are found in all vertebrates, some placodes with a more narrow phylogenetic distribution have been described. In frog embryos, hypobranchial placodes develop ventral to the pharyngeal pouches and give rise to small hypobranchial ganglia of unknown function (Schlosser and Northcutt 2000; Schlosser 2003). Because hypobranchial placodes appear to develop from the same placodal area as epibranchial placodes, it has been proposed that they may be functionally equivalent to epibranchial placodes (and give rise to viscerosensory neurons) but ventrally displaced due to the intervening pharyngeal pouches (Schlosser and Northcutt 2000). Ventral thickenings that may correspond to hypobranchial placodes, have also been found in mammals, although they subsequently degenerate by apoptosis instead of producing sensory neurons (Washausen et al. 2005).

Another placode, the paratympanic placode, which forms adjacent to the anteriormost epibranchial placode, has recently been described in birds (O'Neill et al. 2012). It gives rise to the paratympanic organ, a pouch in the middle ear, which may detect air pressure, and develops into the hair cell like mechanoreceptors of this organ and the somatosensory neurons innervating it.

3 Origin of Cranial Placodes from a Common Primordium

The brief survey of placode development in Chapter 2 shows that placodes are quite diverse and form different types of sense organs, ganglia, and endocrine glands at different positions in the head. However, underlying this diversity are some important similarities between the development of different placodes. In this chapter I will show that these shared aspects of placode development are due to the origin of all placodes from a common primordium, the so-called pre-placodal ectoderm (PPE). I will then review how the PPE originates in early vertebrate development in parallel to other ectodermal territories such as neural crest, neural plate, and epidermis. I will focus in particular on the role of transcription factors during this process, since these play central roles for the cell fate decisions that step by step convert pluripotent cells into the sensory and neurosecretory cell types that concern us in this book.

3.1 ALL PLACODES DEVELOP FROM A COMMON PRIMORDIUM, THE PRE-PLACODAL ECTODERM (PPE)

3.1.1 Shared Aspects of Placode Development

The shared or generic aspects of placode development include (1) the generation of neurons or sensory cells by most placodes and (2) the importance of cell shape changes and morphogenetic movements (e.g. invagination or cell delamination) during placode development (Fig. 2.2). Neurons, sensory cells or both are formed, among other possible derivatives, from all neurogenic placodes. These comprise all placodes except the adenohypophyseal and lens placode (e.g. Noden 1983; Fode et al. 1998; Ma et al. 1998; Schlosser and Northcutt 2000; Andermann, Ungos, and Raible 2002; Begbie, Ballivet, and Graham 2002). However, both of the latter placodes develop from ectoderm that initially expresses but subsequently downregulates neuronal determination genes such as *Neurog2* (Schlosser and Ahrens 2004). This suggests that the adenohypophyseal and lens placode develop from neurogenic ectoderm by active suppression of neuronal fates.

Apart from being neurogenic, most placodes also undergo cell shape changes and/or morphogenetic movements (reviewed in Webb and Noden 1993, Baker and Bronner-Fraser 2001; Schlosser 2010; Breau and Schneider-Maunoury 2015). This is first evident in the thickened appearance of many placodes due to the formation of elongated epithelial cells (columnar epithelia). The adenohypophyseal, olfactory, otic, and lens placode subsequently invaginate (Figs. 1.3A, 2.2) and this is followed by further complex morphogenetic movements for the olfactory pit and otic vesicle (Figs. 1.8, 2.2). The lateral line placodes form primordia, which elongate or migrate large distances to form the various lateral lines (Figs. 1.7, 2.2). And last but not least, neuronal progenitors delaminate from all of the neurogenic placodes, break through the underlying basal lamina and migrate away before congregating with neural crest cells to form the various cranial ganglia (Fig. 2.2).

These generic aspects of placode development suggest that different placodes rely to some extent on shared developmental programs either due to the independent activation of similar gene regulatory networks or due to their origin from a common developmental precursor. A common origin of different placodes was proposed in many classical embryological studies based on the presence of early ectodermal thickenings at the border of the neural plate, from which different placodes subsequently emerge (e.g. Platt 1894; Landacre 1910; Knouff 1935; Nieuwkoop 1963; Nieuwkoop, Johnen, and Albers 1985; Schilling and Kimmel 1994; Miyake, von Herbing, and Hall 1997; reviewed in Baker and Bronner-Fraser 2001; Schlosser 2002). However, while such thickenings have been described in some vertebrate groups such as teleosts, frogs, and mammals; they are absent in other groups (e.g. some salamanders, birds). Embryological studies of these latter taxa have, therefore, often questioned the existence of a common precursor for all placodes (Stone 1922; Northcutt and Brändle 1995; Graham and Begbie 2000; Begbie and Graham 2001). Moreover, in taxa, where such early thickenings of the neural plate border are present, these typically are observed at cranial and trunk levels and give rise to both neural crest and placodes (Knouff 1935; Schilling and Kimmel 1994; Miyake et al. 1997; Schlosser and Northcutt 2000) rather than to placodes alone. Taken together, this indicates that the presence or absence of thickenings at the neural plate border does not provide a reliable criterion to judge whether there is a common precursor for all cranial placodes or not.

I have argued before that if there is anything like a common precursor for all placodes or a pan-placodal primordium, it must fulfill two conditions (Schlosser 2002, 2005; Schlosser and Ahrens 2004). First, all placodes must develop from a common region of origin or pre-placodal region. And second, this pre-placodal region must be biased toward the development of cranial placodes. This means that it must have a distinct regulatory state conferring an autonomous tendency to develop generic placodal properties.

We still do not understand well, how fate decisions are made in development and, thus, it is not easy to define precisely what a "regulatory state" is. I will use this term to refer to relatively stable, non-transitory states or "attractors" of gene regulatory networks (Graf and Enver 2009) (for more detail see Chapter 2 in Schlosser 2021). Thus defined, a state, in which mutually cross-repressive transcription factors A and B are transitorily co-expressed in a cell would not be considered a regulatory state. However, each of the two possible outcomes of such an interaction, that is, states, in which either A or B but not both are stably maintained would qualify as a regulatory state. Thus, a regulatory state will be typically defined by the competence of a cell (based on the location of genes in accessible chromatin and the expression of competence factors – transcription factors required relatively unspecifically for multiple fates) together with the stable expression of a combination of specific transcription factors and cofactors. When cells make fate decisions in development, they may first pass through more general regulatory states with broader competence (e.g. ectoderm), before adopting more specific fates (e.g. neuron).

In the following sections, I will first discuss evidence from fate mapping studies that indeed all cranial placodes originate from a common precursor, the PPE. I will then proceed to show that the PPE is a region with a distinct regulatory state conferring placodal bias and is defined by the expression of transcription factors of the Six1/2 and Six4/5 family and their cofactors belonging to the Eya family. These transcriptional regulators promote shared programs of neurogenesis, morphogenesis and possibly other developmental processes in all placodes.

3.1.2 Origin of all Cranial Placodes from a Common Region

The origin of placodes has been studied in many vertebrates using tissue transplantations or applications of dyes. This allowed to generate fate maps of gastrula or neurula stage embryos, which identify the regions of ectoderm that subsequently give rise to placodes in teleosts (Kozlowski et al. 1997; Whitlock and Westerfield 2000; Dutta et al. 2005), amphibians (e.g. Vogt 1929; Röhlich 1931; Carpenter 1937; Fautrez 1942; Jacobson 1959; Eagleson and Harris 1990; Eagleson, Ferreiro, and Harris 1995; Pieper et al. 2011; Steventon, Mayor, and Streit 2012) and amniotes (Couly and Le Douarin 1985; Couly and Le Douarin 1987; Couly and Le Douarin 1990; Streit 2002; Bhattacharyya et al. 2004; Xu, Dude, and Baker 2008; McCabe, Sechrist, and Bronner-Fraser 2009; Sánchez-Guardado, Puelles, and Hidalgo-Sánchez 2014; Sánchez-Arrones et al. 2017). In all species, a common region of origin for all cranial placodes was identified. This PPE comprises a crescent-shaped region in the outer neural folds and the adjacent ectoderm, which is immediately juxtaposed to the anterior neural plate rostrally and to the cranial neural crest laterally (Figs. 3.1, 3.2).

At gastrula and early neurula stages, the regions of origin of different placodes within the PPE were found to overlap extensively in many of these fate mapping studies and resolve into distinct placodal territories only later. However, due to methodological limitations the degree of overlap is necessarily overestimated in these fate mapping studies and a detailed analysis in *Xenopus* suggested that territories of origin for individual placodes may be segregated within the PPE earlier than fate maps suggest (Pieper et al. 2011).

3.1.3 The Pre-Placodal Ectoderm is Defined by a Ground State Conferring Bias for Placode Development

Classical embryological studies in amphibians revealed that pieces of ectoderm taken from different parts of the embryo and grafted into the PPE region differed in their capacity to develop into placodes or placode-derived structures. For example, ectoderm was more likely to adopt an otic fate when grafted into the region of the PPE fated to develop into otic placode when taken from other parts of the PPE than when taken from outside the PPE (Yntema 1933; Ikeda 1937, 1938; Haggis 1956;

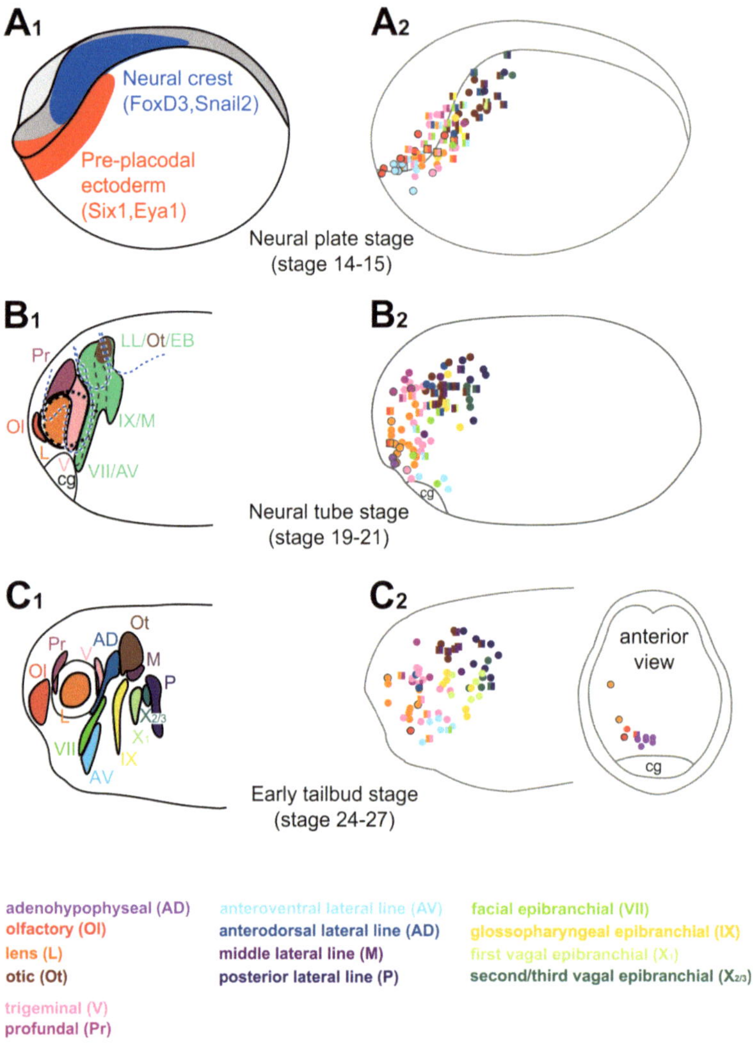

FIGURE 3.1 Development of cranial placodes in *Xenopus*. Development of placodal territories (**A_1-C_1**) and fatemaps (**A_2-C_2**) are shown in lateral views of neural plate stage (**A**), early neural tube stage (**B**), and early tailbud stage embryos (**C**) of *Xenopus laevis*. An anterior view is also shown for the fate map of an early tailbud stage embryo. (See text for details. [A_1-C_1] modified from Schlosser and Ahrens, 2004. [A_2-C_2] modified from Pieper et al. 2011.)

Reyer 1958; Jacobson 1963; Henry and Grainger 1987; Gallagher, Henry, and Grainger 1996; Grainger et al. 1997). Furthermore, experimental mis-expression of various transcription factors can induce supernumerary lens placodes only in cranial ectoderm (which includes the PPE) but not at trunk levels although supernumerary otic placodes can sometimes be induced in the dorsal trunk (Oliver et al. 1996;

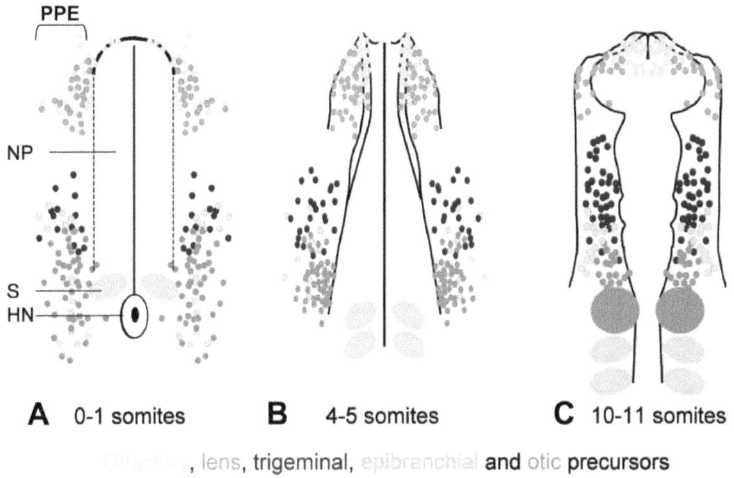

FIGURE 3.2 Development of cranial placodes in the chick embryo. Fate maps of chick embryos at neural plate (**A**), neural fold (**B**) and neural tube stages (**C**). HN, Hensen's node; NP, neural plate; PPE, pre-placodal ectoderm; S, somite. (Reprinted with modification with permission from Streit 2008).

Altmann et al. 1997; Chow et al. 1999; Loosli, Winkler, and Wittbrodt 1999; Köster, Kühnlein, and Wittbrodt 2000; Lagutin et al. 2001; Solomon et al. 2003; Bailey et al. 2006). Recently, a study in chick embryo showed that only the PPE can be induced to express otic markers after treatment with fibroblast growth factors (FGFs) in vitro (Martin and Groves 2006). Other regions of ectoderm only acquired this capacity after transplantation into the PPE region and upregulation of the PPE markers Eya2 and Dlx (as revealed with a pan-Dlx antibody; Martin and Groves 2006). Taken together, these studies suggest that the induction of placodes is a multi-step process and that the PPE is biased toward placodal development because it adopts a pan-placodal ground state, which is a prerequisite for the subsequent induction of various individual placodes.

Experiments in the chick embryo further show that explanted pieces of the PPE continue to express PPE markers (Six1, Eya2), upregulate the anterior placodal marker Pax6 and differentiate as lens regardless of whether these pieces were taken from anterior or posterior parts of the PPE (Bailey et al. 2006). This corroborates that the entire PPE shares a common ground state and suggest that it is initially specified for anterior placodal fates.

The common regulatory state of the PPE is expected to be mediated by transcription factors (or combinations thereof) that are specifically expressed in the PPE. As will be discussed in more detail below, there are in fact many ectodermal transcription factors that overlap in their expression with the PPE or are expressed in some of its parts. However, only very few transcription factors are expressed throughout the entire PPE without extending into adjacent territories (reviewed in Schlosser 2006, 2010; Streit 2007, 2018; Saint-Jeannet and Moody 2014; Moody and LaMantia 2015).

Comparisons between gene expression domains and placodal fate maps so far indicate that only transcription factors of the Six1/2 and Six4/5 family (Oliver et al. 1995; Seo, Drivenes, and Fjose 1998; Ohto et al. 1998; Esteve and Bovolenta 1999; Kobayashi et al. 2000; Pandur and Moody 2000; Ghanbari et al. 2001; Ozaki et al. 2001 Ozaki et al. 2004; Laclef, Souil et al. 2003; Li et al. 2003; Zheng et al. 2003; McLarren, Litsiou, and Streit 2003; Schlosser and Ahrens 2004; Bessarab, Chong, and Korzh 2004; Zou et al. 2004; Pieper et al. 2011) and their cofactors belonging to the Eya family (Abdelhak et al. 1997; Xu et al. 1997, 1999, 2002; Duncan et al. 1997; Kalatzis et al. 1998; Mishima and Tomarev 1998; Sahly, Andermann, and Petit 1999; David et al. 2001; McLarren et al. 2003; Schlosser and Ahrens 2004; Zou et al. 2004) are specifically expressed in the PPE and subsequently continue to be expressed in most cranial placodes (Fig. 3.3). Although several transcription factors are enriched in the PPE, no further transcription factors that are exclusively expressed in the PPE could be identified in recent screens in chick and *Xenopus* embryos (Hintze et al. 2017; Plouhinec et al. 2017). This suggests that Six1/2, Six4/5, and their Eya cofactors indeed play a unique role for defining a common regulatory state (or bias) for the PPE.

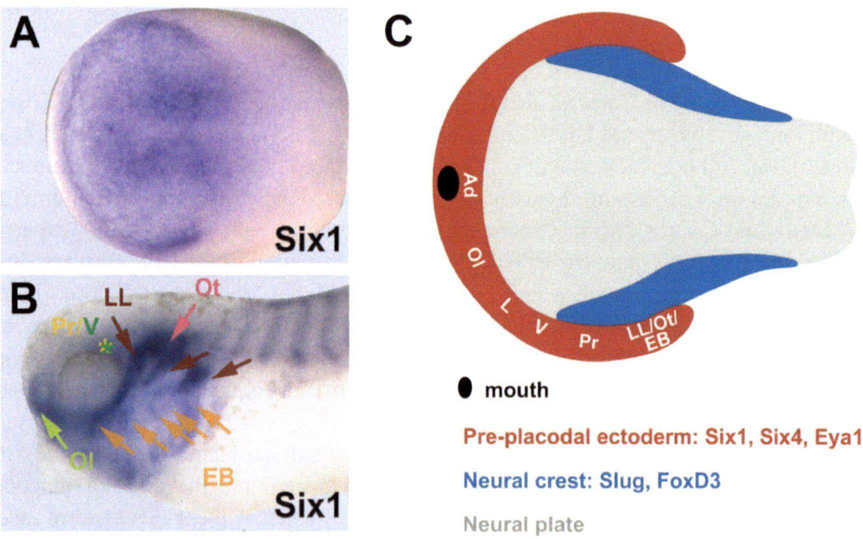

FIGURE 3.3 Origin of placodes from the pre-placodal ectoderm. Six1, Six4, and Eya1 define the pre-placodal ectoderm as illustrated for *Xenopus*. (**A, B**) *Six1* mRNA expression as revealed by in situ hybridization. **A**: *Six1* is first expressed in a crescent-shaped domain around the neural plate. (**B**) At tailbud stages, *Six1* expression continues in all placodes. The position of the fused profundal and trigeminal ganglia is indicated by a double asterisk (the respective placodes are already disappearing at this stage). (**C**) Position of the pre-placodal ectoderm (red) relative to the neural plate (gray) and neural crest (blue). Ad, adenohypophyseal placode; EB, epibranchial placodes; LL, lateral line placodes; Ol, olfactory placode; Ot, otic placode; Pr, profundal placode; V, trigeminal placode. (Reprinted with modification from Schlosser 2006).

3.1.4 Role of Six1/2, Six4/5 and Eya in Pre-Placodal Ectoderm and Placodes

Six proteins are transcription factors that bind DNA with their homeodomain, while interacting with other proteins with their Six domain (Kawakami et al. 2000; Silver and Rebay 2005; Christensen et al. 2008; Kumar 2009). These protein–protein interactions may involve binding of coactivators such as Eya proteins or corepressors such as Groucho proteins (Pignoni et al. 1997; Ohto et al. 1999; Kobayashi et al. 2001; Ikeda et al. 2002; Zhu et al. 2002; López-Rios et al. 2003; Silver et al. 2003; Brugmann et al. 2004; Weasner, Salzer, and Kumar 2007; Patrick et al. 2013). Depending on the cofactors involved, Six proteins may act either as a transcriptional activator or repressor. There are six *Six* genes in vertebrates. These belong to three subfamilies with two members each. Only members of the *Six1/2* and *Six4/5* subfamilies are widely expressed in the PPE, whereas *Six3/6* subfamily members are confined to the anterior part of the PPE and the adjacent neural plate and are involved in the development of anterior placodes and forebrain (Zuber et al. 2003; Cvekl and Duncan 2007; Schlosser 2006; Kumar 2009).

Eya proteins are multifunctional proteins characterized by a N-terminal transactivation domain and an C-terminal Eya domain (Rebay, Silver, and Tootle 2005; Silver and Rebay 2005; Jemc and Rebay 2007; Tadjuidje and Hegde 2013; Rebay 2016). The latter functions as a protein-protein binding domain and also has protein phosphatase activity. After binding to Six proteins, Eya translocates to the nucleus and acts as a transcriptional coactivator (Ohto et al. 1999; Zhang, Knosp et al. 2004). In addition, Eya proteins also form complexes with other proteins and probably serve separate functions in the cytoplasm, which are still poorly understood (Fan et al. 2000; Li et al. 2003; Embry et al. 2004; Xiong, Dabbouseh, and Rebay 2009; Ahmed et al. 2012; Ahmed, Xu, and Xu 2012). The phosphatase activity of Eya proteins may modulate its role as transcriptional coactivator in some contexts (Li et al. 2003). Vertebrates have four *Eya* genes, which originated by duplications of the single invertebrate *Eya* gene. While there are some interspecific differences, at least some members of the *Eya* family such as *Eya1* in *Xenopus* or *Eya2* in the chick show expression throughout the PPE (David et al. 2001; McLarren et al. 2003; Schlosser and Ahrens 2004).

The function of these PPE-specific members of the *Six* and *Eya* families have been studied in several model organisms (mice, chick, zebrafish, and the frog *Xenopus*) after mutations of these genes (knockout) or after inhibiting the translation of the encoded proteins (knockdown). The outcomes of these studies have been reviewed extensively elsewhere (Silver and Rebay 2005; Schlosser 2006, 2010; Jemc and Rebay 2007; Christensen et al. 2008; Wong, Ahmed, and Xu 2013; Xu 2012; Zhou et al. 2018) and will only be summarized here briefly.

Loss of function of Six1 and Eya1 proteins and some other members of these families compromises the development of many placodally-derived sense organs and ganglia as well as the pituitary. Double mutants of Six1 and Six4 in mice show stronger deficits indicating that different Six family members have partly overlapping functions (Grifone et al. 2005; Konishi et al. 2006; Chen, Kim, and Xu 2009). The same is true for different Eya family members (Grifone et al. 2007). In human patients, several types of congenital hearing disorders including Branchio-Oto-Renal (BOR) syndrome have been attributed to mutations in Eya1 or Six1 (Kochhar et al. 2007; Moody and LaMantia 2015).

Detailed characterization of Eya or Six losses of function phenotypes have given us some insights into their mode of action although many open questions remain. It is becoming increasingly clear that these proteins affect placodal development in several independent and context-dependent ways. First, Six1 and Eya1 or Eya2 expression in the PPE is required for promoting or maintaining the expression of other PPE genes and to delimit the adjacent neural crest territory (Brugmann et al. 2004; Christophorou et al. 2009; Maharana and Schlosser 2018). Second, Six and Eya factors then have further functions during the development of placodes from the PPE. Some of these functions appear to be specific for particular types of placode. They have, for example been implicated in affecting the patterning of the inner ear into various territories (Ozaki et al. 2004) or regulating cell lineage decisions in the anterior pituitary (Nica et al. 2006). Other effects appear to be more general and affect different placodes in similar ways. It is the role of Six1 and Eya1 in promoting these generic placodal properties together with their function in PPE maintenance that most likely underlies their capacity to bias the PPE for placodal development and endowing the PPE with the competence to differentiate into any of the different placodal fates.

There is evidence for at least four different categories of such generic effects, viz. effects on (1) morphogenesis (2) cell survival, (3) proliferation, and (4) neuronal and sensory differentiation. I will discuss these briefly here and will consider some of their functions in more detail in Chapter 5.

1. After Six1 or Eya1 loss of function, the invagination and subsequent morphogenetic movements of placodes (e.g. formation of cochlea, semicircular canals, and endolymphatic duct from the otic vesicle) as well as the delamination of neurons are compromised (Xu et al. 1999; Johnson et al. 1999; Laclef, Souil et al. 2003; Li et al. 2003; Zheng et al. 2003; Ozaki et al. 2004; Zou et al. 2004; Zou, Silvius, Rodrigo-Blomqvist et al. 2006; Kozlowski et al. 2005; Friedman et al. 2005; Zou et al. 2008; Bosman et al. 2009; Chen et al. 2009). Similar deficiencies in morphogenesis are also observed in other tissues that depend on Eya1 and Six1 function (Xu et al. 1999, 2002, 2003; Laclef, Hamard et al. 2003; Li et al. 2003; Ozaki et al. 2004; Grifone et al. 2005; Zou, Silvius, Davenport et al. 2006). Conversely, epithelial-mesenchymal transitions (EMT), cell movement, and formation of metastases are increased by elevated levels of Six1 or Eya proteins in various cancers (Yu et al. 2004, 2006; McCoy et al. 2009; Micalizzi et al. 2009; Pandey et al. 2010; Farabaugh et al. 2011; Patrick et al. 2013). There are probably many different pathways, by which Six1 or Eya1 modulate morphogenetic movements. On the one hand, these include transcriptional activation of signaling pathways, cell adhesion molecules or proteins interacting with the cytoskeleton (Yu et al. 2004, 2006; Micalizzi et al. 2009; Farabaugh et al. 2011; Riddiford and Schlosser 2016). On the other hand, they probably include pathways, in which Eya1 acts in the cytoplasm, possibly by dephosphorylating proteins involved in the regulation of actin-dependent cell motility (Xiong et al. 2009; Pandey et al. 2010).

2. Levels of cell death (apoptosis) in placodes are increased after knockout or knockdown of Six1 and Eya1 (Xu et al. 1999; Li et al. 2003; Zheng et al. 2003; Zou et al. 2004; Zou, Silvius, Rodrigo-Blomqvist et al. 2006; Ozaki et al. 2004; Friedman et al. 2005; Kozlowski et al. 2005; Konishi et al. 2006; Chen et al. 2009) indicating that Six1 and Eya1 also promote cell survival. The anti-apoptotic action of Eya1 may be due to its ability to dephosphorylate the histone variant H2AX in response to DNA damage, thereby allowing H2AX to recruit protein cofactors promoting DNA repair rather than apoptosis (Cook et al. 2009; Krishnan et al. 2009). However, this mechanism does not involve Six1 suggesting that Six1 and possibly Eya1 prevent apoptosis by additional and still unknown mechanisms.
3. Placodal derivatives such as the inner ear are often strongly reduced in size in Eya1 or Six1 mutants, at least partly due to reduced cell proliferation (Li et al. 2003; Zheng et al. 2003; Ozaki et al. 2004; Zou, Silvius, Rodrigo-Blomqvist et al. 2006; Zou et al. 2008; Schlosser et al. 2008; Chen et al. 2009). In contrast, overexpression of Six1 and/or Eya1 leads to massive proliferation and a tumor-like thickening of the ectoderm (Kriebel, Müller, and Hollemann 2007; Schlosser et al. 2008). In accordance with this, Six1 and Eya1 have also been shown to promote tumor growth in many cancers, where they directly activate cell cycle control genes such as *CyclinA1*, *CyclinD1*, and *c-Myc* (Ford et al. 1998; Li et al. 2003; Coletta et al. 2004, 2008; Zhang et al. 2005; Yu et al. 2006; McCoy et al. 2009; Miller et al. 2010; Pandey et al. 2010). A recent screen for direct transcriptional targets of Six1 and Eya1 in the PPE indicated that these proteins additionally activate a number of transcription factors (e.g. Hes8, Hes9, Sox2, Sox3) known to promote a proliferative neuronal and sensory progenitor state (Riddiford and Schlosser 2016). These transcription factors are strongly upregulated in regions of the ectoderm overexpressing high levels of Six1 and Eya1 where proliferation is enhanced and neuronal differentiation inhibited (Schlosser et al. 2008; Riddiford and Schlosser 2016, 2017). Taken together, these experimental findings strongly support a function of Six1 and Eya1 in promoting proliferation of sensory and neuronal progenitors. As a note of caution, these effects may be context-dependent. In zebrafish, for example, Six1 was shown to promote proliferation in sensory areas of the otic vesicle, but inhibit it in neuronal precursors (Bricaud and Collazo 2006).
4. After knockout or knockdown of Six1 and/or Eya1, expression of transcription factors promoting neuronal determination and differentiation (e.g. Atoh1, Neurog1, Neurog2, NeuroD1) is reduced and neuronal and sensory differentiation is perturbed in most derivatives of placodes (Xu et al. 1999; Zheng et al. 2003; Laclef, Souil et al. 2003; Zou et al. 2004; Friedman et al. 2005; Bricaud and Collazo 2006; Ikeda et al. 2007; Schlosser et al. 2008; Bosman et al. 2009; Chen et al. 2009; Ahmed et al. 2012; Ahmed, Xu, and Xu 2012). Overexpression of low levels of Six1 and Eya1 instead leads to ectopic differentiation of hair cells and sensory neurons (Schlosser et al. 2008; Ahmed et al. 2012; Ahmed, Xu, and Xu 2012). Six1 and Eya1 form cooperative protein complexes with each other and with other cofactors

to promote differentiation and they do so at multiple levels by activating many different target genes (Ahmed et al. 2012; Ahmed, Xu, and Xu 2012; Riddiford and Schlosser 2016; Li et al. 2020). Direct target genes include *Atoh1* and probably *Neurog1* (Ahmed et al. 2012; Riddiford and Schlosser 2016), which play central roles in specifying hair cells and sensory neurons respectively, as well as many other transcription factors promoting sensory or neuronal differentiation (Bermingham et al. 1999; Chen et al. 2002; Fode et al. 1998; Ma et al. 1998; Woods, Montcouquiol, and Kelley 2004; Li et al. 2020). This indicates that Six1 and Eya1 play important roles for normal differentiation of neuronal and sensory cell types from placodes.

Taking the findings described under (3) and (4) together, Six1 and Eya appear to play a dual role during the development of sensory cells and neurons from the cranial placodes. On the one hand, they appear to keep cells in a progenitor state where they keep dividing and do not differentiate. On the other hand, they seem to promote sensory and neuronal differentiation, which requires that cells stop dividing. I will discuss in Chapter 5, how these apparently conflicting roles can be reconciled.

In summary, there is increasing evidence that all placodes arise from a common region, the PPE and that this region is defined by a distinct regulatory state conferring a bias for cranial placode development. Accordingly, the PPE qualifies as a pan-placodal primordium, that is, a common primordium for all cranial placodes. Based on their specific expression in the PPE and their established role for cranial placode development, Six1/2 and Six4/5 transcription factors and their Eya cofactors are currently the best candidates for factors defining the PPE regulatory state. In the next section, I will address the question on how the PPE is established in the early embryo and how it is positioned and delimited from adjacent ectodermal territories.

3.2 ORIGIN OF THE PRE-PLACODAL ECTODERM (PPE) IN THE EARLY VERTEBRATE EMBRYO

To understand how the PPE is formed, we need to elucidate the changes in the gene regulatory network (GRN) in the early embryo that result in the activation of *Six* and *Eya* genes in their crescent shaped domain of expression at the border of the neural plate. We, thus, must consider, how *Six* and *Eya* genes are regulated by other transcription factors and signaling molecules in the early embryo and how the dynamic interplay between these upstream factors causes their proper spatiotemporal expression in the PPE. In all vertebrate embryos, four different ectodermal territories – epidermis, PPE, neural crest, and neural plate – are established during gastrulation in a precisely registered ventral to dorsal sequence. This suggests that the development of these territories is tightly linked. Indeed, it has become increasingly clear that the induction of the neural plate (neural induction) and neural crest are initiated in parallel even before the onset of gastrulation (reviewed in Pla and Monsoro-Burq 2018). Recent evidence suggests that the first steps of PPE induction can also be traced back to blastula stages and match the early steps of neural induction (Hintze et al. 2017; Trevers et al. 2018).

In the following sections, I will provide an overview of how the different ectodermal territories are established. Since, there is a vast literature on the regulatory interactions leading to the separation of ectoderm from the other germ layers and its subdivision into neural and non-neural ectoderm (reviewed in Kimelman 2006; Stern 2006; Rogers, Moody, and Casey 2009; Ozair, Kintner, and Brivanlou 2013), I will summarize this relatively briefly. I will then discuss the formation of the PPE and the neural crest in the neural plate border region (NPB) in some more detail (reviewed in Schlosser 2006, 2010; Grocott, Tambalo, and Streit 2012; Saint-Jeannet and Moody 2014; Groves and LaBonne 2014; Pla and Monsoro-Burq 2018; Streit 2018).

The limited space available here will not allow me to give a comprehensive review of the literature but I will provide some key references with a particular focus on *Xenopus*, where early ectodermal development has been studied in great detail. Taking the amphibian embryo as a model, I will in the following always refer to neural ectoderm being on the dorsal and non-neural ectoderm being on the ventral side. Please bear in mind that in amniotes, where the embryo develops as a disk on top of the yolk sac, the "dorsal", neural ectoderm is in fact located medially, while the "ventral", non-neural ectoderm is located laterally.

3.2.1 From Pluripotency to Initiation of Lineage Decisions

Although details of pre-gastrula development differ significantly between different vertebrate groups, the earliest divisions of the fertilized egg in all vertebrate groups produce a cluster of pluripotent cells, which can give rise to all germ layers. Such pluripotent cells are found, for example, in the animal cap, a layer of small cells located at the animal pole of the blastula in amphibians or in the epiblast, the outer layer of amniote embryos. The pluripotent state is maintained by a small group of transcription factors with proteins of the SoxB1 (e.g. Sox2 in mammals or Sox3 in chick and *Xenopus*), POU5 (e.g. Oct4 in mammals and birds or Oct25, Oct60, and Oct92 in *Xenopus*), Nanog/Ventx, Klf, and Myc families playing a central role. Together, these factors can induce reprogramming of differentiated cells into pluripotent stem cells (Takahashi and Yamanaka 2006, 2016; Adachi and Schöler 2012). Of these, both the POU5 family as well as the family of Nanog and the related Ventx are vertebrate novelties, that originated during genome duplication in the last common ancestor of vertebrates (Scerbo and Monsoro-Burq 2020).

In late blastula stage embryos, most cells will then lose pluripotency and make their first lineage decision for one of the three germ layers, endoderm, mesoderm, or ectoderm. This process has been particularly well studied in *Xenopus*. Nodal signals that are released from cells in the vegetal hemisphere opposite to the animal pole induce the formation of mesoderm in the adjacent animal cap cells, but not in the animal cap cells located further toward the animal pole. This leads to the formation of an equatorial ring of mesoderm intervening between the endoderm on the vegetal pole and ectoderm on the animal pole (Osada and Wright 1999; Agius et al. 2000; Takahashi et al. 2000; Chiu et al. 2014). A higher level of Nodal signaling on the dorsal side of the embryo (in a region of the vegetal hemisphere known as "Nieuwkoop center" in amphibians) contributes to the induction of dorsal mesoderm, which subsequently will become an important signaling center, the organizer.

In contrast, ectoderm will form from cells that are protected from Nodal signaling. Several mechanisms conspire to prevent nodal signaling in the prospective ectoderm. First, the zinc-finger transcription factor Churchill and its downstream target Smad interacting protein 1 (Sip1, also known as ZEB2) have been shown to play crucial roles during this process (Sheng, dos Reis, and Stern 2003; Snir et al. 2006; Londin, Mentzer, and Sirotkin 2007; Linder et al. 2007; Chng et al. 2010). Sip1 interferes with signaling of Nodal and bone morphogenetic proteins (BMP) because it binds to so-called Smad proteins, which serve as intracellular mediators of these signaling pathways and converts them to transcriptional repressors (Verschueren et al. 1999; Postigo et al. 2003). As a consequence, upregulation of Churchill and Sip1 changes the competence of cells, which will now respond with activation of neural ectodermal genes rather than mesodermal genes in response to FGF signals (Sheng et al. 2003; Londin et al. 2007). This illustrates nicely the importance of competence factors in cell lineage decisions as discussed briefly in Chapter 2. In chick and zebrafish, Churchill itself is preferentially upregulated in dorsal ectoderm by FGF signals (Sheng et al. 2003; Londin et al. 2007) and due to its capacity to interfere with Smad-mediated signaling, may synergize with BMP antagonists in neural induction (see below).

In addition, some of the transcription factors that maintain pluripotency in the early embryo are subsequently also involved in the first lineage decision between ectoderm and endomesoderm. Sox2 inhibits endomesoderm formation in mouse embryos, partly by repressing Brachyury (Thomson et al. 2011; Takemoto et al. 2011). In zebrafish, it has been shown that Sox3, which is initially expressed throughout the animal hemisphere becomes downregulated in the equatorial region, which will give rise to mesoderm probably in response to Nodal signals (Bennett et al. 2007). Sox3, in turn, promotes the formation of ectoderm and inhibits formation of endoderm and mesoderm in *Xenopus* and zebrafish by repressing the expression of *Nodal* genes and activating transcription factors such as Ectodermin that interfere with Nodal signaling (Zhang, Basta et al. 2004; Zhang and Klymkowsky 2007).

Interestingly, both Sox3/Sox2 and Sip1, which promote lineage decisions for ectoderm over endomesoderm by blocking Nodal signaling, also promote neural over non-neural ectodermal fates (see Chapter 5 for details). While Sox3 preferentially activates neural ectodermal genes (Bergsland et al. 2011), Sip1 inhibits BMP as well as Nodal target genes thereby biasing ectoderm toward a neural fate (see below). The widespread expression of Sox3 and Sip1 in the early ectoderm may, therefore help to explain, why the early ectoderm adopts a neural default fate in the absence of BMP (and possibly FGF) signaling molecules in vitro (Muñoz-Sanjuán and Brivanlou 2002).

Under certain conditions, the POU5 family of pluripotency factors also promote neural ectoderm formation in *Xenopus* as well as in mammalian embryonic stem cells (Shimozaki et al. 2003; Cao, Siegel, and Knöchel 2006; Snir et al. 2006; Chiu et al. 2014). In *Xenopus*, this has been shown to be due to the capacity of these factors to antagonize Nodal signaling and endomesodermal gene expression, mediated at least in part by the activation of Churchill and Sip1 (Cao et al. 2006; Snir et al. 2006; Chiu et al. 2014). However, the role of POU5 family proteins in lineage decisions remains somewhat puzzling since other studies in mammals and zebrafish suggest that they promote endomesoderm formation instead (Niwa, Miyazaki, and Smith 2000; Lunde, Belting, and Driever 2004; Reim et al. 2004; Thomson et al. 2011). While species-specific differences may explain some of the conflicting findings, there is also evidence for POU5 family proteins acting in a highly context- and dosage-dependent way. This most likely reflects their known ability to form dimeric complexes with several other transcription factors and activate different target genes depending on their binding partners (reviewed in Onichtchouk and Driever 2016).

In summary, in most vertebrate embryos, cells cease to be pluripotent and make their first lineage decision for one of the three germ layers at the blastula stage. Differential expression of competence factors (e.g. Churchill and Sip) in the germ layers now allows the activation of different sets of target genes in response to the same inductive signals. Moreover, some of the transcription factors promoting pluripotency in the early embryo, subsequently adopt more specific roles in particular germ layers, with SoxB1 (and possibly POU5) promoting ectoderm and, in particular, neural ectoderm.

3.2.2 Upregulation of New Ectodermal Transcription Factors at Blastula Stages

While some of the transcription factors promoting ectoderm formation are inherited from the mother (e.g. Sox3 as discussed in the last section), many additional transcription factors become upregulated in the ectoderm at blastula stages (Heasman 2006; Paraiso et al. 2020). In zebrafish and *Xenopus*, the majority of these transcription factors, whether maternally inherited or zygotically activated are initially distributed throughout the ectoderm. Many of these transcription factors will subsequently become either ventrally restricted (e.g. TFAP2a, Msx1, Dlx3/5, Ventx2, FoxI1-4) and excluded from the dorsal-most, neural ectoderm or will become dorsally restricted and excluded from the ventral, non-neural ectoderm (e.g. Sox3, Sox11, Geminin, FoxD4) as will be discussed in more detail below (reviewed in Grocott et al. 2012; Schlosser 2014). At blastula stages, however, these prospective neural and non-neural transcription factors are still overlapping to a large degree.

In chick embryos, a similar set of factors is broadly overlapping in the pre-gastrula epiblast although sometimes the function of a transcription factor in *Xenopus* appears to be taken on by a different member of the same transcription factor family in the chick. For example, the distribution and function of Dlx5 and Pax7 in chick embryos resembles Dlx3 and Pax3, respectively in frogs (e.g. Pera, Stein, and Kessel 1999; Maczkowiak et al. 2010; Pieper et al. 2011; Grocott et al. 2012). Moreover, in the chick unlike in the frog, many ventrally (chick: laterally) and dorsally (chick: medially) restricted transcription factors appear to be already excluded from the dorsal-most and ventral-most ectoderm, respectively, when they begin to be expressed (Pera et al. 1999; Streit and Stern 1999; Khudyakov and Bronner-Fraser 2009).

The mechanisms underlying the activation of new transcription factors at blastula stages are still poorly understood. However, most ventrally restricted transcription factors such as Dlx3/5, Msx1, GATA2/3, TFAP2a, FoxI1-4, and Ventx1/2 are known to be either directly or indirectly activated by BMP (Suzuki, Ueno, and Hemmati-Brivanlou 1997; Mizuseki et al. 1998; Feledy et al. 1999; Friedle and Knöchel 2002;

Kwon et al. 2010). They are, thus, induced throughout the ectoderm, when BMP signaling molecules become upregulated at all levels of the ectoderm and mesoderm from at least blastula stages on (Hemmati-Brivanlou and Thomsen 1995; Hawley et al. 1995; Streit et al. 1998).

In contrast, many of the transcription factors that will subsequently become dorsally restricted are known to be inhibited by BMP (see below). Nevertheless, at least in zebrafish and *Xenopus* some of them appear to be initially activated throughout the early blastula ectoderm by maternally supplied transcription factors before becoming repressed on the ventral side by BMP (reviewed in Rogers et al. 2009; Lee, Lee, and Moody 2014). The SoxB1 factors Sox3 and Sox2, in particular, have been shown to activate transcription of a large number of genes encoding dorsally restricted transcription factors including other SoxB1 proteins, Sox11, Zic1, Zic3, Geminin, and Myc family proteins (Rogers et al. 2009; Okuda et al. 2010; Bergsland et al. 2011). Cross-regulation between SoxB1, Geminin and Zic1 stabilizes their expression in neural ectoderm; subsequently these transcription factors cooperate in the regulation of multiple neural target genes (Sankar et al. 2016).

FGF also contributes to the induction or maintenance of these various transcription factors during late blastula stages although its precise function is still controversial. In chick embryos, FGF signaling prior to gastrulation has been shown to be essential for the induction of Sox3 and other dorsally restricted genes (Streit et al. 2000; Wilson et al. 2000). However, in *Xenopus*, FGF seems to be dispensable for the induction of Sox3 and some early transcription factors (Zic1, Geminin, Myc) but required for the induction of others (e.g. Sox2, Zic3, Ventx2, Id3) (Delaune, Lemaire, and Kodjabachian 2005; Marchal et al. 2009; Rogers, Ferzli, and Casey 2011; Geary and LaBonne 2018). Finally, inhibition of the Wingless-integrated (Wnt) signaling pathway may also play a role for the induction of prospectively dorsally restricted genes at late blastula stages (Wilson et al. 2001; Heeg-Truesdell and LaBonne 2006; Steventon and Mayor 2012).

In summary, at the blastula stage some maternal transcription factors like Sox3 continue to be expressed throughout the early ectoderm and help to activate other prospectively dorsal or neural transcription factors. At the same time, prospectively ventral or non-neural transcription factors are activated by BMP throughout the ectoderm. At blastula stages, prospectively dorsal and ventral transcription factors are, thus expressed in a widely overlapping pattern (Figs. 3.4, 3.5).

3.2.3 Neural Induction by Signals from the Organizer

Once vertebrate embryos enter gastrulation, the initially broad degree of overlap between dorsal and ventral transcription factors will become increasingly smaller due to signals from the organizer in the dorsal mesoderm. This signaling center occupies the dorsal blastopore lip in fish and amphibian embryos and a structure known as the node (or Hensen's node) in amniote embryos. During gastrulation, the organizer cells move inside into the embryo, where they form dorsal mesodermal structures such as the prechordal plate and notochord, which underlie the developing neural plate. Changes in the regulatory state of the ectoderm in response to signals from the organizer during gastrulation have been traditionally discussed under the

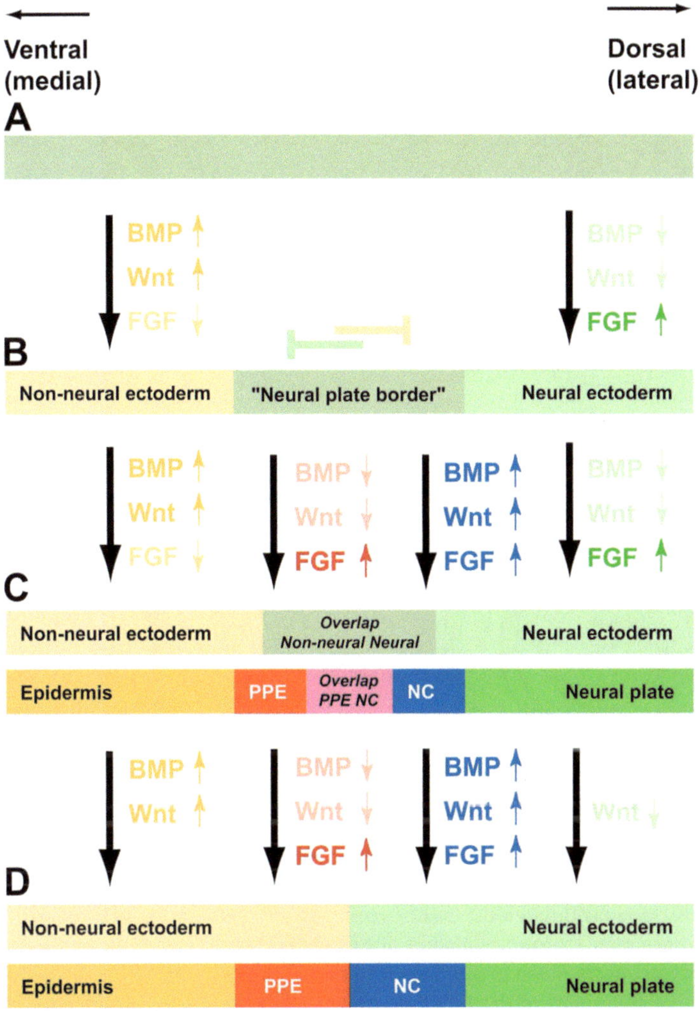

FIGURE 3.4 Overview of neural plate border development. Summary diagram of how different ectodermal territories (epidermis, PPE, neural crest, neural plate) are established during gastrulation in vertebrates. (**A**) At blastula stages there is a uniform ectodermal territory, in which neural and non-neural transcription factors are co-expressed. (**B**) During early gastrula stages, neural transcription factors become dorsally restricted in response to FGF signals and inhibition of BMP and Wnt signaling, whereas non-neural transcription factors become ventrally restricted in response to BMP and Wnt. Both categories of transcription factors overlap at the neural plate border, where additional transcription factors start to be expressed soon (not shown). The neural plate border becomes smaller over time due to cross-repressive interactions between neural and non-neural transcription factors. (**C**) During late gastrula stages, BMP, Wnt, and FGF signals induce neural crest (NC) specifiers in neural ectoderm, while FGF and inhibitors of Wnt and BMP induce PPE specifiers in non-neural ectoderm at the neural plate border. (**D**) After gastrulation, PPE and NC resolve into mutually exclusive territories due to ongoing cross-repressive interactions between neural and non-neural transcription factors and/or between PPE and NC transcription factors.

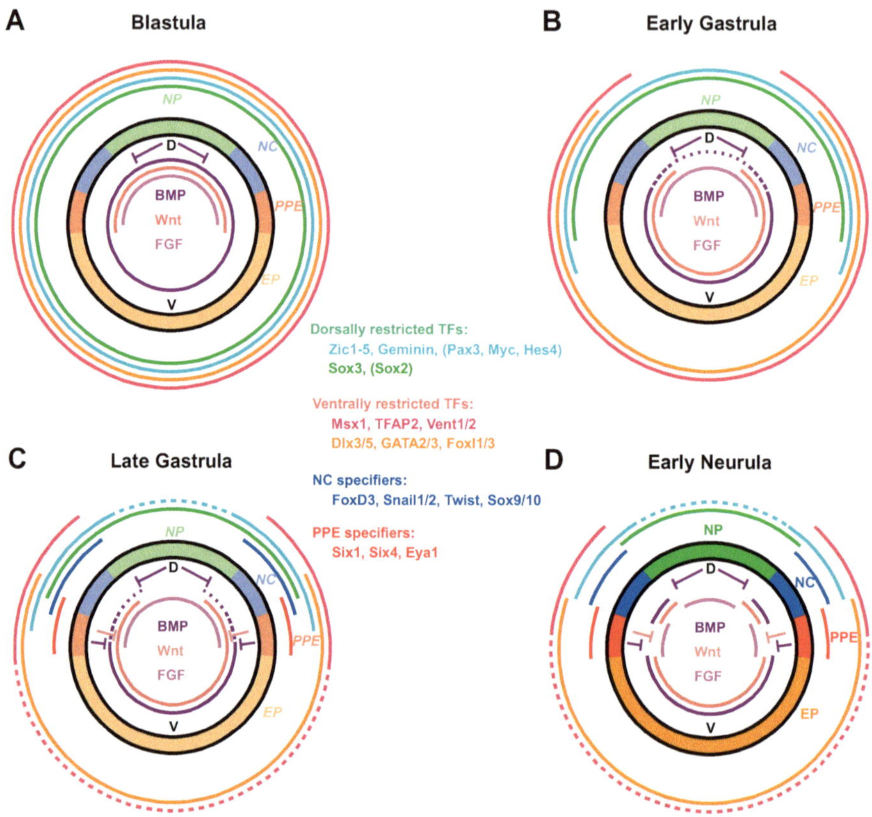

FIGURE 3.5 Establishment of ectodermal territories during early *Xenopus* development. Expression domains of transcription factors are shown as colored lines outside of schematic cross sections through the cranial region of *Xenopus* embryos (D, dorsal; V, ventral). Hatched lines refer to downregulation of expression. Transcription factors listed in parentheses only get upregulated in the respective domains during gastrulation (Pax3, c-Myc and Hes4 in the lateral part of the turquoise domain; Sox2 in the green domain). The regions destined to give rise to neural plate (NP), neural crest (NC), pre-placodal ectoderm (PPE), and epidermis (EP) are shown as fate map for blastula and gastrula stages (**A–C**: faint colors) and as specified territories for early neurula stages (**D**: strong colors). Changes in extent and position of the different territories due to gastrulation movements are disregarded here. Signaling activities of the BMP, Wnt, and FGF pathways along the dorsoventral axis are shown by colored lines inside the schematized embryo, with graded activity of BMP and approximate position of sources of BMP inhibitors and Wnt inhibitors indicated (bars; tissue sources are not depicted). Note that many transcription factors become increasingly dorsally (turquoise and green) or ventrally (orange and pink) restricted, and the region of overlap decreases with gastrulation. Some ventral transcription factors (pink) extend further dorsal than others (orange). Due to cross-repressive interactions, transcription factors defining neural (turquoise) and non-neural (orange) competence become confined to non-overlapping territories at the end of gastrulation and so do neural crest specifiers (blue) and transcription factors defining the PPE (red). (Modified from Schlosser 2014.)

heading of "neural induction" (reviewed in Muñoz-Sanjuán and Brivanlou 2002; De Robertis and Kuroda 2004; Stern 2006; Rogers et al. 2009; Ozair et al. 2013; Monsoro-Burq 2020). However, they are now known to mainly involve the repression of dorsal transcription factors in ventral ectoderm and of ventral transcription factors in dorsal ectoderm (Figs. 3.4, 3.5).

The organizer secretes a cocktail of signaling molecules, which travel between the mesodermal and ectodermal cell layers (and possibly to some extent in the plane of the ectoderm) to change ectodermal gene expression. Organizer signals include various antagonists of BMP (e.g. noggin, chordin, follistatin) and Wnt signaling (e.g. dickkopf, frzb) as well as FGF (reviewed in Stern 2006; Rogers et al. 2009; Ozair et al. 2013; Martinez Arias and Steventon 2018; Monsoro-Burq 2020). Most BMP and Wnt antagonists are proteins that bind to secreted BMP and Wnt proteins, respectively, thereby preventing them from binding to their respective receptors (Balemans and Van Hul 2002; Miyazono, Maeda, and Imamura 2005; Clevers and Nusse 2012; Niehrs 2012). The strength of these diffusible signals declines with increasing distance from the organizer. The resulting concentration gradients of signaling molecules promote the activation or repression of different target genes in different parts of the ectoderm. BMP antagonists are secreted throughout the dorsal organizer, whereas Wnt antagonists are secreted mostly from the part of the organizer that forms the anterior-most dorsal mesoderm (Kiecker and Niehrs 2001; De Robertis and Kuroda 2004; Niehrs 2010; Schohl and Fagotto 2002). Consequently, a gradient of BMP signaling is established along the dorsoventral axis of the embryo (with high BMP levels ventrally). Wnt is repressed dorsally at the early stages of gastrulation and anteriorly at the late stages of gastrulation resulting in the superimposition of an anteroposterior and dorsoventral gradient of Wnt signaling (with low Wnt levels anteriorly and dorsally but high levels posteriorly and ventrally).

The dorsal to ventral recession of BMP signaling during gastrulation is accompanied by progressive exclusion of prospectively ventral or non-neural transcription factors that depend on activation by BMP such as Dlx3/5, Msx1, or TFAP2a from the dorsal ectoderm, while they are maintained ventrally (reviewed in Grocott et al. 2012; Schlosser 2014; Saint-Jeannet and Moody 2014; Simoes-Costa and Bronner 2015; Pla and Monsoro-Burq 2018; Streit 2018). In *Xenopus*, for example, BMP-dependent transcription factors that were initially expressed throughout the blastula ectoderm (e.g. Dlx3, TFAP2a, FoxI1-4) and those that are secondarily upregulated in the ventral ectoderm during gastrulation (e.g. Msx1, GATA2), recede from the dorsal side. Some of them (Dlx3, Dlx5, GATA2, FoxI1-4) will become confined to the prospective epidermis and PPE and excluded from neural plate and neural crest in this process. Others will not recede quite as far ventrally (TFAP2a, Msx1, Ventx) and will continue to be expressed in epidermis, PPE and neural crest, being excluded only from the neural plate (Suzuki et al. 1997; Luo et al. 2002; Schlosser and Ahrens 2004; Matsuo-Takasaki, Matsumura, and Sasai 2005; Rogers et al. 2009; Pieper et al. 2012; Buitrago-Delgado et al. 2015; Alkobtawi and Monsoro-Buro 2020).

Conversely, prospectively dorsal or neural transcription factors such as Sox3, or Zic1/3 are increasingly repressed by persistent BMP signaling in ventral ectoderm but maintained or upregulated in the BMP depleted ectoderm on the dorsal side (reviewed in Grocott et al. 2012; Schlosser 2014; Saint-Jeannet and Moody 2014;

Simoes-Costa and Bronner 2015; Pla and Monsoro-Burq 2018; Streit 2018). Considering again *Xenopus* as an example, transcription factors, that were initially expressed at least weakly throughout the blastula ectoderm (e.g. Sox3, Sox11, Geminin, FoxD4) as well as those that are newly induced by signals from the organizer (e.g. Zic1, Zic3, Sox2), recede from the ventral side until their expression is confined to the neural plate (Sox3, Sox2) or to both neural plate and neural crest domains (Zic1-3, Geminin) (Kroll et al. 1998; Mizuseki et al. 1998; Nakata et al. 1998; Pieper et al. 2012; Buitrago-Delgado et al. 2015). These dynamic changes in the expression of transcription factors result in the establishment of three ectodermal domains; one domain that exclusively expresses neural transcription factors on the dorsal side; another domain that exclusively expresses non-neural transcription factors on the ventral side; and an intervening "neural plate border", where both categories of transcription factors overlap.

A recent study in chick embryos has monitored gene expression profiles in response to signals from the organizer (i.e. Hensen's node) in great detail and has identified many additional transcription factors that change expression during neural induction (Trevers et al. 2018). Transcription factors that were repressed by the node (including Dlx5, GATA2, Msx1, TFAP2a) were typically restricted to the peripheral epiblast (corresponding to ventral, i.e. non-neural ectoderm in frogs). In contrast, the majority of transcription factors that were induced by the node (including Zic1/3, Sox3, Otx2, Etv4/5, Mycn, Trim24, Znf462, Pdlim4, Irx1/2, Cited2, Rybp) are expressed in the central epiblast (corresponding to dorsal, i.e. neural ectoderm in frogs). A largely identical set of transcription factors was shown to be up- or down-regulated during the earliest steps of PPE induction by signals from the underlying head mesoderm (Hintze et al. 2017). Further experiments confirmed that PPE induction and neural induction in the chick pass through a common initial state before transcription factors specific to the neural plate (e.g. Sox1) and PPE (e.g. Six4, Eya2) are induced at a relatively late stage in the process (Hintze et al. 2017; Trevers et al. 2018).

In contrast to these late, dedicated neural plate and PPE markers, many of the transcription factors induced during early steps remain expressed in both neural plate and PPE or in the neural plate border region. Moreover, most of the transcription factors induced during these early steps (including Sox3, Zic3, Mycn) are already expressed in the central epiblast prior to the formation of the organizer in response to signals from the underlying hypoblast (Trevers et al. 2018). This suggests that in chick embryos a step of "pre-neural induction", during which many transcription factors from the neural plate border are induced, precedes the organizer-dependent events of neural induction, during which these and additional transcription factors are specifically upregulated in the neural plate (corresponding to dorsal ectoderm in the frog).

A comparison of these findings with a recent study in *Xenopus* reveals that most transcription factors induced during pre-neural and neural induction in the chick are similarly upregulated during neural induction in frog embryos (Ding et al. 2018). However, some of the transcription factors that are pre-neurally induced in the chick appear to be maternally expressed in *Xenopus* (e.g. Sox3). Taken together, this suggests that in spite of some interspecific differences, mechanisms of neural induction are broadly conserved in vertebrates. In all vertebrate species investigated, neural induction results in a changing expression of transcription factors in the ectoderm during gastrulation in response to signals from the organizer. Whereas some transcription factors become increasingly excluded from the ventral side and restricted to the dorsal, neural ectoderm, others become increasingly excluded from the dorsal side and restricted to the ventral, non-neural ectoderm.

3.2.4 Changes in Gene Expression at the Neural Plate Border During Gastrulation

As gastrulation proceeds, some transcription factors (e.g. Pax3, Myc, Id3, Hes4 – also known as Hairy2) are upregulated at the neural plate border in a domain, where ventrally restricted transcription factors including TFAP2a and Msx1 overlap with Zic1/3 and other dorsal transcription factors (Fig. 3.5); this domain includes the future neural crest region (Bellmeyer et al. 2003; Light et al. 2005; Kee and Bronner-Fraser 2005; Monsoro-Burq, Wang, and Harland 2005; Nichane, de Croze et al. 2008; Aguirre et al. 2013; Alkobtawi and Monsoro-Burq 2020). The expression of TFAP2a, Msx1, and some other ventrally restricted transcription factors also increases in this region, while the expression of some dorsally restricted transcription factors (e.g. Zic1/3) is both amplified in this domain and downregulated in the central neural plate (Nakata et al. 1998; Tribulo et al. 2003; Monsoro-Burq et al. 2005; Sato, Sasai, and Sasai 2005; de Croze, Maczkowiak, and Monsoro-Burq 2011; Buitrago-Delgado et al. 2015).

Many of the transcription factors expressed at high levels at the neural plate border have long been recognized as important upstream regulators of neural crest development, which have been labeled "neural plate border specifiers" (Meulemans and Bronner-Fraser 2004). Apart from transcription factors mainly expressed at the neural plate border (e.g. Pax3, Hes4, Gbx2), neural plate border specifiers include two different categories of transcription factors that establish almost complementary expression patterns during gastrulation, viz. dorsally (Zic1, Zic3, Geminin) and ventrally (TFAP2a, Msx1, Dlx3/5) restricted transcription factors (reviewed in Grocott et al. 2012; Schlosser 2014; Saint-Jeannet and Moody 2014; Simoes-Costa and Bronner 2015; Pla and Monsoro-Burq 2018; Streit 2018). The different expression patterns of these different categories of transcription factors will subsequently be important to establish neural crest and PPE in mutually exclusive territories (see section 3.2.5 below).

The changes in transcription factor expression at the neural plate border region require Wnt and FGF signals from the underlying mesoderm and the ectoderm itself (Saint-Jeannet et al. 1997; La Bonne and Bronner-Fraser 1998; Streit and Stern 1999; Bang et al. 1999; Garcia-Castro, Marcelle, and Bronner-Fraser 2002; Monsoro-Burq, Fletcher, and Harland 2003; Monsoro-Burq et al. 2005; Hong and Saint-Jeannet 2007; Steventon et al. 2009; Patthey, Edlund, and Gunhaga 2009; Li et al. 2009; de Croze et al. 2011; Garnett, Square, and Medeiros 2012; Stuhlmiller and Garcia-Castro 2012). The effects of Wnt signaling appear to be partly mediated by direct Wnt target genes encoding transcription factors such as Gbx2 and TFAP2a. Gbx2 is expressed throughout the ectoderm except for its anterior part, where Wnt signaling is low (Schohl and Fagotto 2002; Li et al. 2009; Borday et al. 2018). Since Gbx2 is required for neural crest formation, its distribution may help to define the anterior boundary of the developing neural crest, although this hypothesis remains to be tested. Whereas both Gbx2 and TFAP2a have been identified as upstream regulators of the neural border fate activated by Wnt signals and critical for mediating its activity (Li et al. 2009; de Croze et al. 2011), the relationships between these two transcription factors remain unexplored.

Moreover, the ventrally restricted transcription factors TFAP2a, Dlx3 and Msx1 were shown to promote upregulation of Pax3, Hes4, Myc and other transcription factors in ectoderm at the neural plate border (Monsoro-Burq et al. 2005; Nikitina, Sauka-Spengler, and Bronner-Fraser 2008; Nichane, de Croze et al. 2008; de Croze et al. 2011; Pieper et al. 2012). However, Pax3, Hes4 and Myc can only be activated in neuralized ectoderm, i.e. in ectoderm, in which BMP signaling is largely inhibited (Monsoro-Burq et al. 2005; Hong and Saint-Jeannet 2007; de Croze et al. 2011; Pieper et al. 2012). This suggests that dorsally restricted transcription factors of unknown identity are crucial for the competence of ectoderm to activate Pax3 and possibly other neural plate border-specific proteins. Cross-regulatory interactions between various transcription factors expressed at the neural plate border subsequently stabilize their expression and render it independent of signaling molecules (Meulemans and Bronner-Fraser 2004; Monsoro-Burq et al. 2005; Nichane, de Croze et al. 2008; Sato et al. 2005; de Croze et al. 2011).

In summary, signals from the organizer not only result in an increasing dorsally or ventrally restricted expression of early ectodermal transcription factors during vertebrate gastrulation (as discussed in the previous section). Additional transcription factors start to be expressed or are enhanced at the neural plate border, where both categories of transcription factors overlap, during gastrulation. This neural plate border region, which ultimately will give rise to both the neural crest and PPE territories, becomes smaller and smaller as gastrulation proceeds.

3.2.5 Induction of Neural Crest and PPE at the Neural Plate Border

At the end of gastrulation, transcriptional regulators involved in the specification of the neural crest (e.g. FoxD3, Snail1/2) and the PPE (e.g. Six1, Eya) will then be first activated by transcription factors expressed in the neural plate border region (neural plate border specifiers) in combination with inducing signals from adjacent tissues. These signals impinge on the neural plate border from the underlying mesoderm and adjacent ectodermal territories (Selleck and Bronner-Fraser 1995; Mancilla and Mayor 1996; Monsoro-Burq et al. 2003; Ahrens and Schlosser 2005; Litsiou, Hanson, and Streit 2005; Hintze et al. 2017) (Figs. 3.4, 3.5). Recent studies suggest that most neural plate border specifiers initially promote not only neural crest but also PPE induction although their subsequent roles for neural crest and PPE specification may diverge (McLarren et al. 2003; Woda et al. 2003; Matsuo-Takasaki et al. 2005; Kwon et al. 2010; Pieper et al. 2012; Maharana and Schlosser 2018). Neural crest and PPE specifiers will first be induced in overlapping territories but will soon come to occupy mutually exclusive territories (see below).

3.2.5.1 Neural Crest Induction

During late gastrulation, the various neural plate border specifiers cooperate in their domain of overlap at the neural plate border to activate the expression of neural crest specifiers such as FoxD3, Snail1/2, Ets1, Sox9, and Sox10 (e.g. Bellmeyer et al. 2003; Tribulo et al. 2003; Kee and Bronner-Fraser 2005; Sato et al. 2005; Monsoro-Burq et al. 2005; Hong and Saint-Jeannet 2007; Nichane, de Croze et al. 2008; Nichane, Ren et al. 2008; de Croze et al. 2011). The latter are specifically expressed in the neural crest, cross-regulate each other's expression and play key roles for regulating neural crest multipotency and migration (reviewed in Meulemans and Bronner-Fraser 2004; Betancur, Bronner-Fraser, and Sauka-Spengler 2010; Simoes-Costa and Bronner 2015; Pla and Monsoro-Burq 2018).

Zic1 and Pax3 play a central role for the activation of neural crest specifiers and the joint activation of these two transcription factors is sufficient to activate a full neural crest developmental program in the ectoderm (Milet et al. 2013; Bae et al. 2014; Plouhinec et al. 2014; Simoes-Costa et al. 2014). However, TFAP2a and Msx1, which had earlier roles in activating Pax3 and other neural plate border specifiers, may also independently contribute to activating neural crest specifiers (Tribulo et al. 2003; Monsoro-Burq et al. 2005; de Croze et al. 2011; Simoes-Costa et al. 2012; Bhat, Kwon, and Riley 2012). Dlx3 and Dlx5 are also involved in the activation of neural crest specifiers, although it is still unclear whether they are cell-autonomously required in the prospective neural crest cells or are instead required for the production of inducing signals in adjacent non-neural ectoderm (McLarren et al. 2003; Woda et al. 2003; Kaji and Artinger 2004; Pieper et al. 2012). At the same time, cross-repressive interactions between some neural plate border or neural crest specifiers (e.g. Ventx1/2, Msx1, Pax3/7, FoxD3) and dedicated neural plate markers (e.g. Sox3, Sox2) help to sharpen the border between the neural crest and the neural plate (Sasai, Mizuseki, and Sasai 2001; Rogers et al. 2008, 2009; Roellig et al. 2017; Maharana and Schlosser 2018; Buitrago-Delgado et al. 2018).

The induction of neural crest specifiers requires the continuation of Wnt and FGF signaling in this region at late gastrulation (Steventon et al. 2009). However, BMP signaling, which had to be repressed to allow upregulation of Pax3 and other neural plate border specifiers during early gastrulation, is now required for induction of neural crest specifiers (Patthey et al. 2009; Steventon et al. 2009). Indeed, BMP signaling increases specifically in the prospective neural crest region during late gastrulation (Faure et al. 2002; Wu et al. 2011; Reichert, Randall, and Hill 2013) possibly due to the upregulation of CV2, a BMP binding protein, and of SNW1, a protein that modulates BMP receptor activation (Wu et al. 2011; Reichert et al. 2013).

3.2.5.2 PPE Induction

Many of the same neural plate border specifiers that are involved in the activation of neural crest specifiers are also required for the induction of dedicated PPE transcriptional regulators (Six1/2, Six4/5, Eya) during late gastrulation, which serve as PPE specifiers. Several studies have provided evidence that ventrally restricted transcription factors (Dlx3/5, GATA2/3, FoxI1/3, Msx1, TFAP2a, Ventx) are required for PPE induction, presumably acting as competence factors in the non-neural ectoderm that promote PPE induction in a signaling-dependent manner (McLarren et al. 2003; Woda et al. 2003; Matsuo-Takasaki et al. 2005; Kwon et al. 2010; Pieper et al. 2012; Maharana and Schlosser 2018). Some of these factors probably activate PPE markers directly as judged by the presence of Dlx/Msx and GATA binding sites in early enhancers of Six1 (Sato, Ishihara, and Kawakami 2005; Sato et al. 2010). The transcriptional activators of the Dlx family are known to recognize identical DNA-binding sites than the transcriptional repressors of the Msx family and functionally antagonize their activity (Zhang et al. 1997). The ability of high level Dlx to outcompete Msx binding in non-neural ectoderm has been suggested to be essential for PPE formation since Msx knockdown in zebrafish is able to rescue the reduction of pre-placodal marker expression after Dlx loss of function (Phillips et al. 2006). However, since Msx1 knockdown interferes with PPE induction in *Xenopus*, while Msx1 overexpression is able to induce PPE markers ectopically in the neural plate (Maharana and Schlosser 2018), it must play additional roles for PPE induction, which are not yet understood.

In addition, several studies suggest that dorsally restricted transcription factors (e.g. Zic1, Irx1/2, Znf462, Pdlim4) as well as transcription factors specifically expressed in the neural plate border region (Pax3, Myc, Hes4) (Bellmeyer et al. 2003; Glavic et al. 2004; Hong and Saint-Jeannet 2007; Pieper et al. 2012; Jaurena et al. 2015; Hintze et al. 2017; Maharana and Schlosser 2018; Sullivan et al. 2019) also promote PPE formation. While Zic1 promotes the production of PPE inducing signals (e.g. retinoic acid) in the neural ectoderm (Jaurena et al. 2015), both Zic1 and Pax3 were shown to be also required cell-autonomously for PPE formation (Maharana and Schlosser, 2018). This suggests that they contribute to the initial activation of Six and Eya during gastrulation, when their territories still overlap with the prospective PPE (see Figs. 3.4, 3.5) although they subsequently repress PPE formation (see below).

Induction of the PPE during late gastrulation requires signals from the underlying mesoderm and the adjacent neural ectoderm (Ahrens and Schlosser 2005; Litsiou et al. 2005; Hintze et al. 2017). While BMP signaling was essential during early gastrulation for the upregulation of ventrally restricted transcription factors (Dlx3/5, Msx1, GATA2/3, TFAP2a, FoxI1-4, and Ventx1/2) that are required for the activation of PPE specifiers, BMP signaling must be repressed during late gastrulation to allow PPE induction (Ahrens and Schlosser 2005; Litsiou et al. 2005; Kwon et al. 2010). A recent study in zebrafish embryos has indeed shown that upregulation of the BMP inhibitor BAMBI in the prospective PPE forming region of the neural plate border at this stage, protects this domain from the increasing BMP signaling in the adjacent prospective dorsal epidermis and neural crest (Reichert et al. 2013). In addition, Wnt inhibitors, FGFs and retinoic acid secreted from the underlying cranial mesoderm and the developing anterior neural plate also contribute to PPE induction (Brugmann et al. 2004; Litsiou et al. 2005; Ahrens and Schlosser 2005; Hong and Saint-Jeannet 2007; Kwon et al. 2010; Jaurena et al. 2015). Wnt and retinoic acid signals from mesoderm and other tissues prevent PPE induction in the trunk (Brugmann et al. 2004; Litsiou et al. 2005; Janesick et al. 2012).

In summary, because of partly shared upstream regulators (e.g. Dlx3/5, Zic, Pax3) and inductive signals (FGF), PPE specifiers and neural crest specifiers will first be induced in overlapping territories at the end of gastrulation but will soon come to occupy mutually exclusive territories, probably mostly due to cross-repressive interactions between a subset of ventrally and dorsally restricted transcription factors serving as competence factors for non-neural and neural ectoderm as discussed in the following section.

3.2.6 Separation of Neural Crest and PPE

Once dedicated neural crest and PPE specifiers have been induced during late gastrulation, expression of some transcription factors expressed at the neural plate border such as the dorsally restricted Pax3, Myc, and Zic1 will recede further dorsally (turquoise in Fig. 3.5). At the same time, a subset of ventrally restricted transcription factors including Dlx3/5, GATA2/3, and FoxI1-4 (orange in Fig. 3.5) will recede further ventrally until they occupy mutually exclusive ectodermal territories with the dorsally receding transcription factors just mentioned. In parallel, the expression of PPE (e.g. Six1/2, Eya) and neural crest specifiers (e.g. FoxD3, Snail1/2) also becomes confined to mutually exclusive territories (reviewed in Grocott et al. 2012; Groves and LaBonne 2014; Schlosser 2014). The time, when a sharp boundary is finally established between these territories may vary somewhat between species and has been suggested to be around the end of gastrulation in *Xenopus* but at later, neurula stages in chick embryos (Pieper et al. 2012; Roellig et al. 2017).

This process of separation may be driven largely by reciprocal repression between these particular subsets of ventrally (Dlx3/5, GATA2/3, and FoxI1-4) and dorsally restricted (e.g. Zic1, Pax3) transcription factors, which probably serve as competence factors for non-neural (PPE and epidermis) and neural ectoderm (neural crest and neural plate), respectively, as will be discussed in more detail in the next section (Pieper et al. 2012; Maharana and Schlosser 2018). It is important to note that not all ventrally restricted transcription factors are involved in the formation of the neural crest – PPE boundary. Some, including TFAP2a, Msx1, Ventx2, and Nkx6.3 extend further dorsally (pink in Fig. 3.5) where they remain overlapping with Pax3, Zic1, and other dorsally restricted transcription factors and play important roles in neural crest specification (Luo et al. 2003; Tribulo et al. 2003; Monsoro-Burq et al. 2005; de Croze et al. 2011; Zhang et al. 2014).

Feedback regulation between Six1, Eya1, or neural crest specifiers (e.g FoxD3, Snail1/2, Sox9/10) and neural plate border specifiers may also contribute to the definition of the PPE – neural crest boundary (e.g. Sasai et al. 2001; Spokony et al. 2002; Aybar, Nieto, and Mayor 2003; Maharana and Schlosser 2018). In addition, non-random cell movements due to expression of different cell surface molecules may play some role (Streit 2002; Ezin, Fraser, and Bronner-Fraser 2009). Finally, positive auto- and cross-regulation between Six1 and Eya1 were shown to stabilize the PPE, while positive cross-regulation between different neural crest specifiers stabilizes the neural crest region (Spokony et al. 2002; Aoki et al. 2003; Aybar et al. 2003; Honoré, Aybar, and Mayor 2003; Brugmann et al. 2004; Schlosser et al. 2008; Christophorou et al. 2009; Maharana and Schlosser 2018).

A recent study indicates that the role of transcription factors for neural crest and PPE formation is strongly dependent on the context of co-expressed transcription factors and, thus, changes when initially overlapping transcription factors become confined to different territories (Maharana and Schlosser 2018). For example, Zic1 and Pax3 promote PPE formation in the presence of Dlx3 but repress it in the absence of Dlx3 (Maharana and Schlosser 2018). Similarly, Dlx3 promotes the expression of dedicated neural crest specifiers in the presence of Zic1 but represses it in the absence of Zic1 (Maharana and Schlosser 2018). This suggests that when Zic1 and Dlx3 (and likely other transcription factors as well) resolve into complementary territories at the end of gastrulation, their role for PPE and neural crest formation, respectively, may switch from activation to repression. This regulatory setup allows the initial upregulation of dedicated neural crest and PPE specifiers in an overlapping fashion in the neural plate border region and their subsequent separation into two mutually exclusive territories.

It deserves to be emphasized here that the dynamically shrinking neural plate border is the region of the ectoderm that retains its immature condition, in which dorsally and ventrally transcription factors overlap, for the longest period of time and is, consequently, the last part of the ectoderm to make lineage decisions. Progenitors at the neural plate border will initially co-express lineage determining transcription factors for neural crest, PPE, and other ectodermal territories before making a decision by downregulating one set and upregulating another set of these transcription factors as has been directly shown in chick embryos (Roellig et al. 2017). Thus, the neural plate border appears to be a region of indecision or of "multi-lineage priming", similar to what has been proposed for the hematopoietic system (Hu et al. 1997; Nimmo, May, and Enver 2015).

It is possible that the combined expression of dorsal and ventral transcription factors may contribute to maintain the expression of some pluripotency factors precisely in this region until the end of gastrulation, as has been recently suggested (Buitrago-Delgado et al. 2015). However, many of the transcription factors that are thought to contribute to the pluripotency network of transcription factors in the animal cap of the *Xenopus* blastula including Snail1 and FoxD3 decline in expression during gastrulation before being upregulated specifically in the neural crest territory at the end of gastrulation (Buitrago-Delgado et al. 2015). In line with this, a recent study demonstrated that Ventx2, a member of the vertebrate-specific Ventx/Nanog family that becomes enriched at the neural border during neurulation, promotes multipotency and ectomesenchymal fate of the cranial neural crest in cooperation with Pax3 and Zic1 (Scerbo and Monsoro-Burq 2020). Taken together, these findings show that many of the transcription factors conferring pluripotency to neural crest cells are induced or upregulated at the end of gastrulation specifically in this territory rather than being maintained from blastula stage animal caps suggesting that neural crest acquires pluripotency downstream of neural plate border formation (Briggs et al. 2018; Scerbo and Monsoro-Burq 2020).

3.2.7 Two Models for Explaining the Origin of Neural Crest and PPE from the Neural Plate Border

To explain how neural crest and PPE arise as two distinct territories from the neural plate border, two different models have been proposed (Fig. 3.6A). The "neural plate border state model" proposes that the neural plate border is a common precursor for

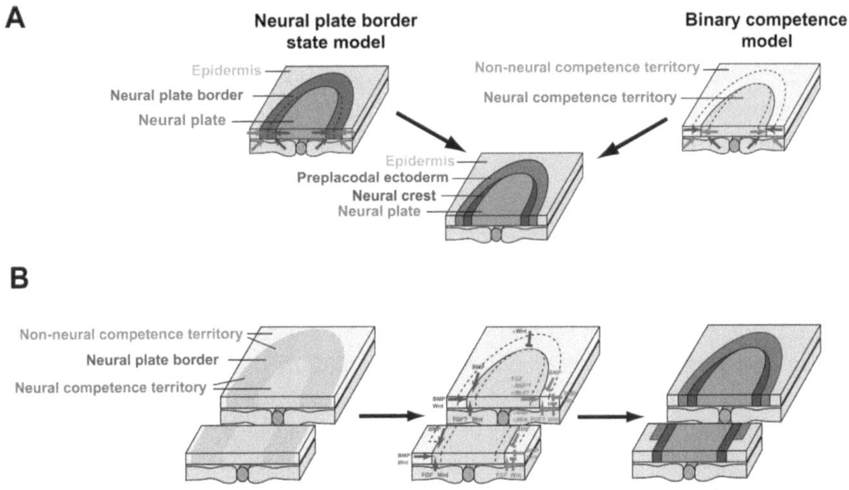

FIGURE 3.6 Models for neural plate border development. (**A**) The 'neural plate border state model' proposes that pre-placodal ectoderm (PPE) and neural crest are induced from a common precursor at the neural plate border, whereas the 'binary competence model' proposes that they are induced from non-neural and neural ectodermal competence territories, respectively. (**B**) Dorsally restricted neural and ventrally restricted non-neural competence factors overlap during gastrulation (left panel) but then resolve into mutually exclusive territories at the end of gastrulation (middle panel). Inducing signals from adjacent tissues induce PPE (FGF, BMP inhibitors, and Wnt inhibitors; shown on right side) and neural crest (FGF, BMP, and Wnt; shown on left side) at the border of non-neural and neural ectoderm, respectively. (Modified from Schlosser 2006.)

neural crest and PPE. Depending on which signals are received, either neural crest or PPE are induced from this region (Brugmann et al. 2004; Meulemans and Bronner-Fraser 2004; Litsiou et al. 2005; Patthey et al. 2009). The "binary competence model" instead proposes that two mutually exclusive competence territories are established in the ectoderm by the end of gastrulation. In the presence of appropriate signals from adjacent tissues, neural crest can be induced from the dorsal, neural competence territory, whereas PPE can be induced from the ventral, non-neural competence territory (Ahrens and Schlosser 2005; Pieper et al. 2012). This is supported by grafting experiments in *Xenopus* that showed that at the end of gastrulation only neural ectoderm is competent to form neural crest and only non-neural ectoderm is competent to form PPE in response to inducing signals (Pieper et al. 2012).

While the emerging picture of a dynamic neural plate border region, which gradually resolves into complementary neural crest and PPE territories, integrates aspects from both models (Figs. 3.4, 3.5, 3.6B), an open question about the hierarchy of lineage decisions at the neural plate border still remains. The binary competence model predicts that a decision for neural vs. non-neural ectoderm should precede the decision between neural plate and neural crest within the neural ectoderm and between epidermis and PPE within the non-neural ectoderm. The neural plate border state model predicts instead that the neural plate border first adopts a regulatory state distinct from neural plate and epidermis followed by a decision between neural crest and PPE within the neural plate border ectoderm. As an alternative to both models, cells at the neural plate border may not follow such a tree-like pattern of binary cell fate decisions at all but instead may adopt neural crest and PPE fates via multiple different pathways from progenitors expressing various combinations of transcription factors.

We currently do not have sufficient information to resolve this question because the importance of many of the transcription factors co-expressed at the neural plate border for directing lineage decisions in the ectoderm is not well understood. However, the dynamic changes of transcription factor expression in single cells documented in several recent studies can be best reconciled with the binary competence model. A recent immunohistochemical study (Roellig et al. 2017) showed that different cells in the neural plate border region of chick embryos express different combinations of transcription factors that ultimately define the neural plate (Sox2), neural crest (Pax7, which is, however, also expressed in the dorsal neural tube) and PPE (Six1) territories; epidermal transcription factors were not analyzed. In spite of this heterogeneity, the majority of cells (44–58%) co-express Sox2 and Pax7 (but not Six1), while only 1–4% co-express Six1 and Pax7 (but not Sox2) and only 0–1% co-express Sox2 and Six1 (but not Pax7). The predominance of cells co-expressing neural plate and neural crest markers conforms to the predictions of the binary competence model. It is further supported by recent reconstructions of cell lineage decisions from single-cell sequencing data in *Xenopus* and zebrafish that indicate that neural crest and neural plate cells arise from a common neural progenitor, while PPE and epidermis arise from a common non-neural progenitor (Briggs et al. 2018; Farrell et al. 2018; Wagner et al. 2018).

There is also increasing evidence from *Xenopus* and zebrafish that the sharp boundary developing between some dorsally (Zic1, Pax3) and ventrally restricted transcription factors (Dlx3/5, GATA2/3, FoxI1/3) separates a dorsal, neural competence territory from a ventral non-neural competence territory. For example, as mentioned in the previous section Zic1 and Pax3 promote PPE formation in the presence of Dlx3 but repress it in the absence of Dlx3 (Maharana and Schlosser 2018). Similarly, Dlx3 promotes the expression of dedicated neural crest specifiers in the presence of Zic1 but represses it in the absence of Zic1

While Zic1 represses PPE in the absence of Dlx3 at the end of gastrulation, Zic1 and Zic3 are required for the induction of both dedicated neural crest and dedicated neural plate transcription factors and promote neural crest over neural plate fate in the presence of Wnt signaling (Sato et al. 2005; Marchal et al. 2009; Maharana and Schlosser 2018). Conversely, while Dlx3 represses neural crest in the absence of Zic1 at the end of gastrulation, Dlx3/5, GATA2/3, and FoxI1/3 are required for epidermal and PPE formation and promote the adoption of epidermal fate in the presence or PPE fate in the absence of BMP (Kwon et al. 2010; Pieper et al. 2012; Bhat et al. 2012; Zhang et al. 2014; Maharana and Schlosser 2018). This suggests that Zic1 and Zic3 act as neural competence factors, which promote the adoption of different fates (neural plate vs. neural crest) in the neural ectoderm, whereas Dlx3/5, GATA2/3, and FoxI1/3 act as non-neural competence factors, which promote the adoption of different fates (epidermis vs. PPE) in the non-neural ectoderm, in a signaling-dependent manner.

Additional ventrally restricted transcription, in particular TFAP2a (and probably Msx1 and Ventx as well) also contribute to non-neural competence and engage in cross-regulatory interactions with Dlx3/5, GATA2/3, and FoxI1/3 that stabilize the non-neural regulatory state (McLarren et al. 2003; Matsuo-Takasaki et al. 2005; Kwon et al. 2010; Bhat et al. 2012; Pieper et al. 2012; Maharana and Schlosser 2018). Although the expression of TFAP2a (and other transcription factors of this category) does not respect the boundary between neural and non-neural competence territories and extends dorsally into the neural crest domain, it can induce epidermal markers only in ectoderm that is not neurally committed, while it promotes neural crest formation in neuralized ectoderm (Luo et al. 2002, 2003; Li and Cornell 2007; de Croze et al. 2011). This suggests that TFAP2a serves distinct functions as a competence factor in the non-neural ectoderm and a neural crest-promoting factor in the neural ectoderm but the molecular basis for its context-dependent function is presently not understood.

3.2.8 Towards a Gene Regulatory Network for Neural Crest and PPE Formation

In conclusion, the PPE and neural crest are established together with adjacent ectodermal territories in several phases during gastrulation (Figs. 3.4, 3.5, 3.6). During early gastrulation, in response to dorsally enriched BMP and Wnt inhibitors together with FGF signals, the expression domains of many ectodermal transcription factors become either restricted to the dorsal (neural) side (Zic1/3, Sox3) or to the ventral (non-neural) side (Dlx3/5, GATA2/3, TFAP2a, Msx1, FoxI1-3) of the ectoderm. However, among the ventrally restricted transcription factors, some (TFAP2a, Msx1, Ventx2) extend further dorsal than others (Dlx3/5, GATA2/3, FoxI1-3). Between the dorsal ectodermal domain, which exclusively expresses neural transcription factors and a ventral ectodermal domain, which exclusively expresses non-neural transcription factors, there is a broad "neural plate border" region, where both categories of transcription factors overlap. This region of overlap changes dynamically and becomes smaller and smaller during gastrulation.

During mid-gastrulation, additional transcription factors (Pax3, Myc, Id3, Hes4) become specifically upregulated in this neural plate border region by the combined activity of ventrally and dorsally restricted transcription factors in response to Wnt and FGF signals. Their expression is subsequently stabilized by positive feedback loops. During late gastrulation, the various transcription factors expressed in the neural plate border region conspire to activate neural crest specifiers (FoxD3, Snail1/2, Ets1, Sox9/10) in response to FGF, BMP, and Wnt signals and PPE specifiers (Six1/2, Six4/5, Eya) in response to FGF and inhibitors of BMP and Wnt signaling. Cross-repressive interactions between a subset of ventrally (Dlx3/5, GATA2/3, FoxI1-3) and dorsally restricted transcription factors (Zic1/3), which probably act as non-neural and neural competence factors, respectively, lead to the ultimately separation of PPE and neural crest to mutually exclusive domains. Subsequently, PPE and neural crest are maintained by positive feedback regulation of PPE and neural crest specifiers, respectively.

The interactions between transcription factors and signaling molecules involved in establishment of the PPE and its adjacent ectodermal territories at the neural plate

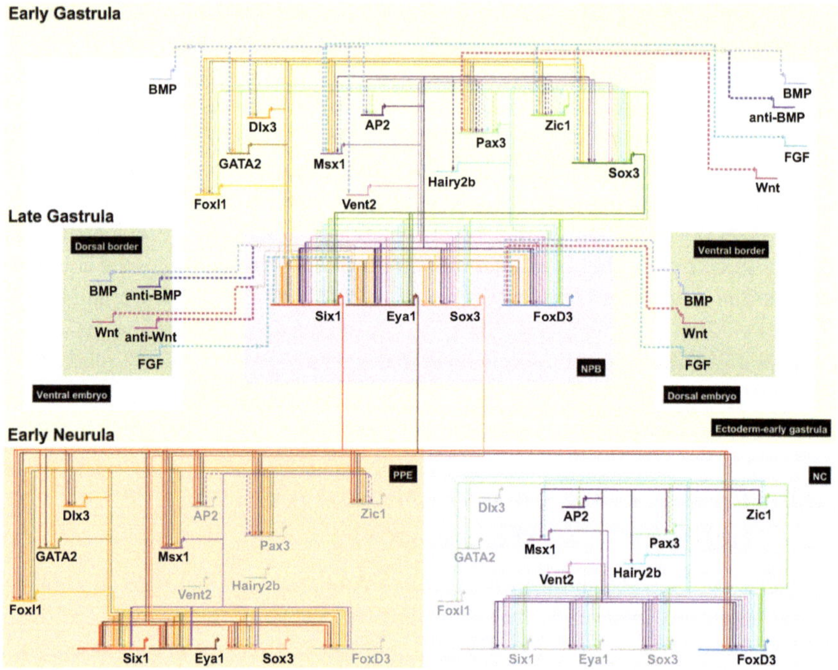

FIGURE 3.7 Gene regulatory network for neural plate border formation in *Xenopus*. Greyed out genes represent those inactive in a particular cell population (e.g. Zic1 in the PPE). All interactions shown are based on functional studies and may be direct or indirect. Solid lines indicate relationships established in (Pieper et al. 2012) and (Maharana and Schlosser 2018), whereas hatched lines indicate relationships established in previous studies (referenced in Maharana and Schlosser 2018). Arrows indicate activation. Bars show repression. Thick lines indicate relationships verified in loss of function (and often also in gain of function) experiments, while thin lines indicate relationships only supported by gain of function experiments. Signaling pathways are shown with extra thick lines. Often there is experimental evidence to support both activation and repression of genes by upstream TFs. Further studies are needed to elucidate the interactions determining these context-dependent effects. In the absence of functional data, temporal changes of regulatory relationships are proposed here based on changing expression patterns. For spatial distribution of transcription factors at the early gastrula, late gastrula, and early neurula stage see Fig. 3.5. (Reprinted from Maharana and Schlosser 2018.)

border allow us to draft a first, tentative GRN of neural plate border development (Fig. 3.7). However, we still have only very limited information, on which of these interactions are direct and which ones are mediated by other factors. Further studies revealing direct transcription factor binding to gene cis-regulatory regions (for example by chromatin immune precipitation followed by sequencing: ChIP-Seq) are, therefore, urgently needed to increase the explanatory and predictive power of this GRN (Davidson 2010; Peter and Davidson 2015).

4 Development of Individual Placodes from their Common Primordium

We have seen in the last chapter that all placodes originate from a common primordium. The latter is defined by the expression of transcriptional regulators of the Six and Eya families in the pre-placodal ectoderm (PPE) which is established as a distinct ectodermal domain during gastrulation. Six and Eya factors bias the PPE to develop into one of the different cranial placodes. During subsequent development, different parts of the PPE will give rise to different types of placodes. Whereas adenohypophyseal, olfactory, and lens placodes will form from its anterior portion, otic, lateral line, and epibranchial placodes will form from its posterior part, with profundal and trigeminal placodes sandwiched in between (Fig. 4.1).

The subdivision of the PPE into individual placodes is a complex process comprising several phases. The first phase involves the multistep regional specification of different placodes within the PPE due to the region-specific induction of different transcription factors. In a second phase, individual placodes separate from each other by a combination of transcriptional cross-repression, apoptosis, and cell movements. In a third phase, placodes undergo invagination or other morphogenetic processes to become internalized and adopt a specific three-dimensional structure. In this chapter, I will review the first two of these three phases, placing particular emphasis on the role of transcription factors during the first phase, which is most relevant for our understanding of how different placodal cell types are generated in a region-specific manner (for previous reviews see Bailey and Streit 2006; Schlosser 2006, 2010; Park and Saint-Jeannet 2010; Grocott, Tambalo, and Streit 2012; Saint-Jeannet and Moody 2014; Singh and Groves 2016). The morphogenetic movements of placodes in the third phase present a fascinating, complex, and still poorly understood topic of its own, which is, however, beyond the scope of this book.

4.1 MULTISTEP INDUCTION OF INDIVIDUAL PLACODES

The subdivision of the PPE into region-specific domains of gene expression is initially linked to general anteroposterior patterning of the ectoderm. Already during gastrulation some transcription factors begin to be expressed only in certain regions along the anteroposterior axis in response to signaling molecules that are unevenly distributed along this axis. These include in particular Wnts, fibroblast growth factor (FGF), and retinoic acid, all of which show low concentrations anteriorly but increase

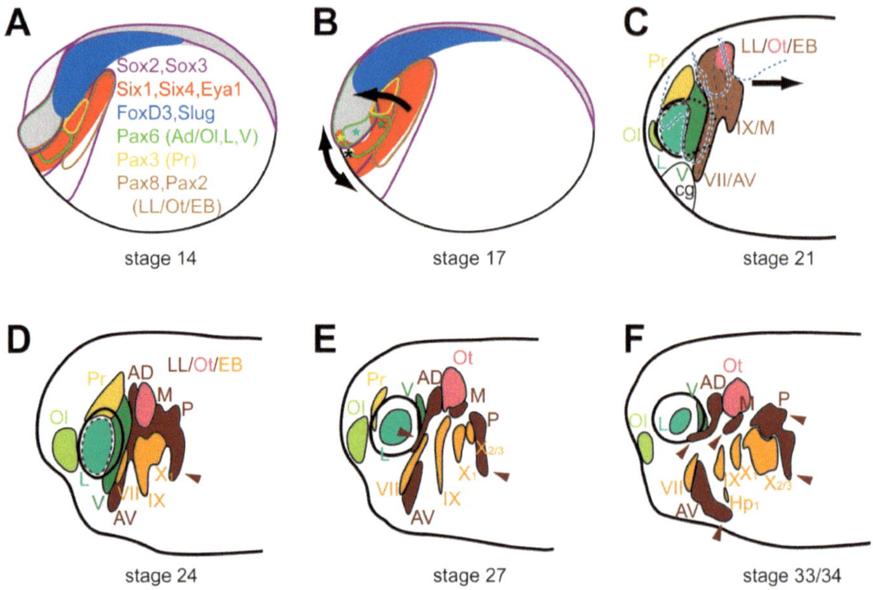

FIGURE 4.1 Summary of placode development in *Xenopus* embryos (lateral views). (**A, B**) Gene expression domains (colored outlines) during neural plate (**A**) and neural fold (**B**) stages with PPE shown in red. Arrows in panel **B** indicate shifts of placodal expression domains due to elevation of the neural folds and formation of the optic vesicles. Green stars identify three areas of Pax6 expression that will contribute to adenohypophyseal and olfactory placode (light green), lens placode (blue green), and trigeminal placode (dark green). Six1 and Eya1 expressions are downregulated at neural fold stages in the regions of prospective lens placode and cement gland (black asterisk). (**C–F**) Development of placodes from end of neurulation until late tailbud stages (adenohypophyseal placode not shown). At stage 21, the posterior placodal area is divided into an anterior and a posterior subregion of thickened ectoderm, separated ventrally by an indentation and dorsally by a region of thinner ectoderm (between broken black lines), while the prospective otic placode (pink) is identifiable as a particularly prominent thickening. The broken blue lines in panel **C** indicate neural crest streams. The arrow in **C** indicates that the posterior placodal area expands posteriorly at early tailbud stages. Brown arrowheads in panels **D–F** indicate developing lateral line primordia. Ad/Ol, anterior placodal area, from which adenohypophyseal (Ad) and olfactory placodes (Ol) develop; AV, anteroventral lateral line placode; cg, cement gland; Hp_1, first hypobranchial placode; L, prospective lens placode (hatched outline), lens placode or lens; LL/Ot/EB, posterior placodal area, from which lateral line (LL), otic (Ot), and epibranchial (EB) placodes develop; M, middle lateral line placode; Ol, olfactory placode; Ot, otic placode or vesicle; P, posterior lateral line placode; Pr, profundal placode; V, trigeminal placode; VII, facial epibranchial placode; IX, glossopharyngeal epibranchial placode; X_1, first vagal epibranchial placode; $X_{2/3}$, second and third vagal epibranchial placodes (fused). (Reprinted from Schlosser 2006; based on Schlosser and Northcutt 2000; Schlosser and Ahrens 2004.)

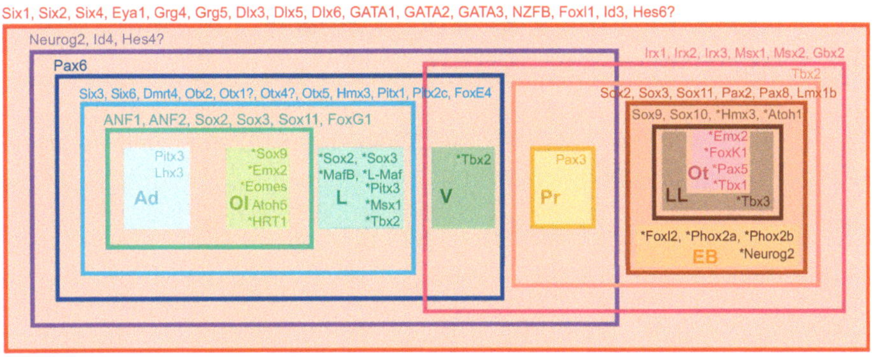

FIGURE 4.2 Transcription factor expression in the PPE during specification of placodes. Schematic summary of expression domains of selected transcription factors in the placodal ectoderm of neural plate stage *Xenopus* embryos. Anterior is to the left. Transcription factors preceded by asterisks refer to expression domains established at later stages. Question marks indicate that precise domain boundaries are not known. Ad, adenohypophyseal placode; Ol, olfactory placode; L, lens placode; V, trigeminal placode; Pr, profundal placode; LL, lateral line placodes; Ot, otic placode; EB, epibranchial placodes. (Reprinted with modification from Schlosser 2006.)

in concentrations toward the posterior. The extent of expression along the anteroposterior axis differs for the transcription factors resulting in a pattern of partly nested, partly overlapping transcription factor domains (Fig. 4.2) (e.g. Schlosser and Ahrens 2004). In response to signals from various signaling centers in the neural plate or tube and the underlying endomesoderm, this pattern of transcription factor expression changes and becomes more complex over time.

The different combinations of transcription factors co-expressed in different domains of the PPE suggest that individual placodes are specified in a stepwise and combinatorial fashion (Torres and Giráldez 1998; Schlosser 2006). However, so far only few of these factors have been confirmed to be functionally required for placode induction experimentally. Many other transcription factors have been implicated in placode induction based on their expression pattern, but we still lack experimental evidence.

Immediately after gastrulation, transcription factors tend to be expressed in a nested pattern with two centers, one anteriorly and one posteriorly (Fig. 4.2). The combination of transcription factors co-expressed anteriorly and posteriorly defines two multiplacodal areas: an extended anterior placodal area giving rise to lens, adenohypophyseal and olfactory placodes (with the latter two arising from a smaller common area, the so-called anterior placodal area), and a posterior placodal area comprising the future otic, lateral line, and epibranchial placodes (Schlosser and Ahrens 2004; Schlosser 2006; Toro and Varga 2007; Ladher, O'Neill, and Begbie 2010; Grocott et al. 2012; Chen and Streit 2013). Trigeminal and profundal placodes originate in between, in an area where some of the anterior and posterior transcription factors overlap (Schlosser 2006). As will be reviewed below, some experimental evidence suggests that different parts of a multiplacodal area initially have equivalent developmental potential and are able to give rise to all of its derivative placodes. Only after new signals induce more localized expression of some transcription factors within each multiplacodal area, is the identity of individual placodes specified. This suggests that PPE cells become specified as individual placodes in multiple steps, first making a lineage decision for a particular multiplacodal area before choosing to adopt a particular placodal fate.

A tissue is said to be specified, when it is able to adopt its fate in isolation without requiring additional signals (Slack 1991). However, the fate of specified tissues is still malleable and may change when exposed to signals promoting alternative fates (e.g. after grafting to a different location in the embryo). When a tissue no longer responds to such contravening signals and has adopted its fate irreversibly, it is said to be committed (Slack 1991). Irreversible commitment of ectodermal regions to a particular placodal fate typically occurs around neural fold stages in amphibians (Yntema 1933, 1939; Schlosser and Northcutt 2001; reviewed in Schlosser and Northcutt 2001) and shortly after neural tube closure in amniotes in an anteroposterior sequence (Baker et al. 1999; Groves and Bronner-Fraser 2000; Bhattacharyya and Bronner-Fraser 2008). However, there are differences between placodes as well as some species-specific differences; for example, commitment occurs earlier in frogs than in salamander embryos (reviewed in Baker and Bronner-Fraser 2001; Schlosser 2002a).

In the following sections, I will first address the initial subdivision of the PPE by anteroposterior patterning mechanisms and then discuss the development of each multiplacodal area.

4.1.1 Initial Subdivision of the PPE along the Anteroposterior Axis

Simultaneous with the establishment of different ectodermal territories (epidermis, PPE, neural crest, and neural plate) along the dorsoventral axis during gastrulation (see Chapter 3), the embryonic ectoderm is also subdivided along the anteroposterior axis. Many classical studies suggest that anteroposterior patterning in vertebrates is linked to neural induction. During this process, cells seem to first adopt anterior identity throughout the neural plate (and the dorsal mesodermal tissues induced by the organizer) followed by a transformation of regional identity in its posterior part. This has been initially proposed in the "activation-transformation" model of Nieuwkoop and colleagues (Nieuwkoop and Nigtevecht 1954; Stern 2001). However, recent studies suggest that this mechanism may only be applicable to anteroposterior patterning in the head region, where hindbrain is initially induced as forebrain followed by posterior transformation, while in the trunk region the spinal cord acquires its posterior identity directly without passing through an anterior state (Metzis et al. 2018; Al Anber and Martin 2019; Polevoy et al. 2019). Here I will only discuss anteroposterior patterning mechanisms in the head since these are relevant for understanding regionalization of the PPE.

Cranial anteroposterior patterning involves the establishment of region-specific domains of transcription factor expression in response to signaling molecules with differential activity along the anteroposterior axis. These signals comprise mostly the posteriorly enriched Wnts, FGFs, and retinoic acid (Altmann and Hemmati-Brivanlou 2001; Schier and Talbot 2005; Niehrs 2010; Hikasa and Sokol 2013;

Carron and Shi 2016). Retinoic acid, which is required for Six1 and Eya1 expression in the PPE (Janesick et al. 2012; Jaurena et al. 2015), also appears to play a central role in setting its posterior boundary. Retinoic acid activates the transcription factor Tbx1, which is expressed in the posterior PPE, as well as the Groucho-associated corepressor Ripply3, which is expressed further posterior but overlapps with Tbx1. Whereas Tbx1 alone promotes FGF8 expression and PPE formation, Ripply3 converts Tbx1 into a transcriptional repressor, thereby preventing expression of FGF8 and induction of the PPE posteriorly (Janesick et al. 2012).

The Wnt gradient plays a particularly important role for the anteroposterior patterning of the PPE as it does for patterning of other tissues along the entire anteroposterior axis. While Wnt signals are secreted throughout the embryo, they are sequestered anteriorly by Wnt antagonists. The localized production of diffusible Wnt antagonists in the anterior portion of the organizer (including the prospective prechordal plate) and other cranial tissues helps to establish a gradient of canonical Wnt signaling with low activity in the anterior and increasing activity toward the posterior. This gradient activates different transcription factors in a concentration-dependent manner along the anteroposterior axis (Kiecker and Niehrs 2001; Petersen and Reddien 2009; Niehrs 2010). Transcription factors encoded by Wnt target genes, such as *Cdx*, *Meis*, *Gbx*, and *Irx*, are activated posteriorly and in turn activate other posterior transcription factors such as Hox proteins. In contrast, transcription factors that are repressed by Wnt signaling including Otx2, FoxG1 (=BF-1), Six3, Fezf1/2 and Hesx1 (=ANF1) are upregulated anteriorly (Hashimoto et al. 2000; Kazanskaya, Glinka, and Niehrs 2000; Gómez-Skarmeta, Calle-Mustienes, and Modolell 2001; Kiecker and Niehrs 2001; Braun et al. 2003; Lecaudey et al. 2005; Nordstrom et al. 2006; Pilon et al. 2006; Li et al. 2009; Elkouby et al. 2010).

However, the expression boundaries of these transcription factors differ depending on their sensitivity for Wnt signaling. The expression of Irx, for example, extends further anterior than that of Gbx, while the expression of Otx2 extends further posterior than that of Six3 (Simeone et al. 1993; Oliver et al. 1995; von Bubnoff, Schmidt, and Kimelman 1996; Bosse et al. 1997; Bellefroid et al. 1998; Glavic, Gómez-Skarmeta, and Mayor 2002; Kobayashi et al. 2002; Steventon, Mayor, and Streit 2012). Anterior and posterior transcription factors are initially activated in an overlapping pattern, but reciprocal repression between them soon establishes sharp boundaries between their expression domains. Whereas the reciprocal transcriptional repression between Otx2 and Gbx2 defines the mid-hindbrain boundary, mutual repression of Six3-Irx3, Fezf-Irx (with equivalent functions of Fexf1 and Fezf2 and the various Irx family members), and/or Arx-Irx positions the zona limitans intrathalamica (ZLI) (Broccoli, Boncinelli, and Wurst 1999; Millet et al. 1999; Martinez-Barbera et al. 2001; Kobayashi et al. 2002; Tour et al. 2002; Braun et al. 2003; Hirata et al. 2006; Jeong et al. 2007; Rhinn et al. 2009; Rodriguez-Seguel, Alarcón, and Gómez-Skarmeta 2009).

The function of many of these anteroposteriorly restricted transcription factors has been best characterized in the neural plate, where they play important roles for specifying different regions of the developing central nervous system (CNS) (reviewed in Schilling and Knight 2001; Wurst and Bally-Cuif 2001; Lichtneckert and Reichert 2005; Holland et al. 2013; Carron and Shi 2016). However, their expression in most cases extends into the PPE or even into more lateral non-neural ectoderm, where their expression boundaries are probably similarly modulated by Wnt signaling (Fig. 4.3). The distribution of transcripts indicates that the boundary between Six3 or Fezf1/2 and Irx runs between the prospective lens placode in the extended anterior placodal area and the profundal/trigeminal placodes (Itoh et al. 2002; Schlosser and Ahrens 2004; Rodriguez-Seguel et al. 2009). Whether reciprocal repression between Fezf and Irx contributes to the sharpening of this boundary, like in the CNS, has not yet been investigated. The boundary between Otx2 and Gbx2, however, is located more posterior between the profundal/trigeminal placodes and the posterior placodal area and is sharpened by mutual repression of Otx2 and Gbx2 (Steventon et al. 2012).

These expression patterns suggest that early anteroposteriorly restricted transcription factors may also contribute to the regionalization of the PPE. Elevation of Wnt signaling during gastrula stage indeed leads to the anterior shift or expansion of trigeminal and otic placodes while Wnt inhibition has the opposite effect (Itoh et al. 2002; Bajoghli et al. 2009; Rhinn et al. 2009; Steventon and Mayor 2012, Steventon et al. 2012). Furthermore, the anteriorly restricted Otx2 and Six3/6 are required for development of the adenohypophyseal, olfactory, and lens placodes (Acampora et al. 1995; Matsuo et al. 1995; Gammill and Sive 2001; Li et al. 2002; Carl, Loosli, and Wittbrodt 2002; Lagutin et al. 2003; Liu et al. 2006). Otx2 but not Six3/6 is also required for the formation of profundal and trigeminal placodes (Steventon et al. 2012). Conversely, the posteriorly restricted Gbx and Irx transcription factors are required for the proper expression of transcription factors specifying the posterior placodal area (including the otic placode) such as Pax8 and Pax2 (Glavic et al. 2004; Feijoo et al. 2009; Steventon et al. 2012). The otic vesicle however, forms in mutants or morphants of particular *Irx* or *Gbx* genes (Wassarman et al. 1997; Lebel et al. 2003; Byrd and Meyers 2005; Feijoo et al. 2009). Whether this reflects functional compensation by other members of the *Irx* or *Gbx* gene families remains to be tested. Irx but not Gbx factors are also required for the development of profundal and trigeminal placodes (Itoh et al. 2002; Feijoo et al. 2009; Rodriguez-Seguel et al. 2009).

Taken together, this indicates that the earliest subdivision of the PPE along the anteroposterior axis is determined by transcription factors that are initially activated in relatively broad domains in anterior or posterior ectoderm by low or high levels of Wnt signaling, respectively, in cooperation with other signals (Figs. 4.2 and 4.3). Co-expression of anteriorly confined transcription factors, such as Otx2, Six3/6, and Fezf1/2 (and absence of Gbx and Irx expression), helps to define the extended anterior placodal area which will form the adenohypophyseal, olfactory, and lens placodes (Fig. 4.3). Co-expression of posteriorly confined Gbx and Irx family members (and absence of Otx2, Six3/6, and Fezf1/2 expression) helps to define the posterior placodal area which will form the otic, lateral line, and epibranchial placodes (Fig. 4.3). In between these two regions, the profundal and trigeminal placodes develop in a part of the PPE, where the anteriorly restricted Otx2 overlaps with posteriorly restricted Irx factors, but which does not express other anteriorly (Six3/6, Fezf1/2) or posteriorly (Gbx) restricted transcription factors (Fig. 4.3).

However, while the transcription factors involved in the early subdivision of the PPE along the anteroposterior axis are required for the formation of these multiplacodal areas, they are not sufficient. In the following sections, I will discuss

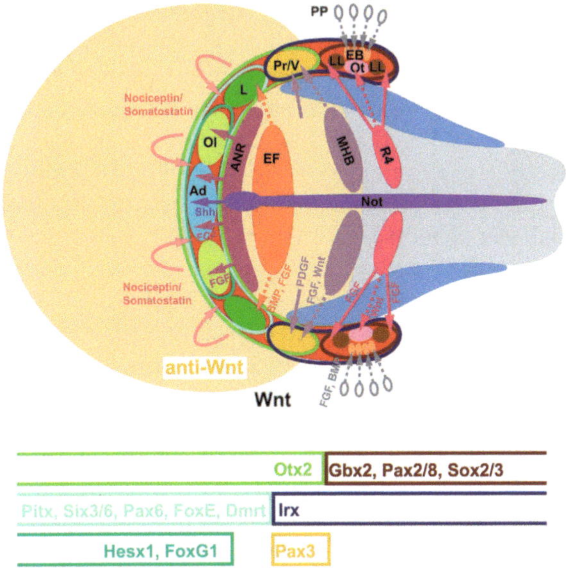

FIGURE 4.3 Signals involved in multistep placode induction. Posteriorly restricted Wnt signals and signaling centers within the mesoderm and neural plate induce transcription factors within the PPE (red) that specify multiplacodal areas (colored outlines) and individual placodes (colored ovals). The extended anterior placodal area (Ad/Ol/L) is indicated by a mint green outline, the anterior placodal area (Ad/Ol) by a dark green outline, and the posterior placodal area (LL/Ot/EB) by a brown outline. Distribution of some transcription factors in corresponding areas is shown below. Hatched arrows indicate signaling events at later (post-neural plate) stages. Ad, adenohypophyseal placode; ANR, anterior neural ridge; EB, epibranchial placodes; EF, eye field; L, lens placode; LL, lateral line placodes; MHB, midbrain–hindbrain boundary; Not, notochord; Ol, olfactory placode; Ot, otic placode; PP, pharyngeal pouches; Pr/V, profundal/trigeminal placodes; R4, rhombomere 4. (Reprinted with modification from Schlosser 2010.)

which additional signals and transcription factors are required for the induction of each multiplacodal area and how the latter subsequently further subdivide into individual placodes.

4.1.2 Origin of Adenohypophyseal, Olfactory, and Lens Placodes

4.1.2.1 Induction of the Extended Anterior Placodal Area

The anterior part of the PPE, which co-expresses Otx2, Fezf1/2, and Six3/6, subsequently gives rise to the adenohypophyseal, olfactory, and lens placodes (Schlosser 2006; Toro and Varga 2007). Additional transcription factors including Pax6, Pitx, and FoxE family members, Lhx2 and DMRT4 are upregulated in this area during gastrulation (Zygar, Cook, and Grainger 1998; Hollemann and Pieler 1999; Kenyon, Moody, and Jamrich 1999; Pommereit, Pieler, and Hollemann 2001; Schweickert, Steinbeisser, and Blum 2001; Zuber et al. 2003; Schlosser and Ahrens 2004; Huang et al. 2005). The expression pattern of these transcription factors varies to some extent between species. Expression of Pax6 in frogs and of Pitx3 in zebrafish, for example, extends further posterior and also includes the trigeminal placode (Schlosser and Ahrens 2004; Zilinski et al. 2005). However, the co-expression of multiple of these transcription factors appears to define a common precursor for adenohypophyseal, olfactory, and lens placodes, which I have termed the extended anterior placodal area (Schlosser 2010), in all vertebrates (Fig. 4.3). Other transcription factors such as FoxG1 or Hesx1 are confined to the more anterior part of this domain, termed the anterior placodal area (Schlosser and Ahrens 2004), which includes the prospective adenohypophyseal and olfactory placodes, but excludes the prospective lens placode.

From neural plate or fold stages on, some of these transcription factors become restricted to individual placodes and additional transcription factors with putative roles in placode specification begin to be expressed in a placode specific pattern. In *Xenopus*, for example, Hesx1, Pitx1, and Pitx2 become confined to (or enriched in) the adenohypophyseal placodes, DMRT4 to the olfactory placode and FoxE3 and Pax6 to the lens placode, while Pitx3 and Lhx3 becomes specifically upregulated in the adenohypophyseal placode, DMRT5, Atoh7 (=Ath5) in the olfactory placodes and Sox2, Tbx2, MafB, and Nrl (=L-Maf) in the lens placode (Taira et al. 1993; Kanekar et al. 1997; Ishibashi and Yasuda 2001; Schlosser and Ahrens 2004; Schlosser 2006; Parlier et al. 2013).

The PPE specifiers Six1 and Eya as well as non-neural competence factors TFAP2a, GATA2/3, Dlx3/5, and FoxI1/3 are required for the expression of many transcription factors expressed in the extended anterior placodal area (Solomon and Fritz 2002; Christophorou et al. 2009; Kwon et al. 2010; Bhat, Kwon, and Riley 2012; Pieper et al. 2012). In addition, the anterior transcription factor Otx2 is an essential upstream regulator for all transcription factors expressed throughout the extended anterior placodal area (Goodyer et al. 2003; Dickinson and Sive 2007; Ogino, Fisher, and Grainger 2008; Steventon et al. 2012; Parlier et al. 2013) and Six3/Six6 may also contribute to their activation (Liu et al. 2006).

However, signals first from the prechordal mesoderm and later from the anterior PPE itself are also required for the induction of Pax6 and Pitx3 in this region (Lleras-Forero et al. 2013; Hintze et al. 2017). The neuropeptides somatostatin and nociceptin have recently been shown to mediate these inductive activities in chick and zebrafish embryos (Lleras-Forero et al. 2013). Somatostatin, secreted from the prechordal plate, induces nociceptin in the overlying anterior PPE and both signals activate Pax6 (and Pitx3) in the anterior PPE. Pax6, subsequently, is required together with Otx2 for the activation of Pitx, DMRT, and other transcription factors (reviewed in Graw 2010; Ogino et al. 2012; Cvekl and Zhang 2017), while FoxE3 is activated downstream of Pitx3 (Ahmad et al. 2013).

Pax6 is excluded from the territory of the prospective profundal placode in the most posterior part of the Otx2 expression domain where Pax3 is expressed. Pax3 is confined to this territory due to its dependence on signals from the adjacent neural plate (see below) in addition to Otx2 (Steventon et al. 2012). Once induced, Pax3 engages in cross-repressive interactions with Pax6 thereby excluding Pax6 from the caudal part of the Otx2 expression domain (Wakamatsu 2011).

4.1.2.2 Induction of Individual Placodes within the Extended Anterior Placodal Area

Several studies indicate that the co-expression of Pax6 with Pitx, DMRT, and FoxE and other transcription factors in the extended anterior placodal area defines a common regulatory state endowing this region with the competence to develop into any of the three anterior-most placodes (adenohypophyseal, olfactory, and lens). Which of these individual placodal fates a particular part of this region will adopt depends on its exposure to signaling molecules from adjacent tissues (Fig. 4.3). Specification of the adenohypophyseal placode requires hedgehog signals from the prechordal plate, notochord, and the floor plate of the developing neural tube (Karlstrom, Talbot, and Schier 1999; Kondoh et al. 2000; Treier et al. 2001; Varga et al. 2001; Sbrogna, Barresi, and Karlstrom 2003; Herzog et al. 2003; Dutta et al. 2005), whereas the olfactory placode requires FGF signaling from the anterior-most neural plate (Bailey et al. 2006) and the lens placode bone morphogenetic protein (BMP) signaling possibly from the adjacent neural plate or the placodal ectoderm itself (Furuta and Hogan 1998; Wawersik et al. 1999; Varga et al. 2001; Sjödal, Edlund, and Gunhaga 2007).

Manipulation of these signaling pathways can expand one anterior placodal fate at the expense of others. Experimental decreases of hedgehog signaling leads to an expansion of olfactory and lens placodes but reduction of the adenohyophyseal placode, while increases of hedgehog signaling have the opposite effect (Kondoh et al. 2000; Varga et al. 2001; Sbrogna, Barresi, and Karlstrom 2003; Cornesse, Pieler, and Hollemann 2005; Dutta et al. 2005; Zilinski et al. 2005). Increasing FGF signaling, in turn, promotes olfactory over lens placodes, while increasing BMP signaling promotes lens over olfactory placodes (Bailey et al. 2006; Sjödal et al. 2007).

While the expression of Pitx, DMRT, FoxE, and Pax6 extends throughout the extended anterior placodal area, FoxG1 and Hesx1 expression are limited to a narrower domain (the anterior placodal area) within this region, which will form the adenohypophyseal and olfactory placodes (Zaraisky et al. 1995; Mathers et al. 1995; Bourguignon, Li, and Papalopulu 1998; Eagleson and Dempewolf 2002). From neural fold stages on, Hesx1 and some Pitx paralogs will then become confined to the adenohypophyseal placode and FoxG1 to the olfactory placode. Another transcription factor Sp8, which is widely expressed in the PPE but excluded from the prospective adenohypophyseal placode, may also help to distinguish between adenohypophyseal and olfactory placode (Bell et al. 2003; Sánchez-Arrones et al. 2017). Hesx1, FoxG1, and Sp8 transcription factors are all required for olfactory placode development; Hesx1 is also essential for development of the adenohypophyseal placode (Xuan et al. 1995; Dattani et al. 1998; Duggan et al. 2008; Kawauchi et al. 2009; Kasberg, Brunskill, and Steven Potter 2013).

Hesx1 is induced in ectoderm expressing Otx2 and Six3/6 by signals from anterior endomesoderm in particular Wnt inhibitors (Hermesz, Mackem, and Mahon 1996; Thomas and Beddington 1996; Martinez-Barbera and Beddington 2001; Matsuda and Kondoh 2014). Due to the localization of these signals, the domain of Hesx1 expression is restricted to the anterior portion of the Otx2 and Six3/6 expressing domain and defines the anterior placodal area. Hesx1 is then required for activation of FGF8 expression in the ectoderm thereby establishing a new signaling center at the anterior border of the neural plate known as the anterior neural ridge (ANR) (Fig. 4.3) (Dattani et al. 1998; Martinez-Barbera and Beddington 2001). FGF8 in the ANR is required for patterning both the non-neural ectoderm and the forebrain in a concentration-dependent manner (Shimamura and Rubenstein 1997; Rubenstein et al. 1998; Shanmugalingam et al. 2000; Eagleson and Dempewolf 2002; Kawauchi et al. 2005; Storm et al. 2006) and is essential for inducing FoxG1 in the anterior placodal area (Shimamura and Rubenstein 1997; Houart, Westerfield, and Wilson 1998; Ye et al. 1998; Eagleson and Dempewolf 2002; Kobayashi et al. 2002; Bailey et al. 2006).

Taken together, this suggests that the extended anterior placodal area develops at the intersection of the PPE and a region where anteriorly restricted transcription factors such as Otx2 and Six3/6 are expressed. Cells in this area probably adopt an individual placodal fate in two steps by first deciding between a lens fate and an anterior (adenohypophyseal/olfactory) placodal fate before taking a decision between adenohypophyseal and olfactory placodes. However, this needs to be confirmed in further studies. Additional hedgehog, BMP, and FGF signals from the diencephalon are subsequently required for the development of the anterior pituitary from the adenohypophyseal placode (Takuma et al. 1998; Treier et al. 1998; Herzog et al. 2003, 2004; Guner et al. 2008), while FGFs, BMPs and Notch signaling from frontonasal mesenchyme, ectoderm, and optic cup are required during multiple steps of both olfactory and lens development (Henry and Grainger 1987, 1990; Furuta and Hogan 1998; Wawersik et al. 1999; La Mantia et al. 2000; Faber et al. 2001; Fisher and Grainger 2004; Kawauchi et al. 2004; Ogino et al. 2008; Maier et al. 2010; Garcia et al. 2011; Pandit, Jidigam, and Gunhaga 2011; Parlier et al. 2013).

4.1.3 Origin of Otic, Lateral Line, and Epibranchial Placodes

4.1.3.1 Induction of the Posterior Placodal Area

The posterior part of the PPE, which is characterized by expression of Gbx and Irx family members but does not express Otx2, Six3/6, and Fezf1/2, gives rise to the otic, lateral line, and epibranchial placodes (reviewed in Schlosser 2006; Baker, O'Neill, and McCole 2008; Ladher et al. 2010; Chen and Streit 2013). Several transcription factors with well-established functions in posterior placode development, including Pax8, Pax2, and Sox2/3 are upregulated in this domain during gastrulation (Heller and Brändli 1999; Schlosser and Ahrens 2004; Gou et al. 2018). The region of co-expression of these transcription factors appears to define a common precursor for the otic, lateral line, and epibranchial placodes termed the posterior placodal area (Fig. 4.3) (Schlosser and Ahrens 2004; Pieper et al. 2011). A recent study in chick embryos has identified Lmx1a, Sox13, and Zbtb16 as additional transcription factors specifically enriched in the otic-epibranchial domain, lateral line placodes being absent in amniotes (Chen et al. 2017). Similar to the extended anterior placodal area discussed above, the posterior placodal area appears to initially share a common regulatory state, endowing this region with the competence to develop into one of the three posterior-most types of placode (otic, lateral line, and epibranchial placodes) at the end of gastrulation. The ultimate fate of each subregion is then specified in a signaling dependent manner as discussed further below.

Somewhat surprisingly, fate mapping studies in *Xenopus* and zebrafish have shown that only the anterior epibranchial (facial and glossopharyngeal) and lateral line placodes (anterodorsal, anteroventral, and middle) are derived from the posterior placodal area as established by the end of gastrulation (Pieper et al. 2011; McCarroll et al. 2012). The more posterior vagal epibranchial and posterior lateral line placodes originate from cranial ectoderm immediately posterior to this area, which only upregulate transcription factors defining the regulatory states for PPE (Six1/2, Six4/5, and Eya) and posterior placodal area (Pax8, Pax2, and Sox2/3) after neural tube closure (Pieper et al. 2011). This shows that development of the posterior-most placodes of these two series is delayed and that they may be established by upstream regulators that are partly different from those activating their anterior cousins.

Additional transcription factors begin to be expressed in the posterior placodal area from neural plate or fold stages on, some of them localized to specific subregions. In *Xenopus*, FoxG1 and Lmx1b are upregulated in the entire posterior placodal area (Bourguignon et al. 1998; Haldin et al. 2003), while Sox9, Sox10, and Hmx3 (=Nkx5.1) become expressed in its dorsal part (prospective otic and lateral line placodes) (Spokony et al. 2002; Honoré, Aybar, and Mayor 2003; Bayramov et al. 2004). After neural tube closure, expression of many transcription factors (Lmx1b, Sox9, Sox10, Irx3, Gbx2) then becomes confined to the otic placode or vesicle (von Bubnoff et al 1996; Bellefroid et al. 1998; Spokony et al. 2002; Honoré et al. 2003; Haldin et al. 2003). At the same time, new transcription factors with putative roles in placode specification and differentiation become specifically upregulated in the otic (Tbx1, Pax5, FoxK1: Heller and Brändli 1999; Pohl and Knöchel 2004; Ataliotis et al. 2005), lateral line (Tbx3: Schlosser and Ahrens 2004) and epibranchial placodes (FoxI2, Phox2a,2b: Talikka et al. 2004; Pohl, Rössner, and Knöchel 2005). Similar schedules of establishing differential expression of transcription factors in placodes derived from the posterior placodal area are observed in other vertebrates (Ladher et al. 2010; Grocott et al. 2012; Chen and Streit 2013; Aguillon, Blader, and Batut 2016; Chen et al. 2017).

The PPE specifiers Six1 and Eya as well as non-neural competence factors TFAP2a, GATA 2/3, Dlx3/5, and FoxI1/3, which were required for development of anterior placodes, are also essential regulators of many transcription factors specifically expressed in the posterior placodal area (Solomon and Fritz 2002; Liu et al. 2003; Solomon et al. 2003; Hans, Liu, and Westerfield 2004; Hans et al. 2007; Christophorou et al. 2009; Kwon et al. 2010; Bhat et al. 2012; Pieper et al. 2012; Khatri, Edlund, and Groves 2014; Edlund, Birol, and Groves 2015; Birol et al. 2016). In addition, the posterior transcription factors of the Gbx and Irx families are required for the expression of Pax8, Pax2 and Sox2/3 in the posterior placodal area (Glavic et al. 2004; Feijoo et al. 2009; Steventon et al. 2012).

The role of FoxI1/3 and Dlx3/5 for the development of the posterior placodal area is particularly well studied. Besides their partly redundant role as general non-neural competence factors, these transcription factors take on distinct and separable functions as competence factors for the posterior placodal area after neural tube closure. Whereas FoxI1 and Dlx3 play central roles in anamniotes, FoxI3 and Dlx5 take over these roles in amniotes (Solomon and Fritz 2002; Liu et al. 2003; Solomon et al. 2003; Hans et al. 2004; Hans et al. 2007; Pieper et al. 2012; Khatri et al. 2014; Edlund et al. 2015; Birol et al. 2016).

FoxI1/3 and Dlx3/5 auto- and cross-regulate each other's expression (Solomon and Fritz 2002; McLarren, Litsiou, and Streit 2003; Solomon et al. 2003; Aghaallaei, Bajoghli, and Czerny 2007; Pieper et al. 2012) and both families of transcription factors act in mediating the competence of ectoderm to respond to FGF (Solomon et al. 2003; Hans et al. 2004, 2007; Padanad et al. 2012; Khatri et al. 2014). However, both transcription factors were shown to be independently required for the formation of the posterior placodal area and the subsequent development of otic, epibranchial, and lateral line placodes (Solomon and Fritz 2002; Liu et al. 2003; Solomon et al. 2003; Hans et al. 2004, 2007; Hans, Irmscher, and Brand 2013; Padanad et al. 2012; Khatri et al. 2014). In zebrafish, FoxI1 has been shown to be necessary for the induction of Pax8 and Sox2/3, whereas Dlx3 is required for the induction of Pax2 in the posterior placodal area in response to FGF signaling (Nissen et al. 2003; Hans et al. 2004; Solomon, Kwak, and Fritz 2004; Mackereth et al. 2005; Sun et al. 2007; Hans et al. 2007). It has been proposed that FoxI1/3 may play a prime role in initiating the development of the posterior placodal area since it appears to serve as a pioneer factor, which binds to regulatory regions of genes silenced by repressive epigenetic marks and makes them accessible to other transcription factors by recruiting histone modifying enzymes and DNA-demethylases (Khatri et al. 2014; Singh and Groves 2016).

While FoxI1/3 and Dlx3/5 are important for the competence of ectoderm to form the posterior placodal area, their expression domains extend further anterior and also cover the regions of the developing profundal/trigeminal (FoxI1/3) or even olfactory (Dlx3/5) placodes (Solomon and Fritz 2002; Solomon, Logsdon, and Fritz 2003; Ohyama and Groves 2004; Schlosser and Ahrens 2004; Brown, Wang, and Groves 2005). FGF signals from the endomesoderm and/or the hindbrain are required to induce Pax8, Pax2 and Sox3/2 expression in a circumscribed field of cells within the wider FoxI-Dlx domain, thereby defining the posterior placodal area (Fig. 4.4), the common precursor of otic, lateral line, and epibranchial placodes (Mansour 1994; Ladher et al. 2000; Phillips, Bolding, and Riley 2001; Leger and Brand 2002; Maroon et al. 2002; Liu et al. 2003; Alvarez et al. 2003; Wright and Mansour 2003; Phillips et al. 2004; Nechiporuk et al. 2007; Nikaido et al. 2007; Sun et al. 2007; Zelarayan et al. 2007; Freter et al. 2008; Yang et al. 2013; Wright et al. 2015; Chen et al. 2017). However, there is some variation between different vertebrates regarding which FGF family members (e.g. FGF3 and FGF8 in zebrafish; FGF3 and FGF19 in chick; FGF3 and FGF10 in mouse) and which tissue sources are involved (reviewed in Schimmang 2007; Ladher et al. 2010; Chen and Streit 2013).

Pax8, Pax2 and Sox3/2 play central roles for the development of the otic and epibranchial placodes derived from the posterior placodal area; their function for lateral line placode development has not been analyzed. Due to partially redundant functions of Pax2 and Pax8, loss of function of each of them individually leads to relatively mild perturbations of otic development in both zebrafish and mouse, while double knockout prevents otic vesicle formation or arrests otic development at early vesicle stages (Favor et al. 1996; Torres, Gómez-Pardo, and Gruss 1996; Christ et al. 2004; Burton et al. 2004; Hans et al. 2004; Mackereth et al. 2005;

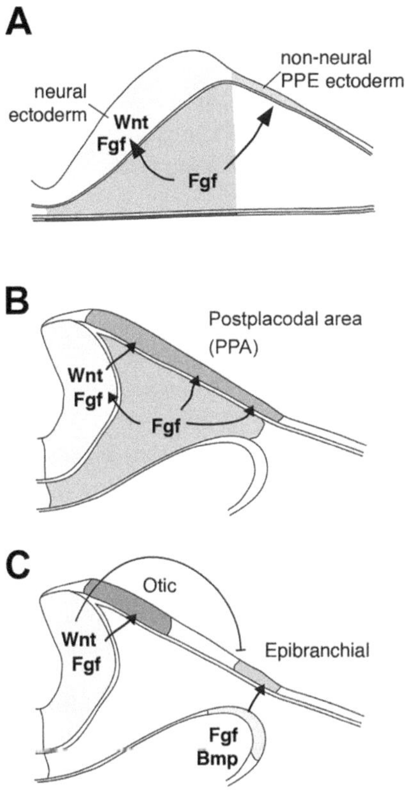

FIGURE 4.4 Multistep induction of the otic and epibranchial placodes. Schematic summary of induction events in the posterior placodal area (PPA) based on data from mouse, chick, and zebrafish. (**A**) During early neurula stages, mesodermal fibroblast growth factor (FGF) induces the PPA in the pre-placodal ectoderm (PPE), and FGF and Wnt signaling in the neural plate. (**B**) During early neural fold stages, FGF signals from mesoderm (and possibly other sources like neural ectoderm and endoderm) consolidate PPA induction and Wnt from the neural plate induces the otic placode in its dorsal part. (**C**) At later neural fold stages, mesodermal FGF signals weaken. Wnt from the neural tube continues to induce otic fate and to repress epibranchial fate, while FGF and BMP signaling from the pharyngeal pouches induce epibranchial placodes. (Reprinted with permission from Sai and Ladher 2015.)

Bouchard et al. 2010). In chick, where the *Pax8* gene has been lost, Pax2 knockdown interferes with the expression of many otic transcription factors and proper otic development (Christophorou et al. 2010; Freter et al. 2012; Chen et al. 2017). Loss of Pax8 and/or Pax2 expression in the posterior placodal area also prevents the activation of transcription factors specific for epibranchial placodes (e.g. Phox2a/b) in both chick and zebrafish (Padanad and Riley 2011; Freter et al. 2012). Taken together, this indicates that Pax2 and Pax8 are required for the development of epibranchial as well as otic placodes. However, they appear to regulate otic and epibranchial fates in a dosage dependent manner with high doses promoting otic over epibranchial development (McCarroll et al. 2012).

The two other early markers of the posterior placodal area, Sox3 and Sox2 (which are also expressed in some anterior placodes) are also required in a redundant fashion for the development of otic, epibranchial and lateral line placodes in zebrafish and mouse embryos and probably act in the same pathway as Pax8 (Rizzoti and Lovell-Badge 2007; Dee et al. 2008; Gou et al. 2018). At later stages, these SoxB1 transcription factors then play additional important roles for regulating neurogenesis in various placodes (see Chapter 5).

Whereas Pax2/8 and Sox2/3 are expressed throughout the entire posterior placodal area, another group of transcription factors including Hmx2/3 and Sox9/10 are confined to the dorsal part of this region which gives rise to the otic and lateral line placodes. However, while mutations in Sox9 or 10 strongly compromise otic development (Liu et al. 2003; Saint-Germain et al. 2004; Yan et al. 2005; Dutton et al. 2009; Breuskin et al. 2009), they do not inhibit the development of neuromasts from lateral line placodes in zebrafish (Grant, Raible, and Piotrowski 2005; López-Schier and Hudspeth 2005; Hans et al. 2013) suggesting that they may be dispensable for the development of lateral line placodes. In contrast, genes of the *Hmx* family including *Hmx2*, *Hmx3*, and possibly *SOHo* are not only required for development of vestibular structures from the dorsal otic vesicle but also for the development of lateral line neuromasts (Hadrys et al. 1998; Wang et al. 2001; Feng and Xu 2010). This suggests that otic and lateral line placodes may possibly share a regulatory state distinguishing them from epibranchial placodes. However, more evidence is needed to support this hypothesis.

4.1.3.2 Induction of Individual Placodes within the Posterior Placodal Area

Once the posterior placodal area is established, individual otic, epibranchial, and lateral line placodes are induced at different locations within this domain. To locally activate transcription factors that define distinct regulatory states for these individual placodes, transcription factors broadly required for the posterior placodal area, such as Pax2, need to cooperate with signaling molecules released from localized signaling centers in adjacent tissues (Figs. 4.3 and 4.4). This has been directly shown for the otic expression of Sall4, which depends on the binding of both Pax2 and the FGF effector Pea3 on its enhancer (Barembaum and Bronner-Fraser 2010). Sall4 together with Sox8 and SoxB1 transcription factors then activates otic enhancers of other transcription factors such as Sox10, Sox3, and Sox2 (Betancur, Sauka-Spengler, and Bronner 2011; Murko and Bronner 2017; Okamoto et al. 2018; Sugahara et al. 2018).

Several studies have investigated the signals required to induce the otic, epibranchial, and lateral line placodes from the posterior placodal area in amniotes and zebrafish (Fig. 4.4). These studies have shown that FGF signaling, which is initially required for the induction of the posterior placodal area, at later stages seems to counteract otic placode specification while promoting epibranchial placode formation (Fig. 4.4) (Nechiporuk et al. 2007; Nikaido et al. 2007; Sun et al. 2007; Freter et al. 2008; Maulding et al. 2014). FGF signals from both endoderm and the developing otic and anterior lateral line placodes itself appear to be specifically involved in the induction of epibranchial placodes in the ventral part of the posterior placodal area (Fig. 4.4) (Padanad and Riley 2011; McCarroll and Nechiporuk 2013). Conversely, Wnt signals from the neural tube are required for the induction of the otic placode dorsally but repress epibranchial placode formation (Fig. 4.4) (Ladher et al. 2000;

Ohyama et al. 2006; Freter et al. 2008; Urness et al. 2010, McCarroll et al. 2012). Notch signaling has also been shown to promote otic at the expense of epibranchial placode formation in mouse but appears to inhibit otic placode formation in chick embryos (Jayasena et al. 2008; Shida et al. 2015). Whether these conflicting findings reflect species-specific differences or differences in the experimental paradigm used in these two studies still needs to be clarified.

The initiation of neurogenesis in the epibranchial placodes subsequently requires additional FGF and BMP signals from the pharyngeal pouches (Fig. 4.4) (Begbie et al. 1999; Holzschuh et al. 2005; Nechiporuk, Linbo, and Raible 2005). These signals activate neurogenesis in ectoderm adjacent to the dorsal (epibranchial placodes) and in some groups also the ventral tip (hypobranchial placodes of frogs) of the pharyngeal pouches. However, PRDC, a BMP inhibitor released by the pharyngeal pouches suppresses neurogenesis in ectoderm directly contacted by the pouches (Kriebitz et al. 2009).

In contrast to the otic and epibranchial placodes, very little is known about the induction of lateral line placodes. A recent study in zebrafish suggests that the signals required for the pre-otic (anterodorsal and anteroventral) lateral line placode and the post-otic posterior lateral line placode differ (another postotic placode, the middle lateral line placode, only develops at very late stages – see Andermann, Ungos, and Raible 2002 – and was not analyzed). Whereas Hmx3 expression in the pre-otic lateral line placodes and the otic placode require FGF signaling, the posterior lateral line placode requires retinoic acid but is inhibited by FGF signaling (Nikaido et al. 2017). This indicates that the posterior lateral line placode, which is not part of the posterior placodal area at the end of gastrulation and only upregulates posterior placodal markers at later stages (Pieper et al. 2011; McCarroll et al. 2012), does not depend on FGF signaling in contrast to the posterior placodal area. It does, however, not clarify, which signals distinguish lateral line placodes from the otic placode within the posterior placodal area. Because otic and epibranchial, but not lateral line placodes, arise next to known signaling centers in the hindbrain and pharyngeal pouches, it was previously proposed that lateral line placodes may represent the default state of the posterior placodal area (Fritzsch, Barald, and Lomax 1998; Schlosser and Ahrens 2004). However, this hypothesis still needs to be tested.

In summary, the posterior placodal area defined by Pax2/8 and Sox2/3 expression is induced at the end of gastrulation (later in amniotes) by FGF signals at the intersection of the competence factors FoxI1/3 and Dlx3/5, the PPE specifiers Six1/2 and Eya and posteriorly restricted transcription factors such as Gbx and Irx. Subsequently, Wnt signaling from the neural tube induces the otic placode dorsally and protracted FGF signaling from endoderm and otic placode induces the epibranchial placodes ventrally, while the signals inducing the lateral line placodes need to be further characterized.

4.1.4 Origin of Profundal and Trigeminal Placodes

The profundal and trigeminal placodes (ophthalmic and maxillomandibular trigeminal placodes of amniotes) arise from a part of the PPE located between the extended anterior and posterior placodal areas, where expression of the anterior marker Otx2 overlaps with the posterior transcription factors of the Irx family (Fig. 4.3). Both placodes produce similar sets of somatosensory neurons and express transcription factors known to be required for somatosensory neuronal differentiation such as DRG11 and PRDM12 (Chen et al. 2001; Jacquin et al. 2008; Chen et al. 2015; Nagy et al. 2015; Patthey et al. 2016). However, it is still not clear whether profundal and trigeminal placodes develop from a common placodal area because currently no transcription factors are known that are exclusively expressed in the ectodermal region giving rise to both developing placodes. Moreover, while the profundal placode is characterized by Pax3 expression, so far no transcription factor uniquely expressed in the trigeminal placode has been identified.

Together with the PPE specifiers (Six1/2, Eya) and non-neural competence factors (TFAP2a, GATA 2/3, Dlx3/5, and FoxI1/3), both Otx2 and Irx are required for the development of profundal and trigeminal placodes (Itoh et al. 2002; Kaji and Artinger 2004; Christophorou et al. 2009; Feijoo et al. 2009; Rodriguez-Seguel et al. 2009; Kwon et al. 2010; Bhat et al. 2012; Pieper et al. 2012; Steventon et al. 2012). Pax3 in the profundal placode subsequently controls its own expression and represses Pax6 anteriorly and Pax2 posteriorly, which helps to delimit the placode from the adjacent extended anterior and posterior placodal areas (Dude et al. 2009; Wakamatsu 2011).

Wnt and FGF from the neural tube, in particular from the midbrain-hindbrain boundary have been shown to be required for the induction of Pax3 in the profundal placode and for trigeminal placode induction (Fig. 4.3) (Stark et al. 1997; Baker, Beddington, and Harland 1999; Lassiter et al. 2007; Canning et al. 2008). Wnt probably acts indirectly via promotion of FGF signaling and FGF signals also play important roles for the subsequent regulation of neurogenesis in the profundal placode (Canning et al. 2008; Lassiter et al. 2009). In addition, PDGF from the neural tube was shown to be required for Pax3 induction in the profundal placode (McCabe and Bronner-Fraser 2008).

4.1.5 Multistep Induction of Placodes – A Brief Summary

In summary, individual placodes are specified in different regions of the PPE in multiple steps (Fig. 4.3). First, during gastrulation an anteroposterior gradient of Wnt signaling together with posteriorly enriched FGF and retinoic acid activate transcription factors in a concentration dependent way in restricted domains along the anteroposterior axis. Cross-repression between transcription factors expressed in adjacent regions (e.g. Otx2-Gbx2 or Six3-Irx) helps to sharpen the boundaries between their expression domains. These transcription factors subsequently are involved in the general anteroposterior patterning of the ectoderm including the PPE.

Second, pre-placodal specifiers (Six1/2 and Eya1) and non-neural competence factors (e.g. FoxI1/3, Dlx3/5) cooperate with anteroposteriorly restricted transcription factors and signaling molecules released from adjacent tissues to induce two multiplacodal areas. The extended anterior placodal area (adenohypophyseal, olfactory, lens) characterized by Pax6, Pitx, DMRT, and FoxE expression is induced in the anterior part of the PPE (expressing Otx2 and Six3/6) by somatostatin and nociceptin signaling from the underlying mesoderm and the ectoderm itself. The posterior placodal area (otic, lateral line, epibranchial placodes) characterized by Pax2/8 and Sox2/3 expression is induced by FGF signals from endomesoderm and hindbrain. The profundal and trigeminal placodes are induced in between these two areas by Wnt, FGF, and PDGF signals from the adjacent mid- and hind-brain.

Third, transcription factors specifying the anterior and posterior placodal areas cooperate with additional signals to further subdivide these two multiplacodal areas. Wnt signaling from the anterior endomesoderm and subsequently FGF8 from the anterior neural ridge defines the anterior placodal area (adenohypophyseal and olfactory placodes) characterized by Hesx1 and FoxG1 expression. In the posterior placodal area, unknown signals activate some transcription factors such as Hmx2/3 only in the dorsal part, which gives rise to the otic and lateral line placodes, possibly defining a common precursor for these two types of placode.

Fourth and finally, cooperation of these transcription factors with localized signals lead to the induction of individual placodes. At the same time, these signals repress alternate placodal fates. Anteriorly, sonic hedgehog from the notochord and floor plate induces the adenohypophyseal placode (and repress olfactory and lens placodes) within the anterior placodal area. Posteriorly, Wnts from the hindbrain induce the otic placode (and repress epibranchial placodes), FGF signals from the endoderm induce the epibranchial placodes (and repress the otic placode), while the signals involved in the induction of lateral line placodes remain unknown.

The regional subdivision of the PPE does not stop here. Due to the interplay of placode-specific transcription factors with locally activated signaling pathways, most individual placodes are further subdivided along the anteroposterior and/or dorsoventral axes with different regions giving rise to different cell types. These patterning events have been extensively reviewed elsewhere for the adenohypophyseal (Scully and Rosenfeld 2002; Rizzoti and Lovell-Badge 2005; Zhu et al. 2007; Zhu, Gleiberman, and Rosenfeld 2007; Kelberman et al. 2009; Pogoda and Hammerschmidt 2009; Davis et al. 2013; Rizzoti 2015), olfactory (Balmer and LaMantia 2005; Maier et al. 2014; Moody and LaMantia 2015), and otic placode (Barald and Kelley 2004; Whitfield and Hammond 2007; Bok, Chang, and Wu 2007; Alsina, Giráldez, and Pujades 2009; Magarinos et al. 2012; Wu and Kelley 2012; Groves and Fekete 2012; Maier et al. 2014; Raft and Groves 2015). I will discuss patterning of individual placodes briefly in the next chapter, but only to the extent relevant for our understanding of how different cell types are specified in placodes.

4.2 SEPARATION OF PLACODES FROM EACH OTHER

Once individual placodes are specified within the PPE by combinations of co-expressed transcription factors, they form stable boundaries and separate from each other. This is achieved by a combination of different mechanisms including transcriptional cross-repression, directed cell movements, and programmed cell death (apoptosis). It has recently been highlighted that the separation of individual placodes occurs in two phases: initial segregation and secondary physical separation (secondary coalescence, compaction) (Breau and Schneider-Maunoury 2014). During the first phase, placodes establish sharp boundaries between them but remain still in close contact with each other, while during the second phase they become physically separated from each other.

4.2.1 Initial Segregation of Placodes

Two major mechanisms have been proposed to contribute to the initial segregation of placodes (reviewed in Schlosser 2006, 2010; Breau and Schneider-Maunoury 2014) (Fig. 4.5). The first mechanism involves boundary formation that accompanies placodal specification. Cells in the border region between two adjacent placodes may initially have not decided their fate and may co-express transcription factors that are activated in response to signals from adjacent signaling centers and direct the specification of two different placodal fates. There may also be random cell movements accompanied by dynamic changes of transcription factor expression profiles when cells come under the influence of different signaling centers. However, mutual cross-repression between these transcription factors will ultimately enforce upregulation of one and downregulation of the other set of transcription factors. This will result in switch-like cell fate decisions and the sharpening of boundaries between adjacent placodes or multiplacodal areas (Fig. 4.5A, B). Several established cases of cross-repression between transcription factors in the PPE such as between Otx2 and Gbx2

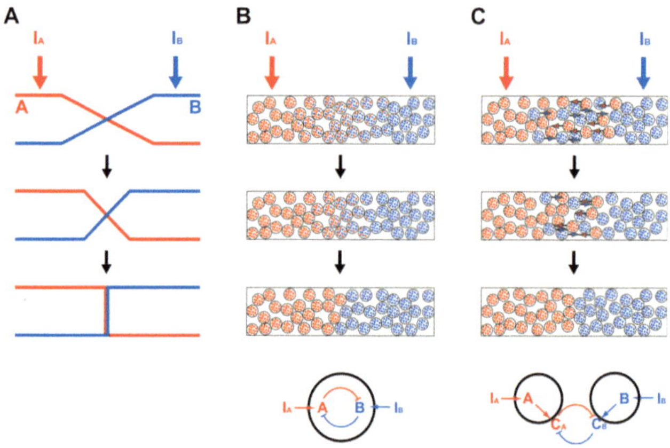

FIGURE 4.5 Two mechanisms for initial placode segregation. (**A**) Transcription factors A and B that are involved in specification of two different individual placodes and are induced by two different inducers (I_A and I_B, respectively), are initially expressed in an overlapping fashion before boundaries between expression domains sharpen over time. Establishment of sharp boundaries may occur by mutual cross-repression of transcription factors or by cell sorting. (**B**) Mutual cross-repression. Overlapping expression domains as depicted in **A** may be due to co-expression of A and B (red and blue dots) in non-specified individual cells (circles). Sharpening of the boundary involves upregulation of one but downregulation of the other transcription factor in individual cells (switch-like cell fate decisions) due to their mutual transcriptional repression (lower panel). There also may be random movements of unspecified cells during this process (not shown). (**C**) Cell sorting. Alternatively, overlapping expression domains as depicted in A may be due to mixing of cells specified early to express different transcription factors. Sharpening of the boundary may then involve sorting of cells by non-random cell movements (arrows), for example mediated by mutually repelling cell adhesion molecules (C_A, C_B), activated by the different transcription factors (lower panel). (Reprinted with modification from Schlosser 2006.)

or Pax6 and Pax3 (Wakamatsu 2011, 2011; Steventon et al. 2012) probably contribute to this process. Other cross-repressive pairs of transcription factors, which have well established roles for boundary formation in the CNS such as Six3–Irx3, Fezf–Irx, and Arx–Irx (Kobayashi et al. 2002; Rodriguez-Seguel et al. 2009), are also likely to play a role in placode segregation.

The second mechanism involves cell movements of pre-specified cells. The border region between two placodes may contain a mixture of cells that are already specified for one or the other placodal fate (due to the stable expression of different core transcriptional regulators). Directional cell movements may then segregate the two cell populations from each other, forming a sharp boundary between them. Known as "cell sorting," this process may be mediated by differential responses of the two cell populations to some chemoattractant (or – repellent) or by differential cell adhesion between the two cell populations (Fig. 4.5A, C). We still know very little about the role of either chemoattractants or differential adhesion during initial placode segregation. However, cells deficient for Pax6 in mouse or overexpressing Dlx5 in chick have been shown to be excluded from the developing lens placode or vesicle (Collinson, Hill, and West 2000; Bhattacharyya et al. 2004). This suggests that these transcription factors may control the expression of different cell adhesion molecules, although Pax6 has also been implicated in the regulation of extracellular matrix molecules (Huang et al. 2011).

The available evidence indicates that both cell-sorting and cross-repressive interactions of transcription factors contribute to the segregation of placodes, although their relative importance is still contentious. A detailed discussion of this topic is beyond the scope of this book and so I will only offer a brief summary here (for reviews see Schlosser 2010; Breau and Schneider-Maunoury 2014, 2015).

In chick embryos, cell movements in the PPE have been documented already at neural plate and fold stages (Streit 2002; Bhattacharyya et al. 2004), before cells are specified and/or committed to particular placodal fates (Baker et al. 1999; Groves and Bronner-Fraser 2000; Baker and Bronner-Fraser 2001; Bhattacharyya and Bronner-Fraser 2008). Tracking of individual cells in the PPE during a comparable time window in *Xenopus* suggests that most of these early cell movements are non-directed and occur locally (Pieper et al. 2011). However, more extensive and directional cell movements have been revealed by time-lapse imaging at later stages of placode development in chick, zebrafish, and frog embryos (Streit 2002; Bhattacharyya et al. 2004; Bhat and Riley 2011; Steventon, Mayor, and Streit 2016), after placodes have become specified and formed clear boundaries of gene expression. Furthermore, coalescence of profundal/trigeminal, otic, and epibranchial placodes was compromised after knockdown of an alpha integrin subunit in zebrafish indicating that interactions of placodal cells with the extracellular matrix are important for these cell movements (Bhat and Riley 2011).

Taken together, this suggests that prior to specification and concomitant segregation of individual placodes, cells in the PPE are mostly subject to random and local cell movements. Subsequently, cell sorting may contribute locally to sharpen the boundaries between adjacent placodes possibly mediated by differential cell adhesion. In the absence of large-scale cell sorting, mutual cross-repression of transcription factors involved in the specification of alternative placodal fates remains the most likely mechanism for initial placode segregation. However, since currently only a few examples of cross-repressing transcription factors (Otx2-Gbx2; Pax6-Pax3) are known and their role in placode specification is poorly understood, more evidence is needed to further support this hypothesis.

4.2.2 Physical Separation of Placodes

At the end of the first phase of placode separation discussed in the last section, relatively sharp molecular boundaries (between expression domains of genes encoding transcription factors and other proteins) have been established between individual placodes. In the next phase of their development, placodes become physically separated from each other and compacted by a combination of several mechanisms. First, the PPE gives rise to epidermal in addition to placodal cells as has been shown in fate mapping studies of chick and *Xenopus* embryos (Streit 2002; Bhattacharyya et al. 2004; Xu, Dude, and Baker 2008; Pieper et al. 2011). This implies that some PPE cells downregulate the expression of PPE specifiers (Six1, Six4, and Eya) and of transcription factors involved in the specification of multiplacodal areas or of individual placodes (e.g. Pax family proteins). These cells may subsequently differentiate as epidermis intervening between individual placodes. Second, programmed cell death by apoptosis has been documented for many developing placodes in mammals and has been proposed to contribute to the separation between the otic and epibranchial placodes (Washausen et al. 2005; Knabe et al. 2009; Washausen and Knabe 2013, 2017, 2018). Third, directional cell movements as discussed in the previous section continue to contribute to the coalescence and compaction of individual placodes (Streit 2002; Bhattacharyya et al. 2004; Bhat and Riley 2011; Kwan et al. 2012; Steventon et al. 2016).

Finally, interactions with adjacent cell populations, most importantly the neural crest, have been implicated during various stages of placode coalescence. A recent study demonstrated that already during neural fold stages in *Xenopus*, distinct neural crest streams interact with epibranchial placodal cells to promote separation of the epibranchial domain into discrete placodes (Theveneau et al. 2013). Neural crest cells were shown to be attracted to epibranchial placodal cells, while the latter are repelled by the neural crest. This "chase and run" behavior is mediated by the secretion of the chemokine SDF1 by the placodes, which attracts neural crest cells by stimulating their CXCR4b receptor (Theveneau et al. 2013).

In conclusion, a number of different mechanisms conspire to physically separate individual placodes from each other after they have been specified for their unique identity. In the next phase of development, various placodes undergo different types of morphogenetic movements resulting in further spatial separation. The adenohypophyseal, olfactory, lens, and otic placodes invaginate partially or fully (Figs. 1.3 and 2.2). The otic placode subsequently forms various pouches ventrally and the semicircular canals dorsally in a complex sequence of epithelial outpocketing and fusion (Figs. 1.8 and 2.2). The lateral line placodes form primordia that elongate (sensory ridges of most lateral lines) or migrate (migratory primordium of the trunk lateral line) along the basal lamina to form the sensory organs of the lateral line system (Figs. 1.7 and 2.2). Moreover, most placodes (i.e. all except adenohypophyseal, olfactory, and lens placodes) give rise to neurons that delaminate from the placode through a breach in the underlying basal lamina and congregate nearby in close interaction with neural crest cells to form the sensory ganglia of the cranial nerves (Fig. 2.2). Discussion of the mechanisms underlying these morphogenetic movements is beyond the scope of this book and I refer the reader to several recent reviews of the topic (reviewed in Mansour and Schoenwolf 2005; Fritzsch, Pauley, and Beisel 2006; Bok et al. 2007; Schlosser 2010; Chitnis, Nogare, and Matsuda 2012; Breau and Schneider-Maunoury 2014; Steventon, Mayor, and Streit 2014; Breau and Schneider-Maunoury 2015; Thomas et al. 2015; Sai and Ladher 2015; Alsina and Whitfield 2017; Cvekl and Zhang 2017; Ladher 2017; Olson and Nechiporuk 2018).

In parallel with these changes in shape, different cell types begin to differentiate from the various placodes. The regional identity of individual placodes acquired during the multistep specification process, discussed in this chapter, is essential for determining which cell types develop from each placode (e.g. lens fiber cells, sensory neurons, hair cells) as well as how they differentiate in a region-specific fashion (e.g. sensory neurons destined to innervate specific targets). It is to these processes of cyto-differentiation that we now turn to in the next chapters.

5 General Mechanisms of Sensory and Neuronal Differentiation

The last two chapters summarized how the pre-placodal ectoderm (PPE), a common primordium of various placodes, is established in the vertebrate embryo and then subdivided into different placodes. Each of these individual placodes then gives rise to a number of different cell types. For example, primary sensory cells (with an axon) develop from the olfactory placode, secondary sensory cells (without an axon) as well as sensory neurons from the otic and lateral line placodes, and sensory neurons but no sensory cells from the profundal/trigeminal and epibranchial placodes. Additional, non-neuronal cell types develop from many of the placodes, in particular, the neurosecretory and lens fiber cells generated by the adenohypophyseal and lens placodes, respectively. In this and the following chapters, I will review how the differentiation of these various placodal cell types is regulated. In this chapter, I will first summarize common mechanisms regulating the formation of neurons and sensory cells in most placodes. In Chapter 6, I will then discuss, how the specific neurons and sensory cells of individual placodes differentiate. In Chapter 7, I will briefly review the differentiation of photoreceptors. While not derived from placodes in vertebrates, photoreceptors will turn out to share a common evolutionary history with sensory cell types derived from placodes (see Schlosser 2021) and, therefore, must be discussed here. In Chapter 8, I will then review how the non-neuronal cell types of the adenohypophyseal and lens placodes differentiate.

As a note of caution, it needs to be pointed out here that besides the specialized placode-derived sensory cell types and retinal photoreceptors, which will be discussed here, there are many other cells in the vertebrate body that are to some extent chemo-, mechano-, or photoresponsive, although not necessarily specialized for such a sensory function. Even though the core regulatory networks (CoRNs) defining the identity of these cell types are often not known, their peculiar developmental origin, transcription factor expression profile, and/or mode of sensory transmission make it likely that they belong to different cell types unrelated to those covered in this book. For example, almost all cells in the developing vertebrate embryo respond to one or the other kind of molecular signal released by other cells and are, thus, in a sense chemoresponsive. Furthermore, many cells other than placode-derived hair cells, in particular, epithelial and cartilage cells, respond to mechanical forces (stretch or pressure) and express a variety of different mechanosensitive ion channels (e.g. Enac channels, some potassium channels, Piezo channels) (Chalfie 2009). Finally, non-visual photosensitive cells have been found in many tissues outside the eyes in vertebrates and other taxa. These cells often appear unspecialized unlike the photoreceptors of the paired eyes and may use different photopigments, although many of these photosensitive cells have not yet been well characterized (Porter 2016).

5.1 COMMON MECHANISMS FOR NEURONAL AND SENSORY DEVELOPMENT

While many organs of the vertebrate body are composed of a rather limited repertoire of different cell types, the central and peripheral nervous systems stand out for the variety of their cell types and the complex patterns of intercellular connections between them. The versatility of receptors, neurotransmitters, and branching patterns observed in the different neurons and sensory cells allows different cells to be wired differently. This highly specific wiring is crucial for the function of the nervous system to coordinate the operation of different organs with each other and with environmental conditions. However, this versatility is superimposed on general features of neuronal and sensory differentiation that are widely shared between different types of neurons or sensory cells (Hobert, Carrera, and Stefanakis 2010; Ernsberger 2015; Burkhardt and Sprecher 2017; Alsina, 2020). These include ion channels and other transmembrane proteins involved in electrical signal transduction, the machinery for synaptic vesicle transport and release, and proteins regulating the outgrowth of dendrites and axons during development and their activity dependent plasticity in the adult.

The gene batteries that encode the proteins mediating these generic neuronal functions appear to be regulated in all neurons and sensory cells by related core networks of transcription factors of the SoxB1 family (Sox1, Sox2, Sox3) and various basic helix loop helix (bHLH) family members (Achaete-Scute-, Atonal-, and Neurogenin-related transcription factors, HES and Id factors). In neurons and sensory cells of placodal origin, Six and Eya family members modulate this generic neuronal core network at various levels. The SoxB1, bHLH, and Six/Eya family members involved as well as details of their regulatory relationships may differ between different neurosensory cell types. These differences probably reflect the divergent evolutionary history of the cell types, as will be discussed in the second volume (Schlosser 2021). However, the overall regulatory logic of these networks appears to be quite conserved for different types of neurons and results in a well-regulated and balanced transition from proliferating stem and progenitor cells to differentiating neurons (Fig. 5.1).

In the following sections, I will first discuss how SoxB1 transcription factors cooperate with other proteins (including HES and Id factors) to keep cells in a proliferating progenitor state, while biasing them toward neuronal or sensory fate. I will then summarize how bHLH transcription factors of the Achaete-Scute- (Ascl), Atonal- (Atoh), and Neurogenin- (Neurog) subfamilies initiate the neuronal and sensory differentiation program in a subset of cells while repressing differentiation and maintaining progenitor states in adjacent cells (lateral inhibition). Finally, I will briefly sketch, how SoxB1 and proneural factors are regulated by Six1 and Eya in cranial placodes.

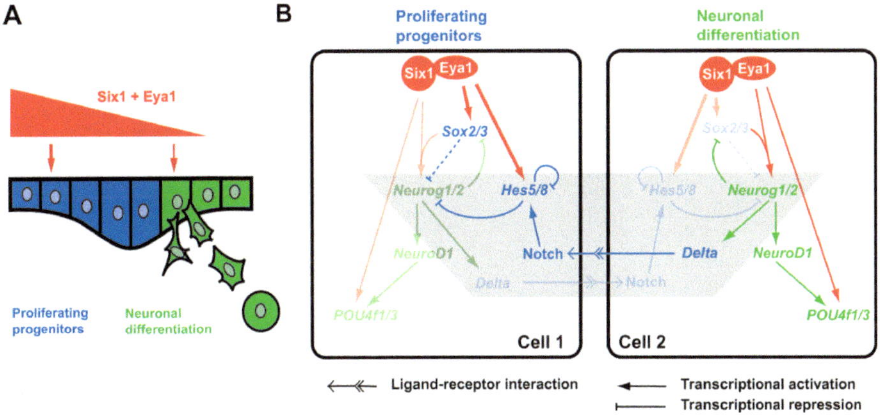

FIGURE 5.1 Model for regulation of progenitor maintenance and neuronal differentiation in placodes. (**A**) High doses of Six1 and Eya1 promote expansion of proliferating progenitors, while declining levels of Six1 and Eya1 (as cells migrate away from placodes) promote neuronal differentiation. (**B**) Interaction of Six1 and Eya1 with genes and proteins involved in progenitor maintenance and neuronal differentiation. In progenitors (blue), neuronal differentiation genes (green) are repressed by Sox2/3 and Hes5/8 transcription factors (hatched line: indirect repression; solid line: direct repression). In differentiating neurons, proneural factors like Neurog1/2 activate NeuroD1 and POU4f1, thereby promoting neuronal differentiation. Moreover, Neurog1/2 also activates transcription of the Notch-ligand Delta, which binds to the Notch receptor in adjacent cells, repressing neuronal differentiation in these neighbors. This lateral inhibition pathway is highlighted by the gray area. High levels of Six1 and Eya1 (thick red arrows) appear to be required for transcriptional activation of target genes that promote progenitor maintenance (e.g. *Sox2/3*, *Hes5/8*, blue), while low levels of Six1 and Eya1 (thin red arrow) appear to be sufficient to activate target genes promoting neuronal differentiation (e.g. *Neurog1/2*, *POU4f1*; green). While Sox2/3 alone inhibits neuronal differentiation upstream and downstream (not shown) of *Neurog1/2*, it poises *Neurog1/2* for activation in conjunction with Six1 and Eya1. A similar network is active in hair cells (with Atoh1 replacing Neurog1/2 and POU4f3 replacing POU4f1). (Modified from Riddiford and Schlosser 2017.)

5.2 SoxB1 TRANSCRIPTION FACTORS: PROGENITOR MAINTENANCE AND NEURONAL OR SENSORY LINEAGE BIAS

The Sox1, Sox2, and Sox3 proteins of vertebrates comprise the SoxB1 subfamily of the HMG-box containing transcription factors. The three SoxB1 factors have partly non-overlapping expression patterns, which also differ between different groups of vertebrates. However, they can typically functionally compensate for each other in areas where they are co-expressed (Miyagi, Kato, and Okuda 2009; Sarkar and Hochedlinger 2013). This indicates that they serve mostly similar functions.

In Chapter 3, I have described how SoxB1 factors contribute to the regulation of pluripotency in the pre-gastrula embryo and how they – Sox2 and Sox3, in particular – subsequently become restricted to the prospective neural plate on the

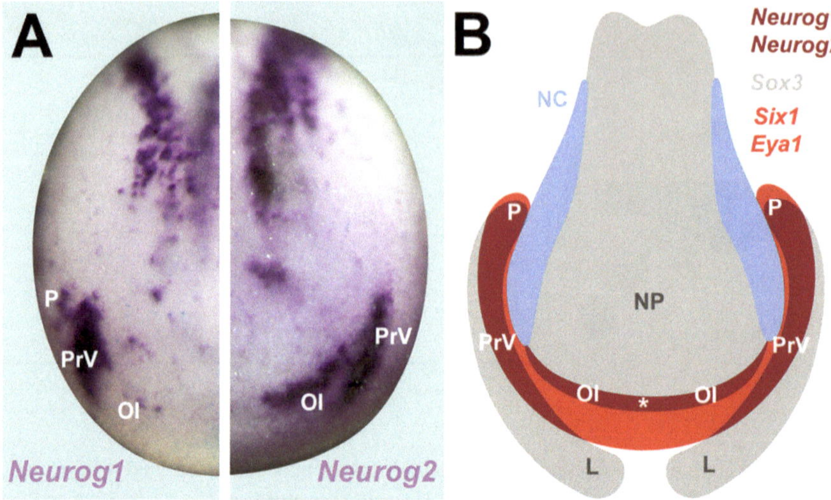

FIGURE 5.2 Expression of *Sox3* and proneural genes *Neurog1/Neurog2* in placodes. (**A**) Placodal expression domains of *Neurog1* and *Neurog2* mRNA in *Xenopus laevis* shown by in situ hybridization. (**B**) Since Six1/Eya1 and Sox3 synergize in activation of *Neurogeningenes*, placodal expression of *Neurog1* and *Neurog2* is activated where Six1/Eya1 and Sox3 expression domains overlap (dark red areas). Asterisk marks area of prospective adenohypophyseal placode, where *Neurog2* is transiently expressed but quickly downregulated (Schlosser and Ahrens, 2004). Expression of *Neurog2* in the posterior placodal (P) area has not yet commenced at this stage. L, lateral domain of Sox3 expression; NC, neural crest; NP, neural plate; Ol, olfactory placode; P, posterior placodal area; PrV, profundal/trigeminal placode. (Modified from Riddiford and Schlosser 2017).

dorsal side of the embryo, where they bias lineage decisions toward a neuroectodermal fate. When the neural plate folds in to form the neural tube during neurulation, the expression of Sox2 and Sox3 is maintained in neural progenitor cells, but downregulated in the differentiating neurons of the CNS. SoxB1 factors are not expressed in the neural crest, but are upregulated in parts of the PPE, even though the latter develops from non-neural ectoderm. In the anterior part of the PPE, Sox2 and Sox3 are expressed in a domain contiguous with the anterior neural plate encompassing the prospective adenohypophyseal and olfactory placodes (Fig. 5.2) (Schlosser and Ahrens 2004; Riddiford and Schlosser 2017). In the posterior PPE, Sox2 and Sox3 are found in a domain encompassing the prospective profundal/trigeminal, otic, lateral line, and epibranchial placodes (Fig. 5.2). Similar to the CNS, both proteins subsequently remain expressed in the proliferating progenitors of placodes, but are downregulated in the differentiating neurons and sensory cells (Rex et al. 1997; Wood and Episkopou 1999; Abu-Elmagd et al. 2001; Ishii, Abu-Elmagd, and Scotting 2001; Kawauchi et al. 2004; Schlosser and Ahrens 2004; Donner, Episkopou, and Maas 2007; Hernandez et al. 2007; Nikaido et al. 2007; Sun et al. 2007; Schlosser et al. 2008).

SoxB1 factors play multiple different roles during development that are strongly context-dependent. However, they have two central and generic functions during neural and placodal development that I will discuss in the following sections. First, they keep progenitor cells in a proliferative and undifferentiated progenitor state, and second, they bias lineage decisions toward neuronal or sensory differentiation (reviewed in Pevny and Placzek 2005; Wegner and Stolt 2005; Miyagi et al. 2009; Kondoh and Kamachi 2010; Wegner 2010; Pevny and Nicolis 2010; Maucksch, Jones, and Connor 2013; Sarkar and Hochedlinger 2013; Shimozaki 2014; Zhang and Cui 2014; Julian, McDonald, and Stanford 2017).

5.2.1 Maintenance of Progenitor States

Studies of chick neural development have shown that SoxB1 factors are both necessary and sufficient for maintaining a neural progenitor state and to prevent neuronal differentiation in the developing neural tube (Bylund et al. 2003; Graham et al. 2003) and this has been confirmed in many studies in other vertebrates (reviewed in Pevny and Placzek 2005; Wegner and Stolt 2005; Pevny and Nicolis 2010; Sarkar and Hochedlinger 2013). To initiate neuronal differentiation, SoxB1 factors, therefore, must be downregulated. Similarly, in placodes *SoxB1* genes are predominantly expressed in proliferating progenitors and downregulated in differentiating neurons or sensory cells reflecting their functions in promoting progenitor states and inhibiting neuronal and sensory differentiation (Abu-Elmagd et al. 2001; Dabdoub et al. 2008; Schlosser et al. 2008; Millimaki, Sweet, and Riley 2010; Evsen et al. 2013; Packard, Lin, and Schwob 2016; Panaliappan et al. 2018)

SoxB1 factors appear to prevent neuronal differentiation in several different ways. On the one hand, SoxB1 factors have been shown to activate the Notch signaling pathway, known to interfere with neuronal differentiation. They may achieve this by directly activating the transcription of the Notch receptor, its ligands and/or Hes proteins (Bani-Yaghoub et al. 2006; Taranova et al. 2006; Agathocleous et al. 2009; Takanaga et al. 2009; Bergsland et al. 2011; Matsushima, Heavner, and Pevny 2011; Surzenko et al. 2013; Zhou et al. 2016; Panaliappan et al. 2018).

On the other hand, SoxB1 factors also prevent neuronal differentiation in Notch-independent ways by directly promoting the transcription of a number of proteins involved in the maintenance of progenitor states (Chen et al. 2008; Holmberg et al. 2008; Rogers et al. 2009; Bergsland et al. 2011; Surzenko et al. 2013). First, they activate the transcription of *CyclinD1*, which mediates the transition between the G1- and S-phases of the cell cycle, thereby promoting cell proliferation (Chen et al. 2008; Hagey and Muhr 2014). Second, they activate components of several signaling pathways (e.g. sonic hedgehog and the epidermal growth factor receptor EGFR) that promote progenitor fates and maintain SoxB1 expression in a positive feedback interaction (Favaro et al. 2009; Takanaga et al. 2009; Aguirre, Rubio, and Gallo 2010; Hu et al. 2010). And third, they transcriptionally activate several other genes implicated in progenitor maintenance. These include genes encoding the transcription factors Geminin (Rogers et al. 2009), Tlx (Shimozaki et al. 2012), and Sox2 itself (Tomioka et al. 2002), as well as the intermediary filament protein Nestin (Tanaka et al. 2004).

5.2.2 Biasing Neuronal and Sensory Cell Fate Decisions

As reviewed in Chapter 3, SoxB1 factors promote a neuroectodermal fate already during the earliest stages of vertebrate development. In line with this, Sox2 has recently been shown to be the major indispensable component of any cocktail of transcription factors capable to directly reprogram mammalian cells into neural precursors, which may differentiate into neurons or glial cells in vitro (Han et al. 2012; Lujan et al. 2012; Ring et al. 2012; Tian et al. 2012). Although Sox2 and Sox3 (but not Sox1) are downregulated in differentiating neurons, many functional studies in different vertebrates have demonstrated that all SoxB1 factors promote neuronal differentiation in the CNS and that they are required for the proper differentiation of various neuronal cell populations (e.g. Mizuseki et al. 1998; Pevny et al. 1998; Kishi et al. 2000; Ferri et al. 2004; Kan et al. 2004, 2007; Zhao et al. 2004; Van Raay et al. 2005; Taranova et al. 2006; Cavallaro et al. 2008; Dee et al. 2008; Miyagi et al. 2008). Similarly, SoxB1 factors were shown to promote the differentiation of both neurons and sensory cells in various cranial placodes (Kishi et al. 2000; Kiernan et al. 2005; Donner et al. 2007; Dabdoub et al. 2008; Dee et al. 2008; Schlosser et al. 2008; Tripathi et al. 2009; Millimaki et al. 2010; Puligilla et al. 2010; Evsen et al. 2013; Steevens et al. 2017; Gou et al. 2018). This indicates that SoxB1 factors provide progenitor cells with a bias toward neuronal or sensory differentiation.

Several mechanisms likely contribute to this capacity of SoxB1 factors to promote neuronal and sensory fates. First, SoxB1 factors preferentially bind to the regulatory regions of neural genes, where they act as pioneer factors opening up closed chromatin making it accessible to other transcription factors as discussed in more detail in section 5.2.3 (Bergsland et al. 2011). Thereby, SoxB1 factors promote competence for neural differentiation. Second, SoxB1 factors can directly activate the transcription of bHLH transcription factors such as Ascl1, Neurog1 or Atoh1, which are the core regulators of the neuronal and sensory differentiation program and they also help to activate some of their downstream genes (Bergsland et al. 2011; Ahmed et al. 2012; Neves et al. 2012; Ahmed, Xu, and Xu 2012; Amador-Arjona et al. 2015; Zhou et al. 2016). Third, SoxB1 factors also promote neuronal and sensory differentiation more indirectly, for example by modulating the activity of signaling pathways. However, these interactions are probably highly context-dependent. For example, Sox2 promotes neuronal differentiation in the adult hippocampus of mice in cooperation with the TCF/LEF transcription factor, a mediator of canonical Wnt signaling (Kuwabara et al. 2009). In contrast, Sox2 promotes neuronal differentiation in neural progenitor cells derived from human embryonic stem cells by transcriptional repression of various components of the canonical Wnt signaling pathway (Zhou et al. 2016).

It must be emphasized here that while SoxB1 factors bias lineage decisions toward neurosensory fates, they are not sufficient to restrict cell fates to a neurosensory lineage or to drive neuronal or sensory differentiation. As expected from a family of transcription factors implicated in the regulation of stem cell pluripotency, SoxB1 expressing cells can give rise to a multitude of cell types other than neurons. In the nervous system, for example, SoxB1 expressing precursors generate glial cells (astrocytes and oligodendrocytes) as well as neurons (Klum et al. 2018). Accordingly, after overexpression of SoxB1 in placodes or non-neural ectoderm of vertebrate embryos, only a subset of the extra SoxB1 expressing cells upregulate neuronal differentiation genes of the bHLH class and differentiate into neurons or sensory cells (Dee et al. 2008; Schlosser et al. 2008; Rogers et al. 2009; Puligilla et al. 2010). Moreover, reprogramming of mammalian cells into neurons in vitro is only possible, when SoxB1 factors are upregulated together with other transcription factors and/or exposed to appropriate signaling molecules (reviewed in Maucksch et al. 2013; Sarkar and Hochedlinger 2013; Julian et al. 2017). This demonstrates that SoxB1 factors can determine lineage decisions for neuronal or sensory cell types only in cooperation with additional factors.

In addition, SoxB1 transcription factors are not always necessary for neuronal differentiation, since neurons can develop from SoxB1-negative progenitors in some lineages such as the neural crest. In the neural crest, Sox10 and other members of the SoxE family (Sox8 and Sox9) may substitute to some extent for SoxB1 factors in supporting neuronal differentiation, although sustained SoxE expression promotes glial and other non-neuronal fates (Carney et al. 2006; Weider and Wegner 2017; Buitrago-Delgado et al. 2018).

5.2.3 Coordination of SoxB1 Functions for Progenitor Maintenance and Neurosensory Lineage Bias

The mechanisms regulating the transition from proliferating progenitor cells to differentiated neurons or sensory cells and the changing functions of SoxB1 factors during this process are still poorly understood. Current evidence suggests that both quantitative as well as qualitative changes of Sox factors together with changing availability of cofactors may play a role.

Evidence for the importance of quantitative changes was provided by studies demonstrating the strongly dosage-dependent effects of SoxB1 factors on proliferation and differentiation in the mouse brain, retina, and otic placode (Taranova et al. 2006, Cavallaro et al. 2008; Dabdoub et al. 2008; Hagey and Muhr 2014). While the highest levels of SoxB1 were shown to promote neural stem cells with relatively low proliferation rates, lower levels of SoxB1 support a highly proliferative progenitor state and favor early stages of differentiation before levels decline further in differentiated cells (Taranova et al. 2006, Cavallaro et al. 2008; Hagey and Muhr 2014). The ability of Sox2 to inhibit proliferation only at high levels was shown to be due to its ability to occupy low affinity binding sites in the *CyclinD1* promoter and recruit co-repressors to inhibit its expression (Hagey and Muhr 2014). If similar differences in binding affinities exist between binding sites in SoxB1 target genes promoting a progenitor state (e.g. Notch pathway genes) and those promoting neuronal and sensory differentiation, declining levels of SoxB1 factors would favor the transition from progenitors to differentiated cells. However, this model requires further experimental validation.

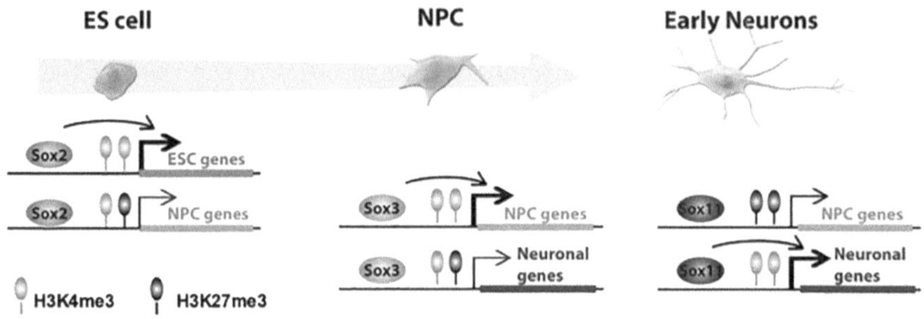

FIGURE 5.3 Changing roles of Sox factors during neurogenesis. Model depicting the sequential binding of different Sox proteins to common downstream genes in stem cells differentiating along the neural lineage. Activating (H3K4me3) and repressive (H3K27me3) histone marks are indicated. Genes carrying both activating and repressive marks are transcriptionally inactive but poised for activation. Replacement of one Sox factor by another leads to the sequential activation of new subsets of genes. ESC, embryonic stem cell; NPC, neural precursor cell. (Reprinted with permission from Bergsland et al. 2011.)

Apart from quantitative changes, qualitative changes of Sox factors have recently also been documented during the transition from pluripotent stem cells to neural progenitor cells and differentiated neurons (Fig. 5.3) (Bergsland et al. 2011). In embryonic stem cells (ESCs), Sox2 binds to the regulatory region of genes actively transcribed in ESCs, which are characterized by the trimethylation of lysine 4 in histone 3 (H3K4me3), a histone modification that supports gene transcription by making chromatin accessible to the basal transcriptional machinery. However, Sox2 in ESCs is also bound to the regulatory regions of genes that will only be activated in neural progenitors and are not actively transcribed in ESC. In ESCs, these carry a repressive histone mark (H3K27me3) in conjunction with the activating H3K4me3 (Fig. 5.3). Such bivalently marked genes are thought to be predisposed or poised for activation but require the binding of additional factors removing the repressive mark for activation. In neural progenitor cells, Sox3 replaces Sox2 at the regulatory regions of the same genes. In parallel, repressive histone marks disappear (by a still unresolved mechanism) and only activating marks remain in the regulatory regions of neural progenitor genes, which become transcriptionally activated. However, bivalent histone modifications persist on the regulatory regions of neuronal differentiation genes, which are still silenced in progenitors (Fig. 5.3). Finally, during neuronal differentiation, replacement of Sox3 by Sox11 binding in the regulatory region of neuronal differentiation genes is likewise accompanied by the disappearance of repressive histone marks and transcriptional activation (Fig. 5.3) (Bergsland et al. 2011).

Taken together, this suggests that the sequential activation of different *Sox* genes may drive the progression of cells from stem and progenitor cells to differentiated neurons by successively making different subsets of target genes accessible for transcription via chromatin modifications. SoxB1 factors have indeed been shown to act as pioneer factors, which are able to occupy their binding sites even in silent, compacted chromatin, and make these regions accessible for other transcription factors

(Bergsland et al. 2011; Soufi et al. 2015; Julian et al. 2017). Some of the proteins with which SoxB1 factors interact to affect chromatin remodeling and deposition of histone marks have been identified (Engelen et al. 2011; Amador-Arjona et al. 2015; Flici et al. 2017), but how these processes acquire specificity to allow the differential regulation of gene transcription in progenitors and differentiating cells remains to be elucidated.

The ability of Sox proteins to bind to DNA in cooperation with various lineage specific cofactors, is probably crucial to account for these context-dependent effects (reviewed in Kondoh and Kamachi 2010; Miyagi et al. 2009; Sarkar and Hochedlinger 2013; Julian et al. 2017). Sox proteins are known to form heterodimers with other transcription factors, in particular of the Pax and POU families, which then bind to composite binding sites in the regulatory regions of target genes. Changes in cofactor availability, thus, may result in the activation of different subsets of target genes. For example, Sox2 forms a DNA binding heterodimer with the POU5 family protein Oct4 in pluripotent stem cells but has been suggested to partner with a POU3 family protein, Brn2, in neural progenitor cells (Tanaka et al. 2004; Lodato et al. 2013). However, no differences in binding specificity between Sox2-POU5 and Sox2-POU3 dimers could be detected (Chang et al. 2017), while POU3 factors were shown to activate neural progenitor genes by binding to a different binding site without dimerizing with Sox2 (Mistri et al. 2015). This suggests that the model, by which replacement of Sox2 binding partners leads to altered DNA binding specificities of the heterodimer and, thus, to the activation of different target genes, is too simple and cannot fully account for the context-specific activity of SoxB1 proteins.

5.3 PRONEURAL bHLH TRANSCRIPTION FACTORS: CORE REGULATORS OF NEURONAL AND SENSORY DIFFERENTIATION

Whereas SoxB1 factors predispose cells to adopt a neural (neuronal, glia or sensory) cell fate, the initiation and coordination of neuronal, glial or sensory differentiation in both central and peripheral nervous system requires the activation of at least one member of a small group of so-called "proneural" transcription factors. First discovered in *Drosophila* by their indispensability for neuronal differentiation and their ability to convert non-neuronal cells into neurons, these proneural proteins belong to two closely related groups of basic helix-loop-helix (bHLH) transcription factors, related to *Drosophila* achaete and scute (Achaete/Scute-like proteins) or atonal (Atonal/Neurogenin/NeuroD-like proteins), respectively (reviewed in Bertrand, Castro, and Guillemot 2002; Guillemot 2007; Huang, Chan, and Schuurmans 2014; Guillemot and Hassan 2017; Baker and Brown 2018; Dennis, Han, and Schuurmans 2019). While in *Drosophila*, proneural transcription factors are first broadly expressed in a larger field of cells before becoming restricted to neural progenitors, this is not always true in vertebrates, where SoxB1 may have taken over some of the early functions of proneural transcription factors Ascl1 and Atoh1 (Hassan and Bellen 2000; Galvez, Abelló, and Giráldez 2017).

Proneural bHLH proteins are members of group A of bHLH family proteins. They bind to so-called E-boxes (CANNTG binding sites) on DNA after heterodimerizing with ubiquitously expressed E-proteins (E12, E47), which are also encoded by group A bHLH genes. The ability of proneural bHLH proteins to transcriptionally activate other genes is antagonized by the Id and Hes/Hey proteins, which belong to groups D and E of bHLH proteins, respectively (reviewed in Bertrand et al. 2002; Kageyama, Ohtsuka, and Kobayashi 2008a; Ling, Kang, and Sun 2014; Wang and Baker 2015; Dhanesh, Subashini, and James 2016). Id proteins can heterodimerize with proneural proteins but lack a DNA-binding domain. They are, thus, able to sequester proneural proteins and interfere with their DNA binding (acting as so-called "dominant-negatives"). Hes/Hey proteins, notably Hes1 and Hes5, can act in a similar fashion, but also can block proneural activity by forming homodimers, which bind to distinct DNA binding sites (so-called N-boxes) and recruit co-repressors of the Groucho protein family, thereby repressing transcription.

5.3.1 Diverse Roles of Proneural Factors in Orchestrating Neuronal and Sensory Differentiation

In vertebrates, as in *Drosophila* and other metazoans (see Schlosser 2021), proneural bHLH genes are both required and sufficient for neuronal and sensory differentiation in both the central and peripheral nervous systems. While different members of the Achaete/Scute-like (e.g. Ascl1) or Atonal/Neurogenin/NeuroD-like (e.g. Atoh1, Neurog1, Neurog2) bHLH proteins function as proneural protein for different types of neurons or sensory cells, these proneural proteins then play a central role in activating neuron- or sensory-specific differentiation gene batteries and initiating cell cycle exit and differentiation. In the cranial placodes, different proneural proteins are expressed in different subsets of placodes in patterns that vary slightly between species (Fig. 5.2A) (e.g. Cau et al. 1997; Fode et al. 1998; Ma et al. 1998; Bermingham et al. 1999; Ma, Anderson, and Fritzsch 2000; Schlosser and Northcutt 2000; Andermann, Ungos, and Raible 2002; Begbie, Ballivet, and Graham 2002; Cau, Casarosa, and Guillemot 2002; Chen et al. 2002; Millimaki et al. 2007; Nieber, Pieler, and Henningfeld 2009). Often several group A bHLH genes are expressed in a single placode with partly divergent, partly redundant functions.

After knockout mutations of proneural genes, neuronal or sensory differentiation is blocked in the cell populations that rely on these particular genes. For example, mice deficient for Ascl1 lack the olfactory sensory neurons derived from the olfactory placode (Guillemot and Joyner 1993; Cau et al. 1997, 2002; Murray et al. 2003); mice and zebrafish deficient for Neurog1 or Neurog2 loose somato- or viscerosensory neurons in the cranial ganglia derived from the profundal/trigeminal, epibranchial, otic, and lateral line placodes (Ma, Kintner, and Anderson 1996; Fode et al. 1998; Ma et al. 1998, 2000; Andermann et al. 2002); and mice and zebrafish deficient for Atoh1 lack hair cells derived from otic or lateral line placodes (Bermingham et al. 1999; Chen et al. 2002; Woods, Montcouquiol, and Kelley 2004; Sarrazin et al. 2006; Millimaki et al. 2007).

Overexpression of proneural proteins, on the other hand, is often sufficient to convert other cell types into neurons or sensory cells. For example, overexpression of Ascl1, Neurog1/2 or NeuroD2 in chick embryos or mammalian cell lines can convert pluripotent cells into neurons (Perez, Rebelo, and Anderson 1999; Farah et al. 2000; Zhang et al. 2013; Aydin et al., 2019), while Ascl1, Neurog1/2, or NeuroD1/4 in *Xenopus* embryos can convert ventral non-neural ectoderm into neurons (Lee et al. 1995; Ma et al. 1996; Perron et al. 1999; Talikka, Perez, and Zimmerman 2002). Furthermore, proneural proteins can even change the fate of differentiated cells. Overexpression of Ascl1, Neurog2, NeuroD1, or NeuroD4 in astrocytes of the mammalian brain can convert these into neurons (Berninger et al. 2007; Guo et al. 2014; Masserdotti et al. 2015; Brulet et al. 2017), whereas overexpression of Atoh1 converts inner ear epithelial cells into hair cells (Zheng and Gao 2000; Woods et al. 2004). Ascl1 and Neurog2 are even sufficient to directly reprogram fibroblasts into neurons (Vierbuchen et al. 2010; Zhang et al. 2013; Chanda et al. 2014). However, the reprogramming efficiency of Ascl1 alone is low and can be greatly increased by the addition of other transcription factors (Brn2, Myt1l) (reviewed in Guillemot and Hassan 2017; Dennis, Han, and Schuurmans 2019). A recent study suggests that a shared generic neuronal program can be activated in mouse fibroblasts by overexpression of 76 different pairwise combinations of transcription factors (Tsunemoto et al., 2018). These typically include a bHLH factor of the Ascl, Neurog, NeuroD or Atoh families in combination with a POU- or nuclear receptor family transcription factor.

Taken together, this suggests that proneural bHLH genes act as selector genes for neuronal and sensory fates in the central and peripheral nervous system. Expression of at least one member of this set appears to be required and sufficient to activate generic aspects of neuronal and sensory differentiation. In other words, proneural bHLH appear to be central members of the CoRN conferring sensory and neuronal identity. The dependence of all neurons and sensory cells on one or the other member of this set of closely related genes probably reflects the fact that all neurons and sensory cells are evolutionarily related cell types, derived from a common type of protoneuron in the metazoan ancestor. We will come back to this hypothesis in Volume 2 (Schlosser 2021).

The central role of proneural bHLH factors for neuronal and sensory differentiation can be broken down into a number of separable functions. First, proneural bHLH factors promote a neuronal or sensory cell fate choice by directly or indirectly activating batteries of differentiation genes that encode the proteins (channels, neurotransmitters, proteins involved in neuronal delamination, and migration) characteristic for the generic neuronal or sensory phenotype. At least some proneural factors (e.g. Ascl1, Neurog2) are able to act as pioneer factors, which bind to their targets in closed chromatin and make them accessible for binding of other transcription factors (Aydin et al., 2019; Alsina, 2020). While different proneural genes initiate neuronal differentiation in each placode (e.g. *Ascl1* in the olfactory placode; *Neurog1* in profundal/trigeminal and otic placodes; *Neurog2* in epibranchial placodes of mice), they all converge on *NeuroD1*, encoding another group A bHLH factor, which then activates neuron-specific gene batteries in postmitotic cells (Ma et al. 1996; Cau et al. 1997; Fode et al. 1998; Ma et al. 1998; Bertrand et al. 2002;

Chae, Stein, and Lee 2004; Cho and Tsai 2004). NeuroD1 also is indispensable in activating neuronal differentiation genes in most parts of the CNS, although other bHLH factors substitute for its function in selected regions such as Ascl1 in the ventral telencephalon (Casarosa, Fode, and Guillemot 1999; Horton et al. 1999). In contrast, the different proneural genes that initiate the differentiation of sensory cells from the otic and lateral line placodes (*Atoh1*) and from the retina (*Atoh7*) appear to activate sensory differentiation genes without first converging on a common generic sensory bHLH transcription factor.

Second, proneural factors and NeuroD1 initiate neuronal or sensory differentiation by promoting cell cycle exit (Farah et al. 2000; Mizuguchi et al. 2001; Novitch, Chen, and Jessell 2001; Politis et al. 2007; Yi et al. 2008; Alvarez-Rodriguez and Pons 2009; Ochocinska and Hitchcock 2009; Lacomme et al. 2012). In addition, they inhibit expression of *SoxB1* genes (Bylund et al. 2003; Dabdoub et al. 2008; Evsen et al. 2013), while upregulating the expression of Sox21, which has been suggested to counteract SoxB1 activity in the regulation of a set of common target genes (Sandberg, Kallstrom, and Muhr 2005; but see Whittington et al. 2015).

Third, proneural factors prevent neuronal or sensory differentiation in adjacent cells. This process is known as lateral inhibition and involves the transcriptional activation of ligands of the Notch signaling pathway such as Delta or Jagged (Fig. 5.4). These ligands are membrane bound proteins that bind to a transmembrane receptor encoded by the *Notch* gene on neighboring cells. Upon binding these ligands, Notch

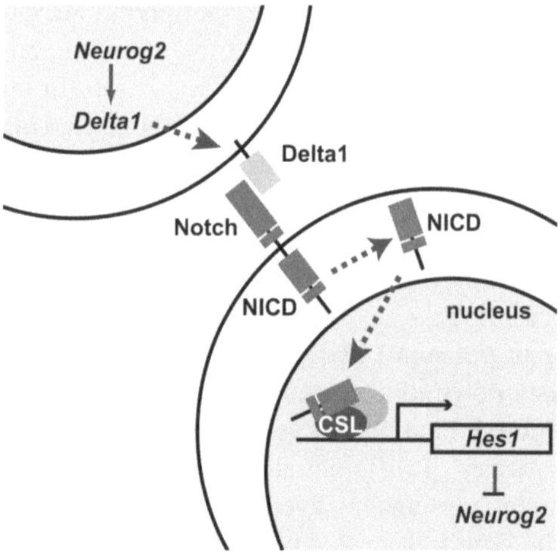

FIGURE 5.4 Notch-mediated lateral inhibition. The Notch receptor is a transmembrane protein, which is activated by binding of Notch ligands such as Delta1, which are expressed downstream of *Neurog2* on the surface of adjacent cells. After ligand binding, the Notch intracellular domain (NICD) is cleaved and enters the nucleus where it acts as a transcriptional co-activator of the CSL transcription factor and helps to activate transcription of target genes such as *Hes1*. The Hes1 protein acts as a transcriptional repressor of multiple target genes including *Neurog2*. (Redrawn and modified from Shimojo et al. 2011.)

undergoes proteolytic cleavage and its intracellular domain moves to the nucleus. There it forms a complex with a transcription factor (Suppressor of hairless; also known as CSL or RBPJ), which then activates multiple target genes (reviewed in Bertrand et al. 2002; Bray 2006; Louvi and Artavanis-Tsakonas 2006; Guruharsha, Kankel, and Artavanis-Tsakonas 2012; Siebel and Lendahl 2017). Among the latter are the genes encoding Hes1 and Hes5, which directly repress transcription of *Ascl1*, *Neurog1/2* or other proneural genes and are, therefore, potent repressors of neuronal or sensory differentiation (Sasai et al. 1992; Ishibashi et al. 1995; Chen et al. 1997; Ohtsuka et al. 1999; Cau et al. 2000; Davis and Turner 2001; Hatakeyama et al. 2004; Kageyama, Ohtsuka, and Kobayashi 2008b; Kobayashi and Kageyama 2014; Dhanesh et al. 2016). Thus, proneural proteins, while activating neuronal or sensory differentiation in the cells, in which they are expressed, repress differentiation and maintain a progenitor state in their neighbors. Importantly, in a region where all cells initially express proneural proteins, these cells interact reciprocally via lateral inhibition. This mechanism will result in the maintenance of high-level proneural protein expression in a subset of cells but extinction of expression in their neighbors, ultimately resulting in a salt and pepper pattern.

Fourth and finally, proneural transcription factors not only activate a generic neuronal or sensory program, but also promote the differentiation of particular subtypes of neurons or sensory cells (reviewed in Bertrand et al. 2002; Huang et al. 2014; Allan and Thor 2015). Recently, the different affinity of Ascl1 and Neurog2 to different E-boxes has been shown to underlie their activation of partly non-overlapping sets of target genes (Aydin et al., 2019). In addition, activation of genes specific for neuronal or sensory subtypes typically requires cooperation of proneural proteins with other transcription factors, often transcription factors of the homeodomain protein superfamily. Proneural proteins are also known to act in combination with other bHLH proteins to specify different cell types in the nervous system. For example, in chick and mouse embryos, Neurog2 is able to specify spinal motor neurons in cooperation with the homeodomain transcription factors Islet1 and Lhx3, but only when expressed in combination with the bHLH protein Olig2 (Mizuguchi et al. 2001; Novitch et al. 2001; Scardigli et al. 2001; Lee and Pfaff 2003; Ma et al. 2008; Lee et al. 2012). Interaction of Neurog2 with Islet1-Phox2a instead promotes formation of cranial motorneurons (Mazzoni et al. 2013). Islet1-Lhx8, in turn, promotes differentiation of a cholinergic population of ventral telencephalic neurons most likely in association with other proneural factors such as Ascl1, since Neurog2 is not expressed in these cells (Cho et al. 2014). While all of these neurons are cholinergic (i.e., they use acetylcholine as a neurotransmitter), Neurog1/2 also promotes the formation of dopaminergic, glutamatergic or peptidergic neurons in other parts of the nervous system in interaction with other cofactors (Ma et al. 1999; Andersson et al. 2007; Jeong et al. 2006; Berninger et al. 2007).

5.3.2 Coordinating the Transition from Progenitors to Differentiating Cells

Although proneural genes promote cell cycle exit and neuronal or sensory differentiation, they are known to be expressed in progenitor cells, which continue dividing.

To resolve this apparent paradox and to explain how proneural transcription factor regulate the transition between progenitors and differentiating cells, several mechanisms have been proposed. First, it has been shown that Neurog2, Ascl1, and Atoh1 are highly unstable proteins, which are quickly degraded by poly-ubiquitination (Vosper et al. 2007; Forget et al. 2014; Urban et al. 2016). The protein stability and DNA binding affinity of Neurog2 and Ascl1 has, however, been shown to be strongly dependent on their phosphorylation status. The proteins have multiple serines, which can be phosphorylated outside of their bHLH domain and the degree of protein stability and DNA binding affinity decreases with increasing numbers of phosphorylated serines (Ali et al. 2011, 2014). Consequently, highly phosphorylated forms of the proteins are only able to bind to target genes in well accessible chromatin such as the *Delta1* gene and presumably other progenitor genes, whereas unphosphorylated forms are able to recruit chromatin remodeling factors and activate neuronal differentiation genes such as *NeuroD1* or *MyT1* in relatively inaccessible chromatin (Ali et al. 2011, 2014; Hindley et al. 2012). It has been further shown that in proliferating cells proneural factors are maintained in a highly phosphorylated form by cyclin-dependent kinases (CDK) suggesting a model, according to which lengthening of the cell cycle and the associated decrease in CDK activity promotes neuronal differentiation by dephosphorylation of proneural proteins (Ali et al. 2011; Hindley et al. 2012; Ali et al. 2014; Hardwick, Azzarelli, and Philpott 2018). In addition, phosphorylation at an evolutionarily conserved residue within the bHLH domain of all proneural proteins can rapidly switch off their proneural activity (Quan et al. 2016).

Second, there is increasing evidence for the proposal that changing dynamics of proneural protein expression govern the transition from progenitors to differentiating cells (Fig. 5.5) (reviewed in Imayoshi and Kageyama 2014; Kobayashi and Kageyama 2014). Proneural proteins such as Neurog2 and Ascl1 were shown to oscillate in neural progenitors in a 2–3 hour period (Imayoshi et al. 2013; Shimojo et al. 2016). These oscillations appear to be driven by out of phase oscillations of Hes1 or related proteins known to repress proneural gene transcription (i.e., Neurog1 levels are low whenever Hes1 levels are high and vice versa) (Shimojo, Ohtsuka, and Kageyama 2008; Imayoshi et al. 2013). The oscillating levels of Hes1 arise because Hes1 represses its own transcription but both the mRNA and protein of Hes1 are degraded very quickly (Fig. 5.5A, B) (Hirata et al. 2002; Bonev, Stanley, and Papalopulu 2012; Tan et al. 2012; Kobayashi et al. 2015).

It has further been observed that mRNA and protein of the Notch ligand Delta1 are oscillating in parallel to Neurog2 in neural progenitor cells (Fig. 5.5C) (Shimojo et al. 2008, 2016). Because progenitors interact with their neighbors via Notch-mediated lateral inhibition, this suggests a model, in which lateral inhibition is actually switching back and forth between neighboring cells (Fig 5.5D, E). Consequently, neighboring cells should show out of phase oscillations in all of these proteins and each cell should alternate between a state (high levels of Neurog2/Delta1 and low levels of Hes1), in which it represses neuronal differentiation in its neighbors and another state (low levels of Neurog2/Delta1 and high levels of Hes1), in which it itself receives a repressive Notch signals from adjacent cells (Kageyama et al. 2008a; Kageyama et al. 2009; Imayoshi and Kageyama 2014). It has recently been shown

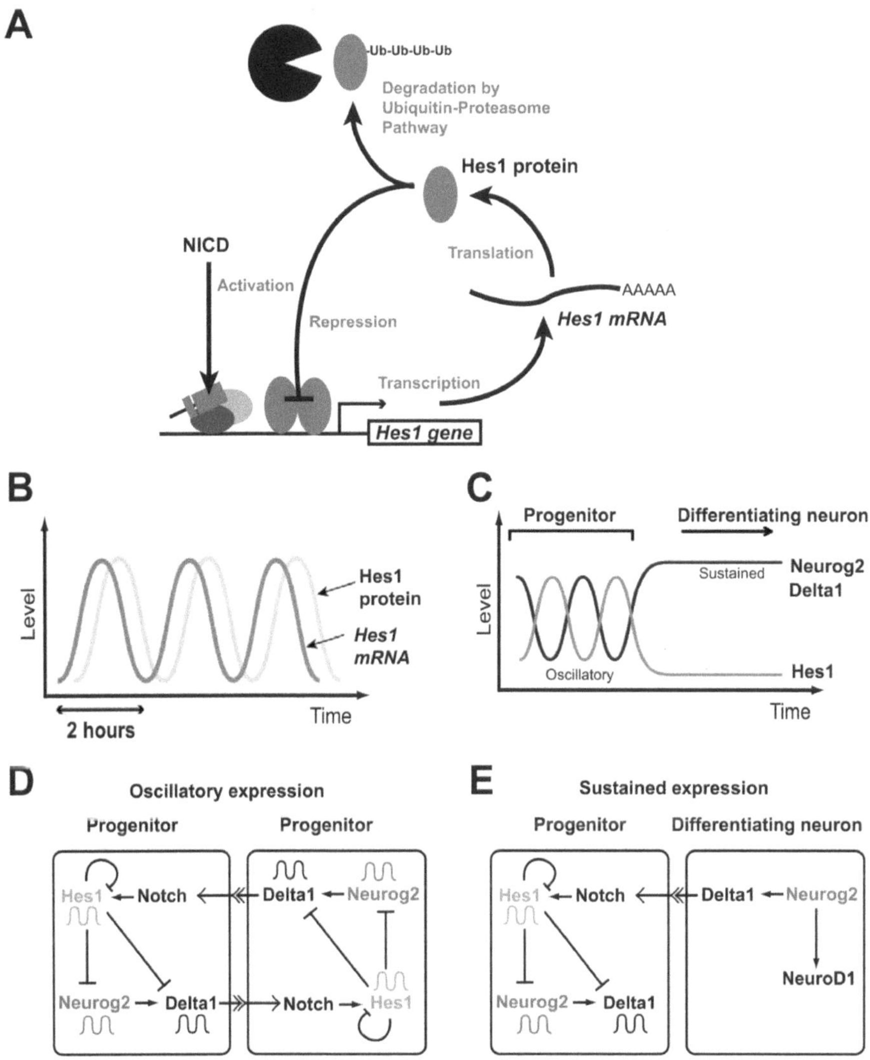

FIGURE 5.5 Oscillations of Hes and proneural proteins drive the transition from progenitors to neuronal differentiation. (**A, B**): Hes1 protein represses transcription of the *Hes1* gene. Both *Hes1* mRNA and Hes1 protein are unstable and quickly degraded (**A**). The combination of auto-repression and mRNA/protein instability leads to oscillatory expression of both *Hes1* mRNA and Hes1 protein with a periodicity of roughly 2 hours (**B**). (**C–E**) Hes1 oscillation in progenitor cells induces the oscillatory expression of Neurog2 and Delta1 by periodic repression (**C, D**). Delta1 activates Notch signaling thereby maintaining a progenitor state and preventing neuronal differentiation in adjacent cells. As long as Delta1 oscillates in both cells (**D**), both cells receive Notch signaling and keep in a progenitor state. When Hes1 is repressed in one cell and stops oscillating, Neurog2 and Delta1 expression become sustained in this cell leading to neuronal differentiation (**C, E**). Delta1 expression in differentiating neurons maintains Notch signaling and progenitor status in adjacent cells (**E**). (Redrawn and modified from Kobayashi and Kageyama 2014.)

that experimental extension of Ascl1 expression from three to six hours promotes cell cycle exit and neuronal differentiation (Imayoshi et al. 2013). This provides strong support for the model that sustained expression of proneural proteins promotes differentiation, while oscillatory expression keeps cells in a progenitor state (Shimojo et al. 2008).

The mechanism regulating the transition between oscillatory and persistent proneural protein expression is still not clear. A reduction of Hes levels due to reduced Notch signaling input may play a role here (Imayoshi et al. 2013) or changes in proneural protein stability or activity due to altered phosphorylation status as discussed above. Gradual accumulation of microRNA miR-9, which dampens Hes1 oscillation, has also been suggested to terminate oscillations, when threshold levels of miR-9 have been reached (Bonev et al. 2012).

Once expression of a certain level of proneural proteins is sustained, they probably activate different sets of target genes that promote neuronal differentiation. Indeed, a recent study in mammalian embryonic forebrain and neural stem cell cultures has shown that Ascl1 activates genes involved in cell cycle progression in proliferating progenitors, but neuronal differentiation genes in cells that stopped dividing (Castro et al. 2011). Again, multiple mechanisms may be involved in this switch. Among the Ascl1 target genes identified in the study mentioned, only genes involved in cell cycle progression but not those promoting neuronal differentiation have binding sites for the Notch mediator CSL (Castro et al. 2011). This suggests that Ascl1 may activate genes involved in cell cycle progression in cooperation with Notch signaling and neuronal differentiation genes, when Notch signaling subsides. It is also possible that target genes promoting neuronal differentiation require longer residence times of proneural transcription factors for their activation. The phosphorylation status of proneural proteins may also alter target gene choice as discussed earlier (Hindley et al. 2012). Prolonged expression of proneural proteins also makes cells unresponsive to Notch-mediated lateral inhibition preventing reversion to a progenitor state. This is at least partly due to their ability to activate the transcription factor MyT1, which prevents activation of *Hes1* and other Notch target genes (Bellefroid et al. 1996; Vasconcelos et al. 2016).

5.4 REGULATION OF SoxB1 AND PRONEURAL PROTEINS BY SIX1 AND EYA IN CRANIAL PLACODES

As far as current evidence indicates, general aspects of neuronal and sensory differentiation from cranial placodes, including its dependency on SoxB1 and proneural proteins, are regulated in a very similar manner to the central nervous system. In contrast to the CNS, however, expression of both *SoxB1* genes and proneural genes in placodes is under control of the pan-placodal transcription factor Six1 and its cofactors Eya1 or Eya2 (Figs. 5.1, 7.2) (see Chapter 3; reviewed in Schlosser 2010; Grocott, Tambalo, and Streit 2012; Wong, Ahmed, and Xu 2013; Xu 2013; Moody and LaMantia 2015; Singh and Groves 2016; Streit 2018). Six4 is acting in partly redundant pathways for at least some placodes (Grifone et al. 2005; Konishi et al. 2006; Zou, Silvius, Rodrigo-Blomqvist, et al. 2006; Chen, Kim, and Xu 2009).

While the neural plate maintains SoxB1 expression from the earliest embryonic stages, expression in the PPE is established independently and regulated by distinct cis-regulatory elements (Uchikawa et al. 2003; Saigou et al. 2010; Iwafuchi-Doi et al. 2011). Although Six1 binding sites have not yet been confirmed in enhancers of *SoxB1* genes, SoxB1 expression in placodes is strongly dependent on Six1 and Eya1 (Fig. 5.1) (Schlosser et al. 2008; Zou et al. 2008; Bosman et al. 2009; Chen et al. 2009; Riddiford and Schlosser 2016). Since SoxB1 expression is not observed in the entire PPE but is confined to two distinct anterior (adenohypophyseal, olfactory) and posterior (profundal/trigeminal, otic, lateral line, epibranchial) subregions of the PPE (Fig. 5.2B), Six1 and Eya1 probably require additional yet unidentified cofactors to activate SoxB1 factors in the PPE (Schlosser and Ahrens 2004).

SoxB1 genes (*Sox2*, *Sox3*) along with proneural genes (*Atoh1*, *Neurog1*), genes promoting neuronal or sensory differentiation (*POU4f1*, *Islet2*, *Gfi1*, *Tlx1*), and genes encoding Hes repressors (*Hes8*, *Hes2*, *Hes9*) have been identified as putatively direct Six1 and Eya1 target genes in a recent screen (Riddiford and Schlosser 2016), suggesting that placodal expression of these genes may be directly regulated by Six1 and Eya1 (Fig. 5.1). *Atoh1*, *Gfi1*, *POU4f3*, *GATA3*, *Pbx1*, and *Six1* were confirmed as direct Six1 targets by enhancer binding and/or ChIP-Seq studies of mouse cochleae (Ahmed et al. 2012; Li et al. 2020). In the cochlea, Six1 also targets genes specific for different subtypes of hair cells and supporting cells as well as a wide range of terminal differentiation genes regulating hair bundle formation (Li et al. 2020). This indicates that Six1 targets both genes expressed in sensory or neuronal progenitors as well as those promoting neuronal differentiation.

Interestingly, the effects of Six1 and Eya1 on progenitor maintenance and differentiation were shown to be strongly dosage dependent (Friedman et al. 2005; Schlosser et al. 2008; Zou et al. 2008; Bosman et al. 2009; Riddiford and Schlosser 2017). In *Xenopus* embryos, high levels of Six1 and Eya1 promoted expression of progenitor genes (*SoxB1*, *Hes*) and proliferation, while lower levels of Six1 and Eya1 instead promoted activation of proneural genes and neuronal differentiation (Fig. 5.1) (Schlosser et al. 2008; Riddiford and Schlosser 2017). Similar dosage effects were observed during mouse inner ear development, where cell proliferation is already compromised in Eya1 mutants with mildly reduced protein levels (hypomorphs), whereas the differentiation of hair cells is only perturbed in mutants with strong reductions in protein levels (Zou et al. 2008). In agreement with these observations, during normal placode development, Eya1 and Six1 are expressed at high levels in proliferating placodal cells that express progenitor markers such as SoxB1 with declining levels in cells that express neuronal differentiation genes (Schlosser et al. 2008; Zou et al. 2008; Chen et al. 2009).

Taken together these observations suggest that Six1 and Eya1 play dual roles during the development of neuronal and sensory cells from placodes promoting both progenitor maintenance and the differentiation of various sensory and neuronal cell types (Fig. 5.1). Whereas high levels of Six1 and Eya1 promote a progenitor state, low levels promote differentiation. The declining levels of Six1 and Eya1 observed as cells migrate away from placodes, thus, may play important roles in regulating the transition from proliferating progenitors to differentiating cells. However, the molecular mechanisms underlying these dosage effects are presently obscure.

Like SoxB1 factors, Six1 has been shown to already occupy enhancers of its target genes before these are expressed (Li et al. 2020). Taken together with its ability to recruit SWI/SNF chromatin remodeling complexes (Ahmed, Xu, and Xu 2012), this raises the possibility that Six1 may act as a pioneer factor, which opens up chromatin and makes enhancers of its target genes accessible for other transcription factors. However, more evidence is needed to confirm this hypothesis. Pioneer factor or not, the binding of Six1 to enhancers of differentiation genes for several different placodal cell types prior to their expression suggests that it acts as a competence factor, which allows cells to differentiate into multiple placode derived sensory or neuronal cell types. Availability of different cofactors may subsequently determine, which particular cell type is adopted.

There is now substantial evidence that Six1 and Eya1 interact with different cofactors to specify different cell fates. Physical interactions between Six1, Eya1, and the Sox2 protein were shown to promote the activation of Atoh1 and otic hair cell development (Ahmed et al. 2012). Protein-protein interactions between Six1, Eya1, Sox2, and the SWI/SNF chromatin remodeling complex, in turn, activate transcription of the neuronal genes *Neurog1* and *NeuroD1*, thereby promoting neuronal differentiation in the inner ear (Ahmed, Xu and Xu 2012). Six1 also forms protein complexes with many other proteins, including many proteins encoded by its own target genes (e.g. Atoh1, POU4f3, Gfi1, GATA3, Rfx) to cooperatively regulate gene expression (Li et al. 2020). Since its target genes include differentiation genes specific for different cell types such as hair cells and supporting cells, this suggests that Six1 may specify different cell types in cooperation with different cofactors.

Apart from their roles in the specification of placodally derived neuronal and sensory cells, Six1 and Eya may also contribute to the differentiation of some neurosecretory cell types of the anterior pituitary derived from the adenohypophyseal placode (see Chapter 8). The differentiation of gonadotropes, corticotropes, and melanotropes was strongly compromised in Eya1 mutants of zebrafish (Nica et al. 2006), and the size of the pituitary was severely reduced in mouse Eya1/Six1 double mutants (Li et al. 2003). This indicates an important role of Six1 and Eya1 in the specification of at least some placodally derived neurosecretory cells, which is still very poorly understood.

Interestingly, however, Six1/2 and Eya may also be required during development of other types of neurosecretory cells derived not from the PPE but from the endoderm. In the pharyngeal endoderm, Six1 and Eya are expressed in the anlagen of the thymus and parathyroid glands, the ultimobranchial bodies (producing the calcitonin secreting C-cells of the thyroid), and in pulmonary neuroendocrine cells (Xu et al. 2002; Zou, Silvius, Davenport, et al. 2006; El Hashash et al. 2011; Travaglini et al. 2020). Moreover, the development of these glands is compromised in mouse mutants of Six1 or Eya1 due to defects in morphogenesis and cell differentiation (Xu et al. 2002; Zou et al. 2006). This has interesting implications for the evolution of cranial placodes as will be discussed in more detail in Chapters 5 and 6 of the second volume (Schlosser 2021). The ability of Six1/2 and Eya to promote the differentiation of different types of sensory or neurosecretory cells in the endoderm and in the non-neural ectoderm (PPE), likely reflects their cooperation with different competence factors in the two tissues, allowing the activation of partially different sets of target genes. However, this needs to be confirmed in further studies.

In summary, many general features of neuronal and sensory differentiation are widely shared between different types of neurons or sensory cells including those derived from cranial placodes. SoxB1 transcription factors maintain cells in a progenitor state and prime them for neural or sensory differentiation. After downregulation of SoxB1 factors, various proneural bHLH transcription factors (e.g. Ascl1, Neurog1/2 or Atoh1) then initiate neuronal or sensory differentiation. While different bHLH factors are active in different cell populations, they appear to play largely overlapping roles in regulating generic aspects of neuronal differentiation. For placode-derived neurons and sensory cells, *SoxB1* genes together with other progenitor promoting genes and proneural bHLH genes are regulated by the pan-placodal transcriptional regulators Six1 and Eya. However, the latter also help to control the specification and differentiation of other placodally derived cell types such as neurosecretory cells.

To establish identities of particular subtypes of neurons or sensory cells, proneural bHLH factors need to cooperate with other transcription factors. These include LIM-type (e.g., Islet1), Paired-type (e.g., Phox2a, Phox2b, DRG11), and POU-type (e.g., Brn3a and Brn3c) homeodomain proteins, COE-type bHLH proteins, Runx proteins and Fox proteins (e.g., FoxG1) as we will see in the next chapters. These other transcription factors are either narrowly expressed in subsets of cells expressing a particular bHLH factor allowing to define subpopulations of these cells (e.g. LIM or Runx transcription factors, which are expressed only in some members of a group of cells), or are expressed in a broad but regionally confined manner, giving region-specific identity to cell types in many different tissues (e.g. Hox transcription factors, which are expressed in defined positions along the anteroposterior axis).

6 Differentiation of Sensory and Neuronal Cell Types from Neurogenic Placodes

After reviewing the common mechanisms regulating the formation of placodal neurons and sensory cells in the last chapter, I will now discuss how these mechanisms are modulated to produce specific neurons and sensory cells from individual placodes. For each placode, I will first provide an overview of structure and function of the various cell types derived from it, and will then review what we currently know about the transcription factors regulating specification and differentiation of these cell types.

6.1 PROFUNDAL AND TRIGEMINAL PLACODES

6.1.1 Cell Types Derived from the Profundal and Trigeminal Placodes

The profundal and trigeminal placodes give rise to a subpopulation of general somatosensory neurons in the profundal/trigeminal ganglion, while another subpopulation as well as all glial cells derive from the neural crest. The general somatosensory neurons residing in the profundal and trigeminal ganglion mediate temperature, touch, and pain sensation from the mouth cavity and the anterior head region (Darian-Smith 1973; Butler and Hodos 2005). They send their axons into the principal nucleus of the trigeminal nerve in the hindbrain (Fig. 1.12). Another class of general somatosensory neurons residing in the mesencephalic nucleus of the trigeminal nerve (located, as the name says, in the midbrain and probably derived from the neural plate) relay information on the position of the jaw muscles (propioception) (Butler and Hodos, 2005).

General somatosensory neurons can be classified into different types according to their morphology, function, and gene expression profile (reviewed in Marmigere and Ernfors 2007; Zagami, Zusso, and Stifani 2009; Liu and Ma 2011). These different types of general somatosensory neurons are found not only in the profundal and trigeminal ganglia but also in other somatosensory ganglia that are completely neural crest derived, including the dorsal ganglia of the glossopharyngeal, and vagal nerves (the facial nerve has only a very small population of somatosensory neurons), and the dorsal root ganglia (DRG) of the spinal cord (Eng et al. 2007; Nguyen et al. 2017; Kupari et al. 2019). In the DRG, the main cell types have been classified by their expression of different Trk-family neurotrophin receptors – these respond to molecules that stimulate neuron survival and axon formation. Glossing over some of the complexities here, there are large propioceptors expressing TrkC, smaller mechanoreceptor expressing TrkB, and very small cells that express TrkA and may mediate pain (nociceptors), temperature (thermoreceptors), itch or weak mechanical stimuli. Recent single-cell sequencing studies have revealed the existence of an even larger diversity of subtypes of somatosensory neurons in the DRG (Chiu et al. 2014; Usoskin et al. 2015; Li, et al. 2016; Gatto et al. 2019) as well as in the profundal/trigeminal ganglion (Nguyen et al. 2017), but nothing is currently known about which transcription factors specify these different types.

6.1.2 Transcription Factors Involved in Specifying Identity of General Somatosensory Neurons

All general somatosensory neurons, whether placode or neural crest derived, appear to rely on the same core regulatory network (CoRN) of transcription factors, which includes one of the Neurogenins (either Neurog1 or Neurog2) as a proneural factor. In most vertebrates, Neurog1 expression predominates over Neurog2 in the profundal/trigeminal, otic, and lateral line placodes, all of which give rise to somatosensory neurons, whereas Neurog2 is more strongly expressed in the epibranchial placodes, which generate viscerosensory neurons (Fode et al. 1998; Ma et al. 1998; Andermann, Ungos, and Raible 2002; Begbie, Ballivet, and Graham 2002; Nieber, Pieler, and Henningfeld 2009). In the DRG, most early born (typically TrkB/C+) neurons are derived from Neurog2 expressing precursors, while later born (typically TrkA+) neurons are mostly derived from Neurog1 expressing precursors (Fig. 6.1) (Marmigere and Ernfors 2007; Liu and Ma 2011). However, there are exceptions to these rules (e.g. in birds: Begbie et al. 2002) suggesting that the functions of Neurog1 and Neurog2 for the specification of sensory neurons subtypes was not conserved during evolution of different vertebrate lineages.

Apart from Neurog1/2, several homeodomain transcription factors are co-expressed in somatosensory neurons of the PNS. This includes Six1 and Six4, which are expressed in general somatosensory neurons derived from either placodes or neural crest (Laclef et al. 2003; Ando et al. 2005; Konishi et al. 2006; Yajima et al. 2014) as well as Islet1, Tlx1 or Tlx3, and POU4f1 (= Brn3a), (Korzh et al. 1993; Fedtsova and Turner 1995; Xiang et al. 1995; Logan et al. 1998; Qian et al. 2001, 2002; Begbie et al. 2002; D'Autreaux et al. 2011; Quina et al. 2012). While Six1, Six4, Islet1, and Tlx1/3 are very widely expressed in both somato- and viscero-sensory neurons of the PNS, POU4f1 expression is limited to somatosensory neurons (Fig. 6.1). In the viscerosensory neurons derived from the epibranchial placodes, it is repressed by Phox2b (D'Autreaux et al. 2011).

In single mutants of Six1 or Six4 or their cofactor Eya1, DRGs and profundal/trigeminal ganglion show relatively minor defects (Laclef et al. 2003; Zou et al. 2004). However, in double mutants of Six1 and Six4, sensory neurons of the DRG and profundal/trigeminal ganglion initially form but their subsequent migration, survival, and terminal differentiation is compromised indicating that Six1 and Six4 act redundantly in promoting specification and survival of general somatosensory neurons probably in conjunction with Eya cofactors (Ando et al. 2005; Konishi et al. 2006; Yajima et al. 2014).

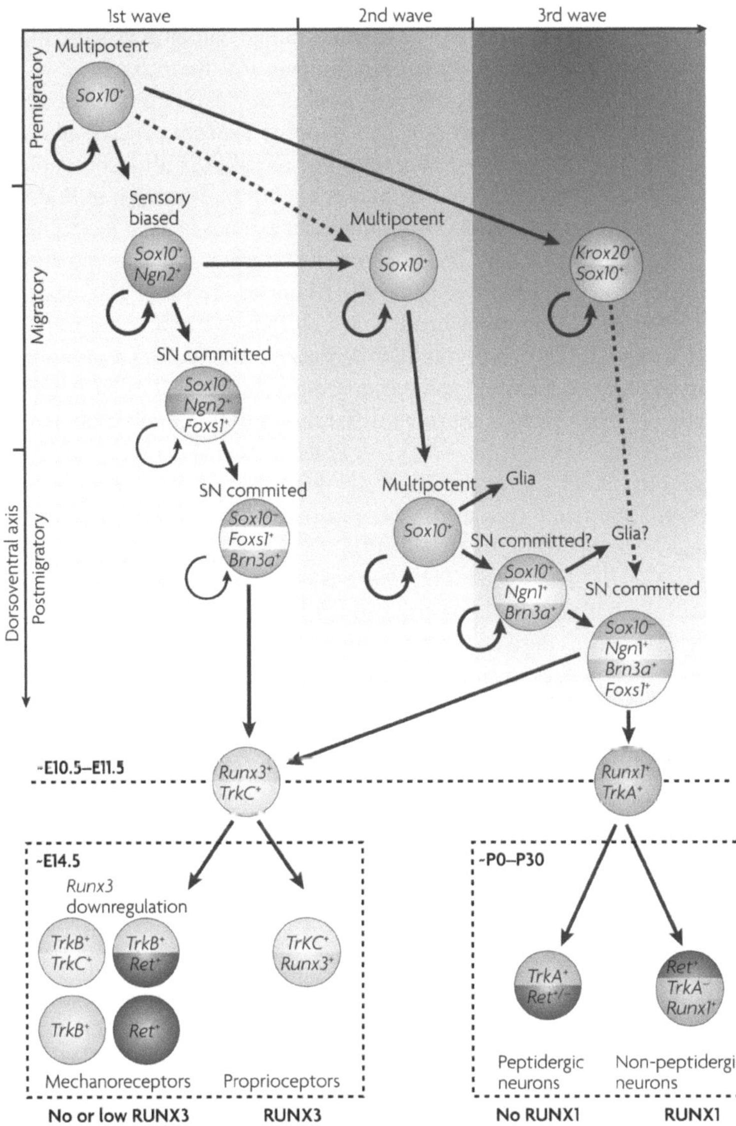

FIGURE 6.1 Specification of somatosensory neurons and their subtypes. Somatosensory neurons originate in several waves regulated by partly different regulatory networks. Progenitors of the first wave transiently express Sox10 followed by Neurog2 (=Ngn2) and POU4f1 (=Brn3a) expression and mostly differentiate into large, TrkB/C+ neurons (mechano- and proprioceptors). Maintained Runx3 promotes the formation of proprioceptors expressing TrkC in this lineage. Progenitors of the second and third wave transiently express Sox10 followed by Neurog1 (=Ngn1) and POU4f1 (=Brn3a) expression and mostly differentiate into small, TrkA+ neurons (both peptidergic and non-peptidergic neurons). Runx1 promotes the formation of non-peptidergic neurons and downregulation of TrkA in this lineage. (Reprinted with permission from Marmigere and Ernfors 2007).

Both Islet1 and POU4f1 transcription factors are upregulated by the Neurog1/2 target gene *NeuroD1* (Lee et al. 1995; Hutcheson and Vetter 2001) and play central roles in specifying somatosensory neuronal identity in the placode or neural crest derived neurons of the peripheral nervous system downstream of Neurog1/2. After knocking out Islet1 in precursors of somatosensory neurons, TrkA+ neurons fail to differentiate, and somatosensory innervation of the skin is almost completely lost, while propioceptors develop normally (Sun et al., 2008). Mutation in POU4f1, similarly prevent proper specification of somatosensory neurons as reflected in pathfinding defects, failure to generate TrkC+ neurons and progressive apoptosis of TrkA+ and TrkB+ neurons (McEvilly et al. 1996; Xiang et al. 1996; Huang et al. 1999; Eng et al. 2001; Marmigere and Ernfors 2007). POU4f1 and Islet1 were shown to co-regulate a similar set of differentiation genes in somatosensory neurons, while suppressing alternative neuronal and mesodermal fates as well as Neurog1/2 expression, thereby helping to make neuronal differentiation irreversible (Sun et al. 2008; Lanier et al. 2009; Dykes et al. 2011). As POU and LIM proteins frequently bind cooperatively to enhancers, this probably involves direct protein-protein interactions between POU4 and Islet1 similar to what has been shown for retinal ganglion cells (Hobert and Westphal 2000; Li et al. 2014).

Because POU4f1, Islet1, and Tlx1/3 are expressed not only in general somatosensory neurons but also in the special somatosensory (all) and/or viscerosensory (Islet1, Tlx1/3) neurons derived from the otic, lateral line, and epibranchial placodes (see below), other factors need to contribute to specifying general somatosensory identity in the profundal and trigeminal placodes. Indeed grafting experiments in the chick have suggested that Pax3 helps to endow cells in the profundal (ophthalmic) placode with general somatosensory identity (Baker, Stark, and Bronner-Fraser 2002). However, other and still unknown transcription factors must substitute for its role in the trigeminal (maxillomandibular) placode, which does not express Pax3 (see Chapter 4) (Baker et al. 1999; Schlosser and Ahrens 2004).

6.1.3 Subtypes of General Somatosensory Neurons

POU4f1 and Islet1 are both required together with Tlx3 for activating the expression of DRG11 (= Prrxl1), a paired type homeodomain transcription factor, in a subset of general somatosensory neurons (Saito et al. 1995; Qian et al. 2002; Sun et al. 2008; Dykes et al. 2011; Regadas et al. 2014; Dvoryanchikov et al. 2017). However, DRG11 is also activated by Phox2b in some viscerosensory neurons and is, thus, unlike POU4f1 not specific for somatosensory neurons. In mouse mutants of DRG11, projections of somatosensory neurons from the DRG and the profundal/trigeminal ganglia to their central targets are perturbed and most small, TrkA+ somatosensory neurons, many of which are involved in pain reception, die. However, the larger mechano- and propioceptors appear largely normal (Chen et al. 2001; Ding et al. 2003; Rebelo et al. 2006). This indicates that DRG11 is required for the specification of a subset of somatosensory neurons. Other transcription factors with partly overlapping expression patterns such as Klf7, PRDM12, and Hmx1 have been shown to also be required for the same subset of neurons characterized by TrkA+ expression (Lei et al. 2001; Laub et al. 2001; Lei et al. 2005, 2006; Quina et al. 2012; Chen et al. 2015; Nagy et al. 2015).

Another pair of transcription factors, Runx1 and Runx3, which can act as either activators or repressors, is also upregulated in subsets of somatosensory neurons downstream of POU4f1 and Islet1 (Dykes et al. 2010, 2011). These transcription factors have been shown to promote switch like cell fate decisions between particular subsets of somatosensory neurons by activating batteries of cell type specific downstream genes (including G-protein coupled receptors (GPCRs), ion channels, and molecules involved in axonal pathfinding), while repressing those of alternative cell types (Fig. 6.1) (Chen et al. 2006; Abdel Samad et al. 2010; Lopes et al. 2012; reviewed in Marmigere and Ernfors 2007; Zagami et al 2009; Liu and Ma 2011). Runx1 is initially activated in small, TrkA+ sensory neurons, but is subsequently maintained only in a subset of these cells, which downregulate TrkA again and form non-peptidergic, cutaneous neurons; the other subset with persistent TrkA forms peptidergic neurons with more widespread projections (Fig. 6.1) (Chen et al. 2006; Kramer et al. 2006). Runx3 is initially activated in relatively large, TrkC+ somatosensory neurons and is subsequently maintained together with TrkC only in a subset of these cells, which develop into propioceptors in the DRG and into a special type of whisker-associated mechanoreceptors in the trigeminal ganglion (Fig. 6.1). Another subset of large somatosensory neurons downregulates Runx3, upregulates TrkB and develops into mechanoreceptors (Kramer et al. 2006; Senzaki et al. 2010).

Whereas Runx1 deficiency leads to an increase of peptidergic, TrkA+ neurons in mice, an overexpression of Runx1 in the DRG results in the opposite phenotype suggesting that persistent Runx1 expression selects a non-peptidergic over a peptidergic neuron type (Chen et al. 2006; Kramer et al. 2006). Similarly, knockout of Runx3 in mice leads to a reduction in TrkC+ propioceptors or mechanoreceptors and increase in TrkB+ mechanoreceptors, while overexpression has the opposite effect suggesting that persistent Runx3 expression selects TrkC+ over TrkB+ neurons (Inoue et al. 2002, 2007; Levanon et al. 2002; Kramer et al. 2006; Senzaki et al. 2010). Interestingly, in zebrafish, Runx3 appears to serve a conserved function in promoting TrkC+ somatosensory neurons, even though these mediate nociceptive rather than propioceptive stimuli (Gau et al. 2017). This indicates that the sensory modalities mediated by a particular cell type may change in the course of evolution.

A recent single-cell RNA-Seq study of mouse DRGs has identified 14 different subtypes of somatosensory neurons, each characterized by expression of a unique combination of subtype-specific transcription factors (Sharma et al., 2020). Interestingly, many of these subtype specific transcription factors, including Runx1/3 and POU4f2/3 are initially co-expressed in a common postmitotic embryonic progenitor and become differentially regulated only when this progenitor differentiates into the different subtypes from late embryonic stages on (Sharma et al., 2020). This close developmental relationship of the different subtypes may reflect their recent evolutionary divergence from a common ancestral general somatosensory cell type.

While Runx transcription factors help to specify a diversity of subtypes of general somatosensory neurons in various placode and neural crest derived ganglia along the body axis, other transcription factors modulate the identity of these subtypes – reflected for example in their projection pattern to the CNS – in a region-specific manner. There is some evidence that transcription factors like Otx2, Irx1/2, and Pax3 that are only co-expressed in the profundal and/or trigeminal placodes within the PPE (see Chapter 4) subsequently contribute to the regulation of subtype-specific properties and/or the establishment of proper projection patterns of the somatosensory neurons derived from these placodes (e.g. Matsuo et al. 1995; Baker et al. 2002; Itoh et al. 2002; Dude et al. 2009; Steventon, Mayor, and Streit 2012). It is, thus, tempting to speculate that the cooperation of region-specific transcription factors (e.g. of the Pax or Hox families) with functional subtype-specific factors (e.g. Runx1/3) may result in the transcriptional activation of guidance molecules that enable the region-specific wiring of functional neuronal subtypes (e.g. mechano- and nociceptors). However, our current knowledge in this area is extremely limited and further empirical data are needed to further substantiate this hypothesis.

In summary, the general somatosensory neurons developing from the profundal and trigeminal placodes rely on a CoRN of POU4f1, Islet1, Tlx1/3, and Pax3 and/or other transcription factors, which are required together with the proneural Neurog1/2 proteins to activate somatosensory differentiation gene batteries. Additional transcription factors including DRG11, PRDM12, Runx1, and Runx3 are required for defining the identity of particular subtypes of somatosensory neurons. All of these transcription factors contribute directly or indirectly to the transcriptional regulation of genes encoding the neurotrophin receptors (reviewed in Lei and Parada 2007), ion channels, guidance molecules, etc., that comprise the phenotype and determine the wiring of a particular subtype.

6.2 EPIBRANCHIAL PLACODES

The epibranchial placodes give rise to viscerosensory neurons residing in the ganglion of the facial nerve (geniculate ganglion) and the distal ganglia of the glossopharyngeal (petrosal ganglion) and vagal nerves (nodose ganglion). These include gustatory (= special viscerosensory) neurons innervating taste buds and general viscerosensory neurons. The latter mediate information on oxygen and carbon dioxide levels and pH from the carotid and aortic bodies as well as information on nutrient levels and stretch or pain from the intestines (Robinson and Gebhart 2008; Hockman et al. 2017; Waise, Dranse, and Lam 2018). All of these neurons send their axons into the nucleus of the solitary tract in the hindbrain (Figs. 1.12, 6.2).

Differentiation of viscerosensory neurons and somatosensory neurons are similar in their dependence on one of the Neurogenins (Neurog2 in most vertebrates) as well as Six1, Islet1, and Tlx1/3 (Korzh et al. 1993; Saito et al. 1995; Fode et al. 1998; Logan et al. 1998; Qian et al. 2001; Begbie et al. 2002; Qian et al. 2002; Zou et al. 2004; Nieber et al. 2009). However, in contrast to somatosensory neurons, viscerosensory neurons express the paired homeodomain proteins Phox2a and Phox2b (Fig. 6.2) (Tiveron, Hirsch, and Brunet 1996; Pattyn et al. 1997; Begbie et al. 2002; Dvoryanchikov et al. 2017; reviewed in Brunet and Pattyn 2002) and a number of other transcription factors (e.g. Prox2, Shox; Patthey et al. 2016). Phox2b inhibits the expression of the somatosensory transcription factor POU4f1 in these neurons (D'Autreaux et al. 2011).

Either Phox2b or Phox2a or both are also expressed in the relay neurons of the viscerosensory neurons in the nucleus of the solitary tract as well as in the visceromotor neurons of the parasympathetic and enteric nervous system (Fig. 6.2). They are, thus, circuit-specific transcription factors, which are present in all neurons

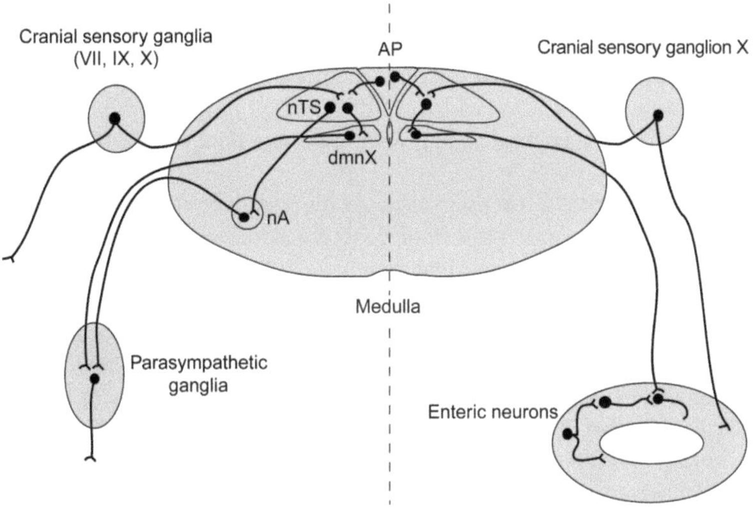

FIGURE 6.2 Phox2b expression in viscerosensory neurons and other neurons of autonomic reflex circuits. Simplified scheme depicting neurons involved in the medullary reflex circuits of the parasympathetic (left) and enteric (right) nervous systems. All neurons shown are Phox2b-dependent. Viscerosensory neurons in the VIIth, IXth, and Xth nerves transmit sensory information (e.g. from carotid body, gut, taste buds) to the nucleus of the solitary tract (nTS). Neurons in this nucleus then activate visceromotor neurons (preganglionic parasympathetic and enteric neurons) in the dorsal motor nucleus of the vagus nerve (dmnX) and the nucleus ambiguus (nA). These, in turn, activate postganglionic motor neurons in the parasympathetic and enteric ganglia. Enteric ganglia also contain intrinsic sensory neurons and interneurons forming local circuits. The area postrema (AP) mediates taste aversion and vomiting. (Modified with permission from Brunet and Pattyn 2002).

comprising autonomic (parasympathetic) reflex circuits. In mutants of Phox2b, most neurons of these circuits are missing and both the first- and second-order viscerosensory neurons adopt a somatosensory identity (Pattyn et al. 1999; Pattyn et al. 2000; Dauger et al. 2003; D'Autreaux et al. 2011). Phox2b, like proneural bHLH factors, promotes cell cycle exit and the adoption of generic neuronal properties due to its ability to promote Neurog2 and Ascl1 expression and inhibit Hes5 and Id2 (Pattyn et al. 1999; Dubreuil et al. 2000, 2002). In addition, Phox2b activates different subtype-specific target genes in the various types of neurons, in which it is expressed. For example, in all noradrenergic neurons including the viscerosensory neurons, Phox2b activates dopamine-b-hydroxylase, the key enzyme of norepinephrine biosynthesis (Kim et al. 1998; Yang et al. 1998; Pattyn, Goridis, and Brunet 2000; Dubreuil et al. 2002; Rychlik et al. 2005; Pla et al. 2008).

There are many different types of viscerosensory neurons (e.g. several types of neurons transmitting gustatory stimuli; chemoreceptors responsive to oxygen, carbon dioxide, pH, or nutrient levels; different types of mechanoreceptors). Recent single-cell RNA-seq studies provided the first insights into the genetic differences between these cells (Dvoryanchikov et al. 2017; Kupari et al. 2019). However, the transcription factors regulating these subtype identities are still largely unknown. In addition, we know very little about how regional differences between viscerosensory neurons in the geniculate, petrosal, and nodose ganglion are controlled. Hox transcription factors, some of which are differentially expressed in the ectoderm of pharyngeal arches, may possibly play some role here (Hunt et al. 1991; Kuratani and Wall 1992).

In summary, the viscerosensory neurons derived from the epibranchial placodes require many of the same transcription factors as the general somatosensory neurons in the profundal and trigeminal ganglia (e.g. Neurog1/2, Islet1, Tlx1/3) but the expression of Phox2b plays a key role in conferring their viscerosensory identity. The shared requirements of somatosensory and viscerosensory neurons for many of the same transcription factors suggests that these neurons may have evolved from a common ancestral type of sensory neuron as will be discussed in more detail in Volume 2 (Schlosser 2021).

Strikingly, several of the transcription factors involved in the specification of subtypes of somato- or viscerosensory neurons are also expressed in the target cells innervated by these neurons in the CNS suggesting a role of these transcription factors in circuit formation. For example, Tlx1 and Tlx3 transcription factors are expressed in many somato- and viscerosensory neurons as well as in the relay neurons in the CNS innervated by them; viz. the nucleus of the trigeminal nerve and dorsal spinal nucleus for the somatosensory pathway; and the solitary nucleus for the viscerosensory pathway (Saito et al. 1995; Tiveron et al. 1996; Logan et al. 1998; Qian et al. 2001, 2002; D'Autreaux et al. 2011). Furthermore, Tlx3 activates DRG11 in the trigeminal nucleus – target of DRG11 positive somatosensory neurons – but Phox2b in the solitary nucleus – target of the Phox2b expressing viscerosensory neurons (Qian et al. 2002). In Tlx1/3, DRG11, and Phox2b mutants, formation of these second order neurons is strongly perturbed (Chen et al. 2001; Qian et al. 2001, 2002; Dauger et al. 2003; Ding et al. 2003; Cheng et al. 2004). The activation of the same homophilic cell surface molecules in both sensory neurons and their relay neurons could provide a possible mechanism for the formation of such circuits but this awaits experimental confirmation.

6.3 OTIC PLACODE

6.3.1 Cell Types Derived from the Otic Placode

The otic placode gives rise to a multitude of cell types, most importantly the hair cells of the vestibular and auditory sensory epithelia and the sensory neurons of the vestibulocochlear ganglion (Figs. 1.8, 1.9). In addition, there are various types of epithelial cells, the secretory cells of the stria vascularis, which produce the endolymph, and several different types of supporting cells.

The mechanosensory hair cell is a highly specialized cell type exhibiting several characteristic features on its apical and basal pole (Fig. 6.3) (reviewed in Fettiplace and Hackney 2006; Goodyear, Kros, and Richardson 2006; Fritzsch et al. 2007; Peng et al. 2011; Fettiplace and Kim 2014; Nicolson 2017; McPherson 2018). On their basal pole, hair cells have peculiar ribbon synapses, which allow these cells to

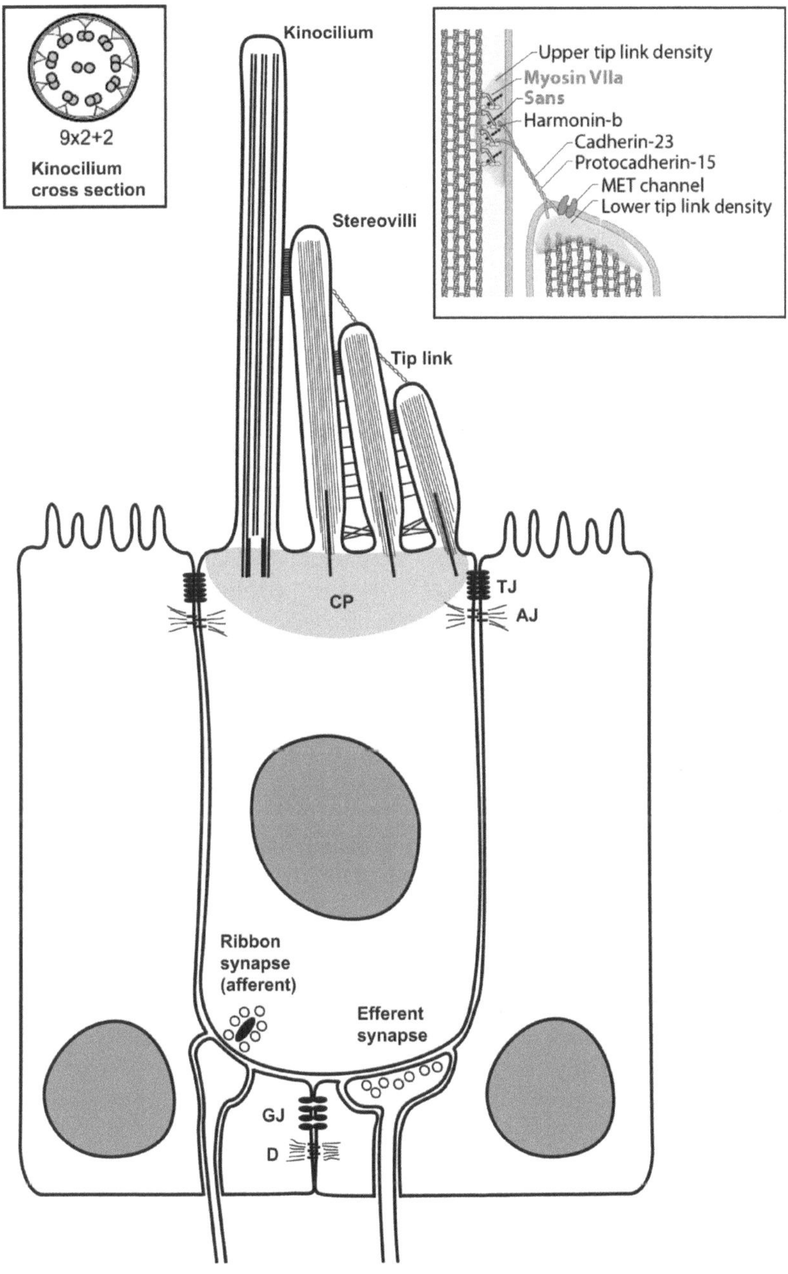

FIGURE 6.3 Diagram of a vertebrate hair cell. See text for details. Left insert: Cross-section through kinocilium. Right insert: Schematic drawing of a tip link between two stereovilli and its associated molecules. AJ, adherens junction; CP, cuticular plate; D, desmosome; GJ, gap junction; TJ, tight junction. (Reprinted with modification with permission from Goodyear, Kros, and Richardson 2006. Left insert: Reprinted with permission from Falk et al. 2015. Right insert: Reprinted with modification with permission from Grati and Kachar 2011).

respond to sustained stimulation with high and constant release rates of synaptic vesicles containing glutamate (Fig. 6.3; see also fig. 2.12 B in Schlosser 2021) (Schmitz, Königstorfer, and Südhof 2000; Matthews and Fuchs 2010; Pangršič, Reisinger, and Moser 2012; Wichmann and Moser 2015). Aggregates of Ribeye proteins provide a docking station for synaptic vesicles and the latter are equipped with a special membrane bound protein, Otoferlin, which promotes vesicle release.

On their apical pole, hair cells carry a "hair bundle" consisting of a single true cilium, the kinocilium, with a 9×2+2 array of microtubules (9 microtubule doublets in the periphery of the cilium; 2 microtubules in the center) as well as a group of 50–100 so-called stereocilia (Fig. 6.3). In spite of their name, the latter are not cilia at all but rather modified microvilli, and thus should be called stereovilli. These are densely packed with actin filaments and anchored in an actin rich cuticular plate at the apex of hair cells. Due to their actin-filled core, stereovilli form stiff rods, which do not bend. When mechanically stimulated, they rather tilt at their constricted base. Stereovilli are increasing in length from one side of the bundle to the other with the longest stereovillum placed next to the kinocilium (Fig. 6.3).

A complex of two proteins, Protocadherin-15 and Cadherin-23 forms a so-called tip link, which connects the mechanotransduction channels located on the tip of a stereovillum with the lateral membrane of its taller neighbors (Fig. 6.3). The tip link is firmly anchored on both sides allowing it to mechanically open the mechanotransduction channels and depolarize the cell when stereovilli bend in the direction of their higher neighbors. Movement of the hair bundle in the opposite direction results in closure of the channel and hyperpolarization (inhibition) of the hair cell. At the upper end of the tip link, Cadherin-23 is coupled via several connecting proteins (MyosinVII, Harmonin-b, Sans, Clarin-1) to the actin cytoskeleton. At the lower end of the tip link, Protocadherin-15 is connected to transmembrane proteins LHFPL5 and TMIE, which together with TMC1/2 form part of the mechanotransduction channel (Fettiplace and Kim 2014; Zhao and Müller 2015; Nicolson 2017). Although the physiological properties of this ion channel have been intensely studied, its molecular nature has long been elusive. Recent evidence suggests that TMC1, a member of the transmembrane channel (TMC)-like family, is one of the pore-forming components of the mechanotransduction channel (Pan et al. 2013). However, the 3D-structure of the pore has still not been elucidated and it is possible that other proteins participate in the pore complex. Several members of the Transient Receptor Protein (TRP) family of ion channels, which play central roles in mechanotransduction of many invertebrate mechanoreceptors (see Chapter 3 in Schlosser 2021), are also expressed in hair cells, but appear to be largely dispensable for mechanotransduction (reviewed in Venkatachalam and Montell 2007; Eijkelkamp, Quick, and Wood 2013; Wu et al. 2016).

Several other proteins (e.g. Whirlin, CIB-2, Epsin8, Myosin XVa) are enriched at the lower end of the tip link and are involved in the regulation of stereovillum length, which is crucial for mechanotransduction (reviewed in McGrath, Roy, and Perrin 2017). The asymmetric, staircase like arrangement of the stereovilli is controlled during development by the movement of the kinocilium from a central to an eccentric position and the direction of these asymmetries is aligned between adjacent hair cells via the Wnt planar cell polarity pathway (reviewed in Goodyear, Kros, and Richardson 2006; Nayak et al. 2007; Sienknecht 2015; Lu and Sipe 2016).

Surprisingly, many of the protein components of the tip link and its upper and lower anchoring complexes (e.g. Protocadherin-15, Cadherin-23, MyosinVII, Harmonin-b, Sans, Whirlin, CIB-2) also play important roles in the photoreceptors of the retina, roles that are unrelated to mechanotransduction and probably involve different functions of these proteins in vesicle trafficking and synapse formation (Cosgrove and Zallocchi 2014). Mutations in these proteins, therefore, cause blindness in addition to deafness, a condition known as Usher syndrome (Cosgrove and Zallocchi 2014). These striking similarities between photo- and mechanoreceptors may point to a deeper evolutionary link between these two cell types as will be discussed in Volume 2 (Schlosser 2021).

In several vertebrate lineages, hair cells have diversified into functionally specialized subtypes (Fig. 6.4). While different subtypes of hair cells have been described in teleosts, sharks, and cyclostomes, their functional significance (e.g. whether any of them preferentially mediate auditory stimuli) is currently not clear (Fay and Popper 2000; Ladich and Popper 2004). In amniotes, where the basilar papilla has become devoted to sound detection, the auditory hair cells of the basilar papilla (cochlea in mammals) can be distinguished from vestibular hair cells of the other sensory areas (reviewed in Eatock and Lysakowski 2006; Elliott, Fritzsch, and Duncan 2018). Moreover, two types of vestibular hair cells can be distinguished by their physiological properties, morphology and the presynaptic terminals formed by the afferent neurons of the vestibulocochlear nerve (Wersäll 1956; Wersäll and Bagger-Sjöbäck 1974; Eatock et al. 1998). Type I hair cells are surrounded by a calyx-like presynaptic ending, whereas type II hair cells are contacted by synaptic boutons (Fig. 6.4 A–C). The vestibular hair cells of the cristae of the semicircular canals (Fig. 1.8) also differ from the hair cells in the utricular and saccular maculae (Fig. 1.8) in that the hair cells of cristae have a very steep staircase of microvilli and an extra-long kinocilium embedded in a gelatinous cupula (Peterson, Cotton, and Grant 1996).

In mammals, the hair cells of the cochlea can be further classified into two distinct types, both of which lose their kinocilium after the apical hair bundle is established (Fig. 6.4D) (Elliott, Fritzsch, and Duncan 2018). The inner hair cells of the cochlea receive mostly afferent innervation from sensory neurons and respond to auditory stimuli. In contrast, the outer hair cells receive mostly efferent innervation and function in stimulus amplification (reviewed in Rubel and Fritzsch 2002; Coate and Kelley 2013; Goodrich 2016). Outer hair cells are depolarized when the underlying basilar membrane rises leading to hair bundle deflection and influx of positively charged ions. This leads to ultra-rapid conformation changes of the membrane protein Prestin resulting in shortening of the cell (Brownell et al. 1985; Zheng et al. 2000; Dallos and Fakler 2002; Dallos 2008; He et al. 2014). The process is reversed and the cell lengthens, when the basilar membrane moves into the other direction. As a consequence, the length of outer hair cells oscillates with the same frequency as the underlying basilar membrane. Because the hair bundle of outer hair cells is fixed to the overlying tectorial membrane, this results in an increased shearing motion over the inner hair cells amplifying the deflection of their hair bundle.

Although inner and outer hair cells are only found in mammals, tall and short hair cells have been described in the basilar papilla of birds and reptiles, which may be their direct evolutionary precursors (Fig. 6.4E). Short hair cells, like the outer

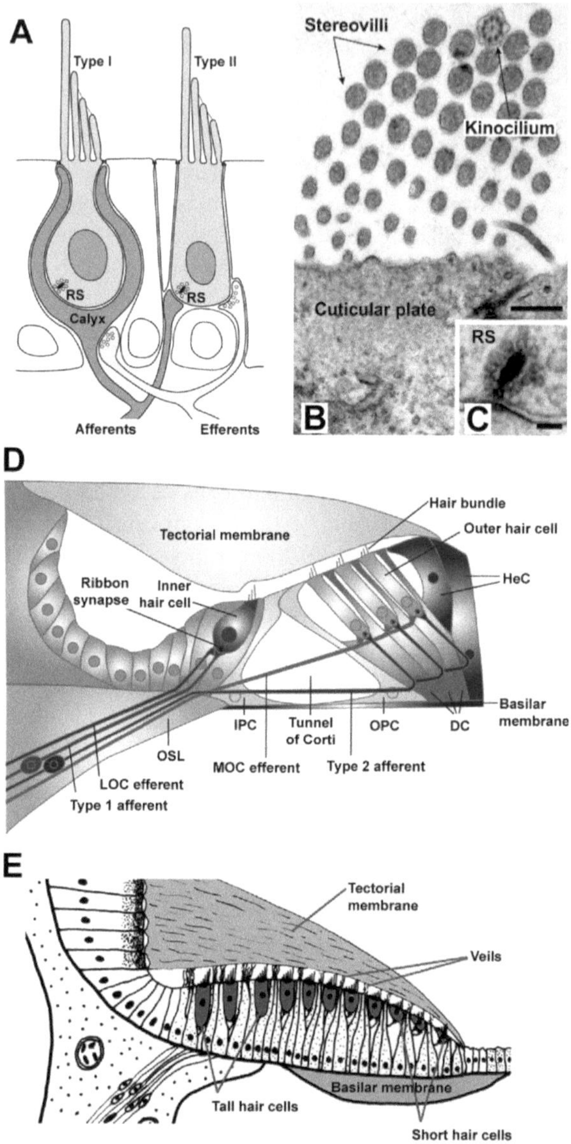

FIGURE 6.4 Diversity of cell types in the inner ear. (**A**) Type I and type II hair cells from a mammalian vestibular organ. RS, ribbon synapse. (**B, C**) Transmission electron micrographs of mouse vestibular hair cell. (**B**) Bundle of stereovilli and kinocilium (scale bar: 500 nm). (**C**) Cross-section of a ribbon synapse (RS) (scale bar: 100 nm). (**D**) Inner and outer hair cells in the mammalian organ of Corti. OSL, osseous spiral lamina; LOC, lateral olivocochlear neuron; MOC, medial olivocochlear neuron; IPC, inner pillar cell; OPC, outer pillar cell; DC, Deiters' cell; HeC, Hensen's cell. Reprinted with modification with permission from Zhang and Coate 2017. (**E**) Tall and short hair cells in the avian basilar papilla. ([A] Redrawn and modified from Eatock et al. 1998; [B,C] Reprinted with permission from Falk et al. 2015; [D] Reprinted with permission from Goodyear and Richardson 2018.)

hair cells of mammals are concentrated on the abneural side of the basilar papilla (away from the ganglion) and are mostly innervated by efferent fibers, whereas tall hair cells, like the inner hair cells of mammals receive afferent and efferent fibers (Hirokawa 1978; Fischer 1992; Chiappe, Kozlov, and Hudspeth 2007; Manley 2011; Beurg, Tan, and Fettiplace 2013). Short hair cells in birds have recently been shown to also use a Prestin motor for stimulus amplification (Beurg, Tan, and Fettiplace 2013) in addition to another mechanism involving active hair bundle motility (Manley and Köppl 1998; Hudspeth et al. 2000; Köppl 2011). Comparison with other vertebrates indicates that Prestin, which belongs to the Slc26A family of anion transporter proteins, evolved a new function as motor protein in amniotes (Schaechinger and Oliver 2007; Tan et al. 2011; Beurg, Tan, and Fettiplace 2013). Taken together, this suggests that individualized hair cells with a distinct function in stimulus amplification and using a Prestin motor evolved already in the ancestor of amniotes, but specialized further into outer hair cells in the mammalian lineage.

In all sensory epithelia, hair cells are surrounded by supporting cells, which are important to maintain ion homeostasis and provide trophic support for the hair cells (reviewed in Monzack and Cunningham 2013; Wan, Corfas, and Stone 2013). They also serve as a reservoir of cells for hair cell regeneration in non-mammalian vertebrates, which may either re-enter the cell cycle and produce new hair cells or directly transdifferentiate into hair cells following damage to the inner ear (reviewed in Brigande and Heller 2009; Brignull, Raible, and Stone 2009; Groves, Zhang, and Fekete 2013; Atkinson et al. 2015; Elliott, Fritzsch, and Duncan 2018). In the mammalian organ of Corti, supporting cells have greatly diversified morphologically and functionally (inner border cell, inner phalangeal cell, inner and outer pillar cells, Deiter's cells, Hensen's cells, Claudius cells, Boettcher cells) (Gale and Jagger 2010) (Fig. 6.4D).

The diversity of hair cells in the inner ear is matched by a corresponding diversity of the special somatosensory neurons of the vestibulocochlear ganglion, which provide the afferent innervation to hair cells and transmit the vestibular and auditory information to the cochlear and vestibular nuclei in the hindbrain (Fig. 1.12). Vestibular hair cells are innervated by neurons residing in the vestibular subdivision of the ganglion, which is clearly separate from the auditory subdivision in mammals. The latter is also known as spiral ganglion because it follows the snail shape of the cochlea. Vestibular neurons may innervate multiple type I or type II hair cells or both and consequently have either calyx- or bouton-like endings or both (Fig. 6.4A) (Wersäll 1956; Fernàndez, Goldberg, and Baird 1990; Eatock and Lysakowski 2006). The auditory ganglion contains two main types of neurons (Rubel and Fritzsch 2002; Nayagam, Muniak, and Ryugo 2011; Bulankina and Moser 2012; Coate and Kelley 2013; Goodrich 2016). Type 1 neurons make up the majority of ganglion cells (about 95%) and innervate the inner hair cells with several type 1 neurons projecting to each hair cell. Type 2 neurons provide sparse afferent innervation to the outer hair cells (Fig. 6.4D). As we will see further in section 6.3.3, type 1 neurons are now known to comprise a heterogeneous assembly of neurons with at least three subtypes (Petitpre et al. 2018; Shrestha et al. 2018). In contrast to ganglia containing general somatosensory neurons, no TrkA+ neurons are found in the vestibulocochlear ganglion, where most neurons express TrkB and/or TrkC (Fritzsch and Beisel 2004).

6.3.2 Development of Neuronal Versus Sensory Cells

The different cell types produced by the otic placode develop in highly stereotyped locations in response to a complex patchwork of transcription factors expressed in regionally restricted domains (Fekete and Wu 2002). These domains are set up in response to signaling molecules secreted from surrounding tissues that pattern the otic vesicle along the anteroposterior and dorsoventral axes (reviewed in Alsina, Giráldez, and Pujades 2009; Groves and Fekete 2012; Wu and Kelley 2012; Maier et al. 2014; Raft and Groves 2015). Proper patterning along the anteroposterior axis of the ear requires high retinoic acid signaling (from the adjacent mesenchyme) posteriorly and FGF (e.g. from rhombomere 4 of the hindbrain) anteriorly, while dorsoventral patterning depends on Wnt signaling from the dorsal hindbrain and sonic hedgehog from notochord and floor plate.

Already before invagination of the otic vesicle is complete, these signaling pathways conspire to define a neurosensory domain expressing Sox2/3 anteroventrally and a complementary non-sensory domain expressing Tbx1 and Lmxa/b transcription factors (Abelló et al. 2007; Schneider-Maunoury and Pujades 2007; Abelló et al. 2010; Bok et al. 2011). Mutual antagonism between Sox2/3 and Tbx1 or Lmx1a/b helps to define the spatial extent of the neurosensory domain (Raft et al. 2004; Xu et al. 2007; Nichols et al. 2008; Koo et al. 2009; Abelló et al. 2010; Radosevic et al. 2011; Gou, Guo, et al. 2018).

The neurosensory domain gives rise to both Neurog1-dependent neurons and Atoh1-dependent sensory cells of the inner ear (Fig. 6.5) (Adam et al. 1998; Ma et al. 1998; Andermann et al. 2002; Alsina et al. 2004; Pujades et al. 2006; Millimaki et al. 2007; Millimaki, Sweet, and Riley 2010) and it has been suggested that hair cells and the sensory neurons innervating them may share a common developmental lineage (Fritzsch et al. 2002). In support of this, a fate mapping study of the chick otic placode suggested that sensory neurons tend to originate from the same region of the otic placode as the sensory areas innervated by them (Bell et al. 2008). Lineage analyses in the chick, mouse, and zebrafish ear confirmed the existence of a common progenitor for sensory neurons and hair cells for the utricular and saccular maculae but not for the cristae of the semicircular canals and the basilar papilla or organ of Corti (Satoh and Fekete 2005; Raft et al. 2007; Sapede, Dyballa, and Pujades 2012). Moreover, Atoh1 expressing hair cells in the maculae were increased at the expense of neurons after knockdown of Neurog1 in mouse and zebrafish suggesting fate conversion of a common progenitor (Ma, Anderson, and Fritzsch 2000; Matei et al. 2005; Raft et al. 2007; Sapede, Dyballa, and Pujades 2012). Taken together, these findings support the existence of a common neuronal-sensory progenitor for the saccular and utricular maculae but suggest a potentially different origin for hair cells from other sensory areas and the sensory neurons innervating them.

Expression of Sox3/2 in the neurosensory domain of the inner ear endows this region with the competence to form both sensory neurons (Neurog1) and hair cells (Atoh1). However, there are some differences between species with respect to timing and function of Sox2 and Sox3 expression. In both zebrafish and chick, Sox3 precedes Sox2 expression and promotes neural differentiation. High levels of Sox2 but not Sox3 are then maintained in the developing prosensory areas (maculae, cristae,

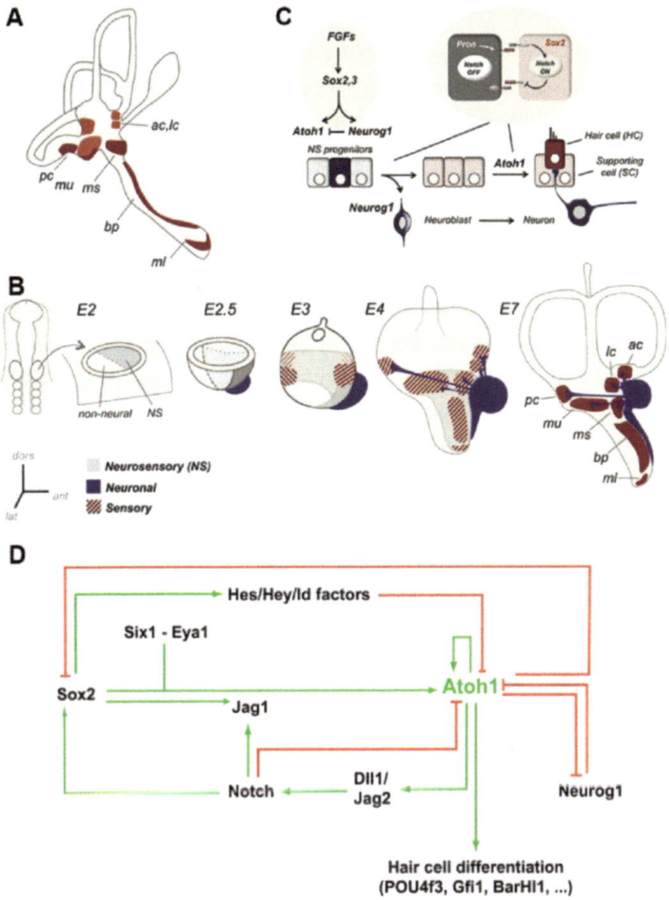

FIGURE 6.5 Development of neurons and sensory cells in the inner ear. (**A**) Diagram of chick inner ear with sensory patches indicated in red: ac, anterior crista; bp, basilar papilla; lc, lateral crista; ml, lagenar macula; ms, saccular macula; mu, utricular macula; pc, posterior crista. (**B**) Development of the inner ear in the chick from embryonic days (E) 2–7. The neurosensory (NS) domain is established early, before invagination. It subsequently gives rise to the hair cells and supporting cells of the sensory patches (red) and the neurons of the vestibulocochlear ganglion (blue). (**C**) Specification of neurons (blue) and hair cells (red) during inner ear development. After induction by FGF, Sox2, and Sox3 in the neurosensory (NS) domain promote expression of both Atoh1 and Neurog1. Subsequently, neurons (specified by Neurog1) differentiate first and delaminate from the neurosensory domain, followed by the differentiation of hair cells (specified by Atoh1). During both steps, cells expressing the proneural gene (Pron) – *Neurog1* or *Atoh1* – repress differentiation in adjacent cells via Notch mediated lateral inhibition (insert). This allows, first, to single out neuronal precursors from cells remaining in the neurosensory domain and, second, to decide between hair cells and supporting cells. (**D**) Central role of Atoh1 in the core regulatory network regulating cell type identity in vertebrate hair cells. ([A,B] Reprinted with permission from Neves, Vachkov, and Giráldez 2013; [C] Reprinted with permission from Galvez, Abelló, and Giraldez 2017; [D] Modified from Raft and Groves 2015.)

and papillae) where they promote sensory differentiation. Ultimately, Sox2 is downregulated in differentiating hair cells but maintained in supporting cells (Neves et al. 2007; Abelló et al. 2010; Evsen et al. 2013; Gou, Vemaraju, et al. 2018). In mammals, only Sox2 is expressed in the neurosensory domain and is required for both neural and sensory differentiation (Kiernan et al. 2005; Puligilla et al. 2010; Steevens et al. 2017). However, subpopulations of neurons and hair cells differentiate in the mammalian ear even in the absence of Sox2, suggesting that other, unknown factors may partly compensate for Sox2 loss of function (Dvorakova et al. 2016, 2020). In contrast to chick and mammalian embryos, a large proportion of sensory neurons and hair cells develops even in double mutants of Sox2 and Sox3 in zebrafish (Millimaki, Sweet, and Riley 2010; Gou, Vemaraju, et al. 2018). This indicates that these transcription factors are not essential for neural and sensory differentiation in zebrafish, suggesting that Atoh1 and Neurog1 expression are initiated by different upstream regulators.

Gain of function experiments in different vertebrates show that Sox2 together with Six1 and Eya1 can promote ectopic Neurog1 and Atoh1 expression in the non-neurosensory domain of the otic vesicle suggesting that these factors help to activate Neurog1 and Atoh1 (Fig. 6.5C) (Puligilla et al. 2010; Jeon et al. 2011; Ahmed et al. 2012; Ahmed, Xu, and Xu 2012; Neves et al. 2012; Evsen et al. 2013; Gou, Vemaraju, et al. 2018). However, like in other placodes, SoxB1 factors prevent terminal differentiation in the inner ear and, thus, must be downregulated to allow neuronal and sensory differentiation to proceed (Dabdoub et al. 2008; Evsen et al. 2013; Puligilla and Kelley 2017). Once the proneural factors Neurog1 (and its downstream target NeuroD1) and Atoh1 are stably activated in neuronal or hair cell progenitors, respectively, they help to initiate differentiation by promoting the downregulation of SoxB1 factors in these progenitors (Dabdoub et al. 2008; Evsen et al. 2013). Neuronal and sensory progenitors also prevent their neighbors from adopting the same fate via Delta1-Notch mediated lateral inhibition (Fig. 6.5C) (Adam et al. 1998; Haddon et al. 1998; Abdolazimi, Stojanova, and Segil 2016).

Given that both neurons and sensory cells develop from the neurosensory domain, how are sensory and neuronal progenitors separated from each other in the otic vesicle? The answer to this question differs for different species. In zebrafish, neuronal and sensory progenitors begin to form simultaneously in adjacent domains of the otic vesicle (Haddon and Lewis 1996). The expression of Pax2a and Dlx3b/4b, which become restricted to the prospective sensory areas, may help to promote sensorigenesis over neurogenesis, while FoxI1 has the opposite effect (Riley et al. 1999; Hans, Irmscher, and Brand 2013; Gou, Guo, et al. 2018). In contrast, neurogenesis precedes sensorigenesis in amniotes (Adam et al. 1998; Pujades et al. 2006; Raft et al. 2007). Neurog1 is first upregulated within the Sox3/2 positive neurosensory domain and Neurog1 expressing cells delaminate and upregulate NeuroD1 and other neuronal differentiation genes. The Neurog1-positive domain then becomes smaller, while Atoh1 is upregulated in a complementary domain, which will form the sensory patches of the inner ear (Raft et al. 2007). Mutual antagonism between Neurog1 and Atoh1 (due to transcriptional cross-repression, which may be direct or mediated by NeuroD) help to maintain the restriction of both factors to complementary regions in amniotes as well as in zebrafish (Matei et al. 2005; Raft et al. 2007; Jahan et al. 2010;

Gou, Vemaraju, et al. 2018). The specification of sensory patches and maintenance of Atoh1 requires Notch signaling via the ligand Jagged1/Serrate1. In contrast to other Notch ligands such as Delta1, Jagged1 signaling activates the expression of Jagged1 in neighboring cells, which helps to maintain Sox2 and, indirectly, Atoh1 expression in the developing sensory areas (Eddison, Leroux, and Lewis 2000; Daudet and Lewis 2005; Daudet, Ariza-McNaughton, and Lewis 2007; Hartman, Reh, and Bermingham-McDonogh 2010; Neves et al. 2011). In this process of lateral induction, Notch signaling acts very differently from lateral inhibition and helps to homogenize the fate of adjacent cells.

It is still not entirely clear, why Neurog1 is upregulated before Atoh1 in the inner ear, but becomes extinguished once Atoh1 expression has been initiated (Fig. 6.5C). Several mechanisms have been suggested to contribute to this sequence of events. First, high levels of Notch signaling in the early neurosensory domain have been shown to directly activate the Neurog1 but not the Atoh1 enhancer and, thus, may activate Neurog1 first (Jeon et al. 2011). Second, while Sox2 activates Atoh1 expression it has also been suggested to activate repressors of Atoh1 such as Id and Hes/Hey transcription factors, thereby delaying the onset of Atoh1 expression (Neves et al. 2012, Neves, Vachkov, and Giráldez 2013; Puligilla and Kelley 2017). Third, posttranscriptional and posttranslational mechanisms may contribute to preferentially destabilize Atoh1 mRNAs or proteins during early otic vesicle stages (reviewed in Galvez, Abelló, and Giráldez 2017). However, once Atoh1 expression has initiated, it transcriptionally activates its own expression in a positive feedback loop, thereby stabilizing Atoh1 expression and repressing Neurog1 (Helms et al. 2000; Raft et al. 2007).

6.3.3 Transcription Factors Involved in Specifying Identity of Special Somatosensory Neurons

Once the Neurog1 expressing neuroblasts delaminate from the otic vesicle they begin to transiently express NeuroD1 and downregulate Neurog1 (Liu et al. 2000; Kim et al. 2001; Bell et al. 2008; Jahan et al. 2010; Evsen et al. 2013). Expression of Islet1 and POU4f1 is initiated next, similar to the general somatosensory neurons derived from the profundal and trigeminal placode (Adam et al. 1998; Li et al. 2004; Radde-Gallwitz et al. 2004; Deng et al. 2014; Durruthy-Durruthy et al. 2014). However, the special somatosensory neurons of the vestibulocochlear ganglion are clearly different from general somatosensory neurons in that they specifically innervate hair cells of the inner ear peripherally and the vestibular and cochlear nuclei in the brainstem centrally. Currently, it is not well understood, how the identity of special somatosensory neurons is specified. Recently, several transcription factors (ESRRG, Hmx3, POU6f2, Runx1, Irx2) were found to be enriched in the vestibulocochlear ganglion (Patthey et al. 2016). However, with exception of Irx2 they are expressed only in subsets of its neurons and Irx2 does not appear to be essential for its formation (Lebel et al. 2003). It is likely that transcription factors that distinguish the otic placode from other placodes such as Pax8 and Pax2 may play a role here (see Chapter 4). Indeed, in Pax2/8 double mutants but not in single mutants the numbers of sensory neurons in the vestibulocochlear ganglion are drastically reduced (Bouchard et al. 2010). This suggests that Pax8 and Pax2 may play a redundant role in specifying the identity of vestibulocochlear neurons but also indicates that other unknown factors are to some extent able to compensate for their loss.

Several additional transcription factors have been suggested to be involved in the specification of different subtypes of neurons within the vestibulocochlear ganglion, although little is known about their mode of action. A few transcription factors are enriched in the vestibular or auditory part, respectively, of the vestibulocochlear ganglion (Lu et al. 2011). For example, Tlx3 is preferentially expressed in vestibular neurons (Koo et al. 2009; Lu et al. 2011). Lmx1a, which is expressed in the auditory but not vestibular part of the neurosensory area, was shown to repress Tlx3 and vestibular neuron formation (Koo et al. 2009). GATA3, in turn, is preferentially maintained in auditory neurons, which are severely reduced and show aberrant wiring in GATA3 knockout mice (Karis et al. 2001; Lawoko-Kerali et al. 2004; Jones and Warchol 2009; Lu et al. 2011; Appler et al. 2013; Duncan and Fritzsch 2013). GATA3 and the transcription factors Prox1 and Mafb are initially co-expressed in all auditory neurons. Subsequently, GATA3 and Mafb decrease in maturing type 1 neurons, while Prox1 decreases in maturing type 2 neurons (Nishimura, Noda, and Dabdoub 2017). Recent single-cell sequencing of the auditory neurons in the mouse spiral ganglion confirmed that GATA3 together with some other transcription factors (POU4f2, Klf7, Mafb) is specifically enriched in the type 2 neurons innervating the outer hair cells (Shrestha et al. 2018; Petitpre et al. 2018). In addition, these studies have revealed three classes of type 1 neurons that innervate the inner hair cells and apart from a shared expression of Prox1 differ in their profile of transcription factor expression.

6.3.4 Transcription Factors Involved in Specifying Identity of Hair Cells

Atoh1 is an essential regulator of hair cell differentiation (Fig. 6.5D). Inner ear hair cells are lost in mice and zebrafish after Atoh1 loss of function during development (Bermingham et al. 1999; Zheng and Gao 2000; Chen et al. 2002; Woods, Montcouquiol, and Kelley 2004; Millimaki et al. 2007; Pan et al. 2011) indicating a central role of Atoh1 for initiation of hair cell differentiation. Conditional knockout of Atoh1 at late stages of hair cell development also revealed a role of Atoh1 for subsequent hair bundle formation and survival of hair cells (Cai et al. 2013; Chonko et al. 2013). Conversely, overexpression of Atoh1 in supporting cells or nonsensory regions promotes ectopic formation of hair cells in the inner ear (Zheng and Gao 2000; Woods et al. 2004; Millimaki et al. 2007; Gubbels et al. 2008; Sweet, Vemaraju, and Riley 2011; Kelly et al. 2012; Liu et al. 2012; Xu et al. 2012; Yang et al. 2012). We still know very little about the mode of action of Atoh1 during hair cell development because the small numbers of hair cells impede the identification of direct Atoh1 target genes. However, 233 of 313 transcripts enriched in hair cells were previously identified as direct target genes of Atoh1 in either cerebellum or intestine by ChIP-Seq (Klisch et al. 2011) and contain Atoh1 binding sites suggesting that they may also be direct transcriptional targets of Atoh1 in hair cells (Cai and Groves 2015). These putative targets include transcription factors implicated in hair cell development (e.g. POU4f3, Gfi1, Pax2), proteins involved in hair bundle development (Elmod1), several components of the mechanotransduction complex (Usherin, Myo7a, Eps8L2, Myo6), and a subunit of the acetylcholine receptor mediating responses to the efferent innervation of hair cells suggesting that Atoh1 broadly controls diverse aspects of hair cell differentiation (Cai and Groves 2015; Scheffer et al. 2015).

However, Atoh1 also has important roles in cell specification in the cerebellum and other parts of the CNS, in secretory cells of the intestine and in the mechanosensory Merkel cells (reviewed in Mulvaney and Dabdoub 2012; Cai and Groves 2015; Jahan, Pan, and Fritzsch 2015; Costa et al. 2017). These context-specific roles indicate that Atoh1 needs to cooperate with other factors to specify hair cell identity. The transcription factors POU4f3, Gfi1, and possibly BarHl1 appear to play a particularly important role here. All three of these transcription factors act downstream of Atoh1, and POU4f3 is directly activated by Atoh1 in conjunction with several cofactors (Fig. 6.5D) (e.g. TCF3, GATA3, ETV4, NMyc, or ETS2) (Hertzano et al. 2004; Masuda et al. 2012; Ikeda et al. 2015). Hair cell differentiation is inhibited in mouse mutants of both POU4f3 and Gfi1, indicating that they are required for hair cell differentiation (Erkman et al. 1996; Xiang et al. 1997, 1998; Wallis et al. 2003). In BarHl1 mutants, hair cells first form but then die progressively, suggesting a role of BarHl1 in hair cell maintenance (Li et al. 2002). Moreover, overexpression of Atoh1 can efficiently induce hair cell fate in the chick inner ear or in vitro only in combination with POU4f3 and Gfi1 (Costa and Henrique 2015; Duran Alonso et al. 2018; Costa et al., 2019). Whereas POU4f3 may act as a pioneer factor which opens up chromatin to allow for Atoh1 binding, Gfi1 has been suggested to switch the role of Atoh1 from a general activator of neuronal genes to a specific activator of hair cell genes (Costa et al., 2019). It has been proposed that the cooperative action of Atoh1, POU4f3, and Gfi1 may be due to direct protein-protein interactions but this needs to be confirmed experimentally (Costa et al. 2017). In addition, Lhx3, which is specifically expressed in differentiated hair cells, is required for hearing, but its function in hair cells remains to be clarified (Hertzano et al., 2007; Rajab et al., 2008).

Since Atoh1, POU4f3, Gfi1, and Lhx3 are also co-expressed in Merkel cells (Wallis et al. 2003; Haeberle et al., 2004; Masuda et al., 2011; Woo et al., 2015), additional transcription factors are probably involved in regulating hair cell identity. Two possible candidates for such factors are GATA3 and Pax2. Mutants in either GATA3 or Pax2 show deficiencies in hair cell differentiation (Favor et al. 1996; Torres, Gómez-Pardo, and Gruss 1996; Karis et al. 2001; Bouchard et al. 2010; Duncan and Fritzsch 2013) and both transcription factors have been shown to synergize with Atoh1 in promoting hair cell formation (Masuda et al. 2012; Chen et al. 2013; Walters et al. 2017). Taken together, this suggests that the CoRN that specifies hair cell identity comprises at least Atoh1, POU4f3, and Gfi1 and probably other transcription factors with region-specific expression in the otic placode such as Pax2 and GATA3.

In addition, at least in mammals hair cell specification requires inhibition of Prox1, which represses Atoh1 and Gfi1, thereby suppressing hair cell fate (Kirjavainen et al. 2008). Indeed Prox1 is initially widely expressed in precursors of hair cells and sensory neurons, but subsequently downregulated in hair cells by an unknown mechanism and maintained only in supporting cells and some sensory neurons (Stone, Shang, and Tomarev 2003).

Differences between auditory and vestibular hair cells depend on transcription factors that differ in expression between the vestibular and auditory part of the inner ear. This has been clearly shown in zebrafish, where Pax5 is required downstream of Hmx2/3 for maintenance of hair cells in the utricular macula and vestibular function (Kwak et al. 2006; Feng and Xu 2010). However, this role of Pax5 may not be conserved throughout vertebrates since no deficits in balance or hearing were reported for Pax5 knockout mice (Urbanek et al. 1994; Reimer et al. 1996).

With the advent of single-cell sequencing, we recently also have gained insights into how transcription factor expression profiles differ between different types of hair cells or hair cells and supporting cells in mammals. These studies revealed several transcription factors differentially expressed between vestibular and cochlear hair cells (Burns et al. 2015; see also Perl, Shamir, and Avraham 2018) as well as the inner and outer hair cells of the cochlea (Liu et al. 2014; Li, Qian, et al. 2016). Another study employing RT-PCR to analyze the expression of 192 genes in single cells from the organ of Corti, revealed unique gene expression profiles for all of its nine major cell types – inner and outer hair cells and seven different types of supporting cells (Waldhaus, Durruthy-Durruthy, and Heller 2015). Unfortunately, studies confirming the function of such factors in subtype specification are currently still lacking. However, recent conditional knockout experiments in mice have established a role of Rfx1/3 transcription factors not for initial specification but for maintenance of outer hair cells and for the regulation of hair bundle differentiation in cooperation with Six1 (Elkon et al. 2015; Li et al. 2020).

In summary, the special somatosensory neurons derived from the otic placode have highly specific functions in innervating the hair cells of the inner ear and transmitting the information to special nuclei in the brain stem. The specification of these neurons depends on a CoRN of Neurog1, Islet1, and POU4f1 transcription factors resembling the CoRN specifying the general somatosensory neurons of other ganglia. Similarly, specification of hair cell identity in the inner ear involves a CoRN of Atoh1, POU4f3, Gfi1, and Prox1 that is not unique to hair cells but also used in other sensory cell types. The CoRNs specifying the sensory neurons and hair cells of the inner ear, therefore, most likely also involve regional transcription factors that are specifically expressed in the otic placode such as Pax2/Pax8 or GATA3. The similarities in transcription factor expression between hair cells (Atoh1, POU4f3) and the sensory neurons that innervate them (Neurog1, POU4f1) have led to suggestions that the secondary sensory cells (hair cells) and their innervating neurons of vertebrates have evolved by duplication and divergence from mechanosensory primary sensory cells as found in many invertebrates (Fritzsch, Beisel, and Bermingham 2000; Fritzsch and Beisel 2004; Fritzsch, Pauley, and Beisel 2006; Fritzsch and Elliott 2017). We will discuss this proposal in more detail in part 3 of the book.

6.4 LATERAL LINE PLACODES

6.4.1 Cell Types Derived from the Lateral Line Placodes

Similar to the otic placode, the lateral line placodes (lost in amniotes) give rise to sensory cells, their innervating sensory neurons and supporting cells (Fig. 1.6, 1.7) (reviewed in Schlosser 2002a; Ghysen and Dambly-Chaudière 2007; Ma and Raible

2009; Chitnis, Nogare, and Matsuda 2012; Piotrowski and Baker 2014; Thomas et al. 2015). In lampreys and gnathostomes these mechanosensory cells are clustered into multicellular neuromasts, whereas hagfishes have only unicellular mechanoreceptors (Northcutt 1989; Braun and Northcutt 1997). The sensory cells of the mechanosensory neuromasts (Fig. 6.6A) are hair cells, which resemble type 2 vestibular hair cells and have a long kinocilium embedded in a gelatinous cupula (Nicolson 2017). They are interspersed with several types of supporting cells (Blaxter 1987; Lush et al. 2019). Each neuromast is composed of two types of hair cells with hair bundles arranged along the same axis (dorsoventral or anteroposterior) but with opposing polarity, so that one type is maximally activated by stimulation in one direction and the other type by stimulation in the opposing direction (Görner 1961; Flock and Wersäll 1962; López-Schier et al. 2004). Pairs of hair cells with opposing polarity are generated by symmetrical divisions during neuromast development (López-Schier et al., 2006). The overall structure of hair cells in the lateral line with tip links and ribbon synapses closely resembles the structure of inner ear hair cells and recent sequencing results in the paddlefish *Polyodon* show that they express many of the proteins typical for hair cell ribbon synapses (Ribeye, Otoferlin, $Ca_v1.3$, Vglut3) (Modrell et al. 2017).

In addition, many gnathostome vertebrates have ampullary electroreceptors (Fig. 6.6B), which have a central cilium surrounded by microvilli on the apical side resembling immature hair cells (Northcutt, Catania, and Criley 1994; Jørgensen 2005). Interspersed with supporting cells, such cells are found inside of jelly-filled

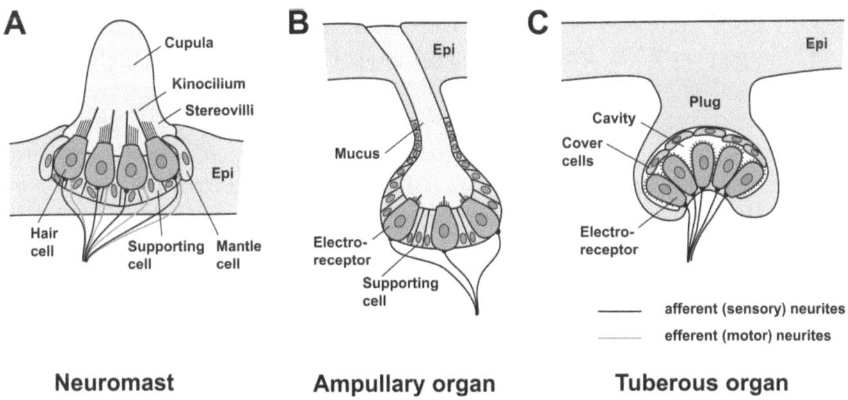

FIGURE 6.6 Diversity of hair cells and electroreceptors in the lateral line system. (**A**) Neuromast with hair cells, supporting cells, and mantle cells in the teleost lateral line. (**B**) Ampullary organ from a cartilaginous fish with electroreceptors responding to cathodal stimuli. Similar ampullary organs are found in actinopterygians including some teleosts. However, electroreceptors in the ampullary organ of teleosts respond to low frequency anodal stimuli. (**C**) Tuberous organ of a gymnotid teleost (gymnomast) with specialized electroreceptors (studded with microvilli) responding to high frequency anodal stimuli. Epi, epidermis. Neuromasts receive both afferent and efferent innervation, whereas electroreceptors receive only afferent innervation. (Adapted from Jørgensen 2005 and Baker, Modrell, and Gillis 2013).

ampullary organs sunk into the epidermis. Fate mapping studies have shown that these ampullary organs, like neuromasts, are derived from lateral line placodes (Northcutt, Brändle, and Fritzsch 1995; Modrell et al. 2011; Gillis et al. 2012). They also form ribbon synapses and express many of the same proteins involved in vesicle release in hair cells (Modrell et al. 2017). These similarities strongly suggest that ampullary electroreceptors have evolved by modification of hair cells (Baker and Modrell 2018). However, in contrast to hair cells they respond to cathodal (external negative) stimuli using voltage gated calcium channels in their apical membrane (Bodznick and Montgomery 2005). Interestingly, the same splice variant of the $Ca_v1.3$ calcium channel that mediates electroreception in the ampullary organs of cartilaginous fishes is expressed in vestibular hair cells of pigeons, where it may play a role in magnetoreception (Bellono, Leitch, and Julius 2017; Nimpf et al. 2019).

Ampullary electroreceptors are not found in cyclostomes, which have either no electroreceptors (hagfishes) or have elongated electroreceptive cells (the so-called end buds of lampreys) on the surface of the epidermis (Baker et al., 2013). Among gnathostomes, electroreceptors and ampullary organs have been lost in frogs and teleosts, but have re-evolved independently in two different groups of teleosts within the osteoglossomorphs and ostariophysans (Bullock, Bodznick, and Northcutt 1983; New 1997; Northcutt 1997; Baker, Modrell, and Gillis 2013). Teleost electroreceptors also resemble hair cells in their apical specializations and, like hair cells, they respond to anodal (external positive) stimuli using voltage-gated calcium channels in their basal membrane (Bodznick and Montgomery 2005). In each of these two electroreceptive teleost clades, one group also evolved electric organs as well as tuberous electroreceptors, which lack ducts and are able to detect the high-frequency electric fields generated by these electric organs (Fig. 6.6C) (Gibbs 2004; Bodznick and Montgomery 2005; Jørgensen 2005; Kawasaki 2005; Baker, Modrell, and Gillis 2013). Based on histological observations, teleost electroreceptors have been suggested to also originate from lateral line placodes (Northcutt 2003), but this is still contentious and needs to be corroborated by fate mapping studies. While the use of a different voltage-sensing mechanism indicates that teleost electroreceptors are not homologous to the electroreceptors of other gnathostomes, their similarities to hair cells suggests that they may also have evolved from lateral line hair cells (Bullock and Heiligenberg 1986; Bodznick and Montgomery 2005). Further information of their transcription factor expression profile and embryonic origin will help to either refute or strengthen this hypothesis.

The special somatosensory neurons of the lateral line are derived from the same lateral line placodes than the hair cells and electroreceptors that they innervate and their outgrowing dendrites track the migrating or elongating primordia of the lateral lines (Fig. 1.7) (Metcalfe 1985; Gilmour et al. 2004). The cell bodies of the lateral line neurons reside in special lateral line ganglia and they send their axons to special lateral line nuclei in the hindbrain (Wullimann and Grothe 2014). It has been shown that there are intrinsic differences between lateral line neurons that send their axons out first and track the primordium ("leaders" with higher levels of NeuroD1 expression) and those that send their axons out later ("followers") (Sato, Koshida, and Takeda 2010; Sato and Takeda 2013). Furthermore, since each lateral line sensory neuron selectively innervates only hair cells of one polarity, lateral line ganglia must contain different subpopulations of sensory cells, which supply hair cells of different polarity (Nagiel, Andor-Ardo, and Hudspeth 2008; Faucherre et al. 2009). Apart from this, little is currently known about the existence of different subtypes of lateral line ganglion cells.

6.4.2 Transcription Factors Involved in Specifying Identity of Lateral Line Placode Derived Cells

Lateral line placodes are patterned relatively early into a proximal Neurog1-positive domain, from which lateral line neurons delaminate, and a distal domain, which gives rise to the lateral line primordia (Northcutt and Brändle 1995; Gompel, Dambly-Chaudière, and Ghysen 2001; Sarrazin et al. 2006; Mizoguchi et al. 2011). The patterning mechanisms underlying the separation of these two domains are only very poorly understood. However, Notch signaling in the distal part of the posterior lateral line placode in zebrafish has been shown to be crucial for promoting the primordium over the neurogenic part (Mizoguchi et al. 2011). The lateral line primordia then elongate (in the head region) or migrate (posterior lateral line placodes) along stereotyped pathways and deposit neuromasts and ampullary organs along the way (Fig. 1.7) (Northcutt, Catania, and Criley 1994; Ledent 2002; Sapede et al. 2002). The mechanisms underlying the patterning and migration of the primordia have recently been studied in great detail for the migrating primordia of the zebrafish posterior lateral line. Discussion of these mechanisms is beyond the scope of this book and I have to refer the interested reader to recent reviews (Schlosser 2002a; Ghysen and Dambly-Chaudière 2007; Ma and Raible 2009; Chitnis, Nogare, and Matsuda 2012; Piotrowski and Baker 2014; Thomas et al. 2015; Dalle Nogare and Chitnis 2017).

Little is known about the transcription factors involved in specification of lateral line sensory neurons. However, the expression of Neurog1, Islet1, and POU4f1 in the developing lateral line ganglia indicates that they are related to other somatosensory neurons (Korzh et al. 1993; Andermann et al. 2002; Riddiford and Schlosser 2016). Similarly, the transcription factor Atoh1, which is essential for hair cell differentiation in the inner ear, also plays a central role for the differentiation of hair cells in the lateral line (Sarrazin et al. 2006; Millimaki et al. 2007; Nechiporuk and Raible 2008). Like in the inner ear, Atoh1 is subject to positive autoregulation in lateral line neuromasts, while suppressing Atoh1 expression in neighboring cells via Notch-mediated lateral inhibition (Itoh and Chitnis 2001; Sarrazin et al. 2006; Matsuda and Chitnis 2010). POU4f3 or the related POU4f2 and Gfi1 are also expressed in lateral line hair cells together with regional markers of the dorsal posterior placodal area such as Pax2/8 and Hmx2/3 (see Chapter 4) (DeCarvalho, Cappendijk, and Fadool 2004; Schlosser and Ahrens 2004; Dufourcq et al. 2006; Feng and Xu 2010; Modrell et al. 2017; Lush et al. 2019). Taken together, this suggests that hair cells of the inner ear and lateral line are regulated by a similar CoRN. However, in contrast to the inner ear, Prox1 expression is maintained in both hair cells and supporting cells of zebrafish lateral line neuromasts, where it is dispensable for initial specification of hair cells, but required for their terminal differentiation (Pistocchi et al. 2009).

A recent single-cell sequencing study of zebrafish neuromasts suggests that there is only one type of hair cells in neuromasts but has uncovered remarkable diversity of supporting cells, presumably reflecting different stages and types of hair cell progenitors (Lush et al. 2019). The transcription factor expression profile of electroreceptors in ampullary organs of the paddlefish, a non-teleost actinopterygian, was recently shown to be similar to neuromast hair cells including expression of Atoh1 and POU4f3, while NeuroD4 is only expressed in the ampullary organs (Modrell et al. 2017). This lends support to the proposal that these electroreceptors evolved as sister cells of neuromast hair cells (Baker and Modrell 2018).

In summary, the limited evidence available suggests that the specification of special somatoensory neurons and hair cells in the lateral line relies on CoRNs similar to those specifying the neurons of the vestibulocochlear ganglion and the hair cells of the inner ear. Although sensory neurons and hair cells derived from lateral line, placodes clearly exhibit lateral line specific specializations (e.g. in ultrastructure and projection patterns), we currently do not know which transcription factors control these differences to neurons and hair cells derived from the otic placode.

6.5 OLFACTORY PLACODES

6.5.1 Cell Types Derived from the Olfactory Placode

The olfactory placode gives rise to several different cell types (Fig. 1.12). These include the chemosensory receptor cells, which are primary sensory cells with an axon. They, thus, serve both as chemoreceptor cells and sensory neurons, which transmit olfactory information to the forebrain (reviewed in Beites et al. 2005; Nicolay, Doucette, and Nazarali 2006; Murdoch and Roskams 2007; Kam, Raja, and Cloutier 2014; Maier et al. 2014; Sokpor et al. 2018). In addition, the olfactory placode gives rise to cells that migrate into the brain, where they differentiate into neurons, which release GnRH or other neuropeptides. Apart from these neuronal cell types, the olfactory placode also produces the mucus producing cells of the Bowman glands, sustentacular or supporting cells, and cells of the respiratory epithelium (Graziadei and Graziadei 1979; Klein and Graziadei 1983; Nomura, Takahashi, and Ushiki 2004; Hansen and Zielinski 2005).

6.5.1.1 Morphology of Olfactory and Vomeronasal Receptor Cells

The chemosensory receptor cells derived from the olfactory placode may respond to general odorants (volatile or water-soluble molecules) or to molecules involved in intra- or interspecific communication (reviewed in Mombaerts 2004; Ache and Young 2005; Breer, Fleischer, and Strotmann 2006; Touhara and Vosshall 2009; Munger, Leinders-Zufall, and Zufall 2009; Kaupp 2010; Suarez, Garcia-Gonzalez, and de Castro 2012). In lungfishes and tetrapods, these chemoreceptor cells are separated into two separate epithelia, the main olfactory and vomeronasal epithelia, which have different projections into the forebrain (Figs. 1.10) (Dulac and Torello 2003; Butler and Hodos 2005; Gonzalez et al. 2010). The receptor cells of the main olfactory epithelium, variously known as olfactory receptor cells or olfactory sensory neurons, send their axons to the main olfactory bulb. In contrast, the receptor cells of the vomeronasal epithelia, known as vomeronasal receptor cells or vomeronasal sensory neurons, instead project to the accessory olfactory bulb, which, in turn, projects to different higher order centers than the main olfactory bulb.

Despite the clear anatomical segregation of both systems in tetrapods, the functional differences between them are still a bit puzzling. Based on early experiments, the olfactory epithelium and main olfactory pathway were long thought to be involved in the detection of general odorants, while the vomeronasal epithelium and accessory olfactory pathway were considered to be specialized for the detection of pheromones (reviewed in Wysocki 1979). Pheromones are signals released from members of the same species with important roles in intra-specific communication (e.g. during mating). However, it has now become clear that the olfactory system can respond to some pheromones, whereas the vomeronasal system can respond to olfactory cues other than pheromones (reviewed in Baxi, Dorries, and Eisthen 2006 and see further discussion below). Alternatively, it has been proposed that the olfactory receptor cells respond to smaller and more volatile molecules, whereas vomeronasal receptor cells respond to larger, and often non-volatile molecules, but even this rule is not without exceptions (Baxi, Dorries, and Eisthen 2006). In spite of the overlap in function between the two systems, the general role of the main olfactory system appears to be linked predominantly to the detection of general odorants, which are often small and volatile molecules. In contrast, the general role of the vomeronasal system appears to be related mainly to the detection of large and often non-volatile molecules. These are often transferred to the receptors by direct contact with a substrate. Many of these large molecules are involved in triggering species-specific and innate behaviors (Silva and Antunes 2017). They include many pheromones used in intra-specific communication as well as signals involved in inter-specific communication, such as odors released by prey animals that are detected by predators or odors released by predators that are used by the prey for predator avoidance (kairomones).

Olfactory and vomeronasal chemosensory receptor cells carry either cilia or microvilli or both on their apical side, and these protrusions contain the proteins involved in the detection of odorants and subsequent signal transduction (Fig. 6.7). The distribution of ciliary and microvillous cells varies between different vertebrates. This has been reviewed extensively before (Eisthen 1992; Eisthen 1997; Elsaesser and Paysan 2007) and will be only briefly summarized here. The vomeronasal receptor cells of tetrapods have microvilli (Fig. 6.7C), whereas the main olfactory epithelium of most tetrapods is dominated by ciliary receptor cells (Fig. 6.7A). In birds and some groups of reptiles, these ciliary receptor cells also carry microvilli. However, the main olfactory epithelium of salamanders and larval frogs contains a mixture of ciliary and microvillous receptor cells, and the latter have also been found as a rare cell type in the main olfactory epithelium of other tetrapods (Rowley, Moran, and Jafek 1989; Eisthen 1992; Elsaesser et al. 2005).

In cyclostomes and bony fishes, the single olfactory epithelium is often composed of both microvillous and ciliary receptor cells. In some actinopterygians (e.g. sturgeons, *Polypterus*) the ciliary receptor cells also carry small microvilli (Hansen and Zielinski 2005). In addition, crypt cells, which display both cilia and microvilli

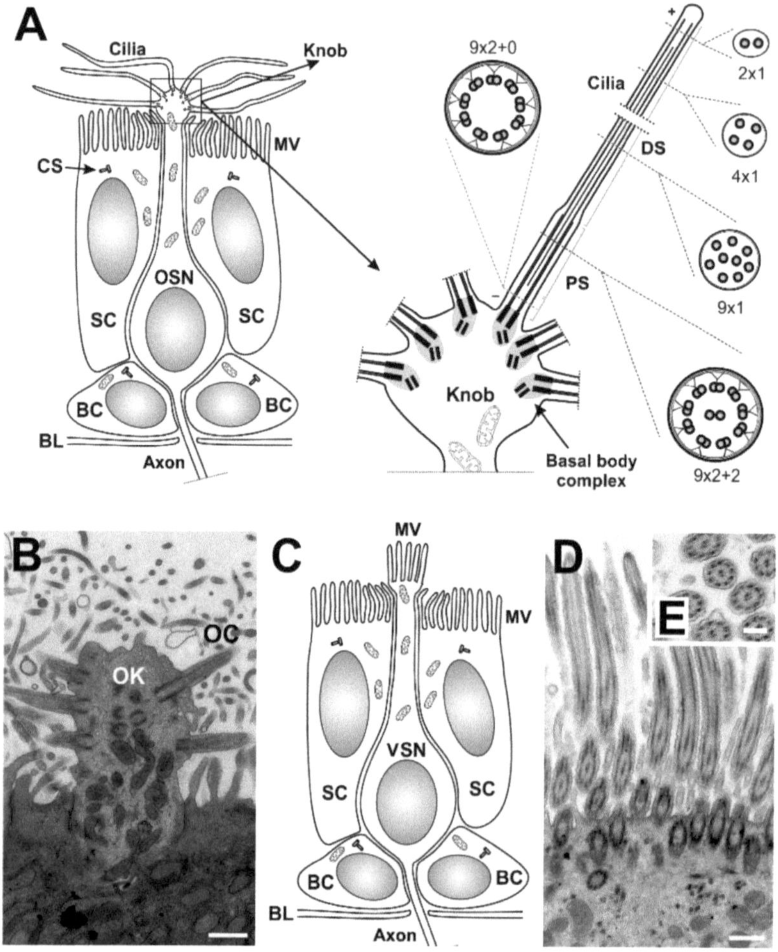

FIGURE 6.7 Olfactory and vomeronasal receptor cells. (**A**) Diagram of a mammalian olfactory sensory neuron (OSN) and its olfactory knob. They are surrounded by supporting cells (SC) with microvilli (MV) on their apical surface and continually replaced by basal cells (BC) throughout life. OSNs are primary sensory neurons with an axon coursing towards the olfactory bulb. The dendrite of OSNs terminates in an olfactory knob carrying specialized sensory cilia responsible for olfaction. Cross sections through various parts of the cilium are shown with numbers of microtubules indicated. BL, basal lamina; CS, centrosomes; DS, distal segment; PS, proximal segment. (**B**) Transmission electron micrographs of olfactory knob (OK) with olfactory cilia (OC) in adult mouse olfactory epithelium (scale bar: 500 nm). (**C**) Diagram of a mammalian vomeronasal sensory neuron (VSN) with microvilli. (**D, E**) Transmission electron micrographs of adult mouse respiratory epithelium with multiple motile cilia on one cell (shown in cross section in E) (scale bars: A, 500nm; B, 200 nm). (Reprinted with modification with permission from Falk et al. 2015).

in an apical recess or "crypt", have been described as a third cell type in teleosts (Hansen and Zeiske 1998; Hansen and Finger 2000). However, only ciliary receptor cells have been found in lampreys and only microvillous receptor cells in cartilaginous fishes (Eisthen 1992; Laframboise et al. 2007; Ferrando and Gallus 2013). This diversity indicates that the relation between cellular morphologies and categories of odors detected is likely to be complex. I will discuss this further below after introducing the different classes of olfactory receptor proteins.

The ciliated olfactory receptor cells of vertebrates have a single apical dendrite ending in a dendritic knob, from which multiple cilia (3–10 in fishes, 10–30 in mammals) extend into the covering mucus (Fig. 6.7A). The cilia are supported by a $9 \times 2 + 2$ array of microtubules but are non-motile in most vertebrates. The long distal segments of the cilia, which harbor the receptor and signal transduction proteins, are oriented parallel to the epithelial surface, where they interdigitate with cilia from adjacent receptor cells and form a dense ciliary mat (Menco 1997; McEwen, Jenkins, and Martens 2008; Falk et al. 2015). Since every olfactory receptor cell expresses only one type of odorant receptor (OR) (see section 6.5.1.3), this ensures that sensitivity to different odorants is relatively evenly spread in the olfactory epithelium. During development of these ciliary cells, the centrioles inherited from the mitotic spindle first multiply before they migrate to the dendritic knob and form the basal bodies of the cilia (Cuschieri and Bannister 1975; Menco and Farbman 1985; Menco and Morrison 2003). Since microvillar receptor cells in several vertebrate groups have multiple centrioles, it has been proposed that these may be derived – possibly several times independently – from the ciliary type (Eisthen 1992). This scenario is also supported by the co-occurrence of microvilli on ciliated olfactory receptor cells in many vertebrates as summarized above.

6.5.1.2 Receptor Proteins and Signal Transduction Pathways in Olfactory and Vomeronasal Receptor Cells

The receptor proteins that interact with odorants in the vertebrate olfactory system belong to several families of seven transmembrane GPCRs and other transmembrane proteins. For most of these families, the number of genes encoding functional family members is highly variable in different species and accompanied by an equally variable set of non-functional pseudogenes (see figs. 2.4 B, 2.12 A in Schlosser 2021). This reflects rapid lineage-specific expansions and contractions due to both selectively neutral and adaptive evolutionary changes (Ache and Young 2005; Nei, Niimura, and Nozawa 2008; Niimura 2009; Suarez et al. 2012; Silva and Antunes 2017). The OR family is the largest of these GPCR families with approximately 150 members in zebrafish, 400 in humans, and more than 1000 in some rodents and elephants (Niimura 2009; Niimura, Matsui, and Touhara 2014). *OR* genes are organized in several clusters occupying different chromosomes. They encode proteins responding to a wide range of small molecules. Related *OR* genes have also been found in other deuterostomes and in cnidarians but not in protostomes indicating that the *OR* family is an ancient metazoan gene family, which was lost in protostomes but was greatly expanded in vertebrates (Churcher and Taylor 2011).

In addition, vertebrates use other receptor gene families in olfaction, all of which are vertebrate specific. The trace amine associated receptors (TAARs) comprise another family of GPCRs with 15 members in mouse, 6 in humans, and 57 in zebrafish (Liberles and Buck 2006), which evolved only in gnathostomes (Hashiguchi and Nishida 2007; Hussain, Saraiva, and Korsching 2009). They specifically respond to amines, which are used as important social cues transmitted, for example, by urine. There are also two GPCR families of vomeronasal receptors, each ranging from a few to 200–300 members depending on the species (Silva and Antunes 2017). The olfactory receptors related to class A (ORA) receptor family, which includes mammalian vomeronasal type 1 receptors (V1R) and the T2R bitter taste receptors of bony fishes and tetrapods, has been found in all vertebrate groups including lampreys and, thus, probably originated with the vertebrates (Dulac and Axel 1995; Del Punta et al. 2002; Saraiva and Korsching 2007; Libants et al. 2009; Grus and Zhang 2009). The olfactory C (OlfC) receptor family, which includes mammalian vomeronasal type 2 receptors (V2R) and is related to the T1R sweet and umami taste receptors, has only evolved in the gnathostomes (Matsunami and Buck 1997; Ryba and Tirindelli 1997; Alioto and Ngai 2006; Grus and Zhang 2009). While V1Rs mostly respond to volatile pheromones, V2Rs respond more strongly to waterborne signals including peptides and major urinary proteins (Touhara and Vosshall 2009). A last GPCRs family of receptors involved in olfaction, the formyl peptide receptors (FPR) with seven members in mouse evolved only in rodents by recruitment of immune-specific receptor proteins to the vomeronasal system (Liberles et al. 2009; Riviere et al. 2009; Dietschi et al. 2017). Similar to their cognates in the immune system, vomeronasal FPRs detect pathogen and inflammation dependent molecules (Riviere et al. 2009; Bufe, Schumann, and Zufall 2012).

Recently, membrane-spanning four-domain proteins (MS4A) have been discovered as a new small family of vertebrate olfactory receptors with 17 members in mouse, which do not belong to GPCRs and have only 4 transmembrane domains (Zuccolo et al. 2010; Greer et al. 2016). They respond to a chemically diverse range of molecules including gases, fatty acids, and peptides, which tend to elicit innate appetitive or aversive reactions (Greer et al. 2016). MS4A receptors are specifically expressed in cells also expressing the transmembrane protein Guanylylcyclase (GC) D or G, which has previously been implicated in odorant detection, but appears to be dispensable for MS4A mediated responses (Kaupp 2010; Greer et al. 2016). Whether and how MS4A and GC interact in the perception of certain odors, thus, requires further clarification.

The signals detected by different families of olfactory receptor proteins are relayed to the cell by different signal transduction pathways. Receptors of the OR family activate adenylylcyclase type III (ACIII) mediated by the olfaction specific G-protein α-subunit $G\alpha_{olf}$. cAMP produced by ACIII then opens channels of the cyclic nucleotide-gated (CNG) family such as CNGA2. This leads to Ca^{2+} influx and depolarization of the cell, which is amplified by the opening of Ca^{2+}-dependent chloride channels (Fig. 6.8A) (reviewed in Touhara and Vosshall 2009; Munger, Leinders-Zufall, and Zufall 2009; Kaupp 2010). TAARs were also reported to use this canonical OR transduction pathway (Zhang et al. 2013). In contrast, receptors of the V1R and V2R families interact with the G-protein α-subunits $G\alpha_i$ or $G\alpha_0$ to

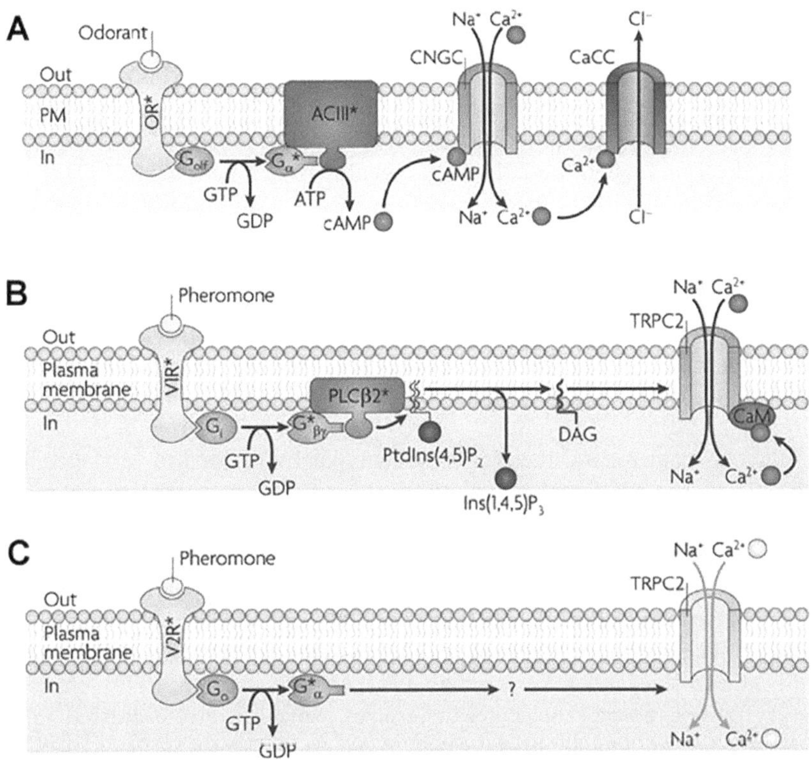

FIGURE 6.8 Signal transduction in olfactory and vomeronasal receptors. (**A**) The binding of an odorant to the olfactory receptor (OR) sequentially activates the olfaction-specific G protein α subunit (G_{olf}), adenylyl cyclase type III (ACIII), the olfactory cyclic nucleotide-gated channel (CNGC), and a Ca^{2+}-activated Cl^- channel (CaCC). Activation of both channel types leads to depolarization. (**B**) The binding of the pheromone to a V1R receptor successively activates the inhibitory G-protein α subunit G_i; phospholipase Cβ2 (PLCβ2), which produces inositol-1,4,5-trisphoshate (Ins(1,4,5)P_3) and diacylglycerol (DAG) from phosphatidylinositol-4,5-bisphoshate (PtdIns(4,5)P_2); and finally the transient receptor potential cation channel C2 (TRPC2). Activation of TRPC2 mediates Na^+ and Ca^{2+} influx, leading to a depolarization. Recovery and adaptation may involve binding of Ca^{2+}–calmodulin (CaM) to TRPC2. (**C**) The binding of pheromones to V2R receptors activates the G-protein α subunit G_o. In some V2R-expressing neurons, TRPC2 may be involved in generating depolarizing currents, but other signaling mechanisms may exist. (Reprinted with permission from Kaupp 2010).

ultimately open ion channels of the TRP family, in particular, TRPC2 in mammals (Fig. 6.8B, C). The proteins transducing the signal from G-proteins to the TRP channel have only been characterized for V1R receptors, where they comprise phospolipase C (PLC), inositol-1, 4, 5-trisphosphate, and diacylglycerol (DAG) (Fig. 6.8B) (reviewed in Munger, Leinders-Zufall, and Zufall 2009; Touhara and Vosshall 2009; Kaupp 2010). How signals of MS4A receptors are transduced is currently unknown (Greer et al. 2016).

6.5.1.3 Mechanisms of Olfactory Receptor Protein Choice

Current evidence suggests that only a single allele of a single *OR* gene is expressed in any olfactory receptor cell (Chess et al. 1994; Shykind et al. 2004; Clowney et al. 2012). Single-cell RNA-Seq confirmed that co-expression of several ORs occurs only transitorily in immature olfactory receptor cells (Hanchate et al. 2015; Saraiva et al. 2015; Tan, Li, and Xie 2015; Scholz et al. 2016; Fletcher et al. 2017). This "one neuron one receptor" rule also appears to apply to TAAR and V1R receptors but not to V2Rs or MS4As (Silvotti et al. 2007; Munger, Leinders-Zufall, and Zufall 2009; Johnson et al. 2012). V2Rs can be classified into four families A–D, and one member of the A, B, or D families is always co-expressed with one C family member in a single cell (Martini et al. 2001; Silvotti et al. 2007; Ishii and Mombaerts 2011). The functional significance of this co-expression is still unclear. In contrast, multiple members of the MS4A receptor family are co-expressed in a single receptor cell (Greer et al. 2016).

Whereas vomeronasal receptor proteins typically respond to very specific molecules (i.e., they are narrowly tuned), there is great variety of response properties among the ORs. While some ORs are narrowly tuned, others are broadly tuned and respond to many different odor molecules (Monahan and Lomvardas 2015). Receptor cells of the main olfactory epithelium expressing the same OR or TAAR receptor project to the same glomerulus in the main olfactory bulb where they form synapses with mitral cells which transmit information further to higher brain centers (Ressler, Sullivan, and Buck 1994; Vassar et al. 1994; Mombaerts et al. 1996; Johnson et al. 2012). The OR receptor molecules themselves, which are also expressed on axonal growth cones, are thought to contribute to the establishment of these specific connections during development (Vassalli et al. 2002; Barnea et al. 2004; Feinstein et al. 2004, Strotmann et al. 2004; Nakashima et al. 2013; Tsai and Barnea 2014). However, details of this process are still poorly understood. The combinatorial activation of different glomeruli by single odor molecules or odorant mixtures permits the discrimination between multitudes of different odors.

The apical and basal receptor cells of the vomeronasal epithelium, which express V1R and V2R receptors, respectively, similarly project to specific glomeruli in the accessory olfactory bulb (Belluscio et al. 1999; Rodriguez, Feinstein, and Mombaerts 1999; Del Punta et al. 2002). In contrast, the MS4A expressing receptor cells, which reside in the main olfactory epithelium, project to a "necklace" of multiple interconnected glomeruli encircling the main olfactory bulb. Each bead of the necklace, thus responds to multiple MS4A stimuli suggesting that the system may be involved more with the detection rather than discrimination of particular cues (Greer et al. 2016).

The mechanisms ensuring the monoallelic and monogenic expression of olfactory receptors have been particularly well studied for the OR family, but similar types of mechanisms have been postulated to underlie receptor choice in the TAAR, V1R, and V2R families (Dalton and Lomvardas 2015). An intricate combination of non-random and random mechanisms is responsible for OR choice. Non-random mechanisms involve the requirement for upstream transcription factors for the transcriptional activation of *OR* genes. *OR* genes have promoters with binding sites for various homeodomain and COE (Collier/Olf/EBF) transcription factors ensuring that *OR* genes are only expressed in differentiating olfactory receptor neurons (Wang, Tsai, and Reed 1997; Vassalli et al. 2002; Young, Luche, and Trask 2011; Plessy et al. 2012). In addition, several enhancer-like elements have been identified in *OR* geneclusters, which appear to control the expression of *OR* genes from this and possibly other clusters (Khan, Vaes, and Mombaerts 2011; Vassalli, Feinstein, and Mombaerts 2011; Markenscoff-Papadimitriou et al. 2014). Activation of such elements by regionally confined transcription factors may lead to the well-documented restriction of expression of particular *OR* genes to different zones in the olfactory epithelium (Miyamichi et al. 2005). However, this hypothesis still awaits experimental verification.

While these non-random mechanisms define a subset of *OR* genes available for expression in a given receptor cell, only one allele of one *OR* gene will ultimately be stably expressed and the choice is due to a stochastic selection process. In olfactory progenitor cells, all *OR* genes are silenced by repressive modifications of histone proteins, viz. H3 lysine 9 trimethylation and histone H3 lysine 20 trimethylation. These lead to compaction of chromatin, rendering *OR* genes inaccessible for transcription (Magklara et al. 2011). When immature olfactory receptor cells start to express the histone demethylase LSD1, it will begin to remove these repressive histone marks in random order from *OR* genes (Lyons et al. 2013). The first *OR* gene activated will then prevent expression of all other *OR* genes by a complex feedback mechanism involving the induction of Activating Transcription Factor 5 (ATF5) and downregulation of LSD1 (Wang et al. 2012; Dalton, Lyons, and Lomvardas 2013; Lyons et al. 2013). As long as demethylation is a slow and stochastic process, followed by rapid feedback inhibition of demethylase activity, this mechanism ensures the stable selection of a random single *OR* allele in each olfactory receptor cell (Tan, Zong, and Xie 2013).

6.5.1.4 Functional Specializations of Olfactory Receptor Cells

Despite the large diversity of receptor proteins expressed in the chemoreceptor cells derived from the olfactory placode, these cells are morphologically relatively uniform, comprising sensory neurons with either cilia or microvilli (and sometimes both) on their apical side, as discussed above. Studies in different vertebrates have found that, generally, ORs, TAARs and the associated signal transduction proteins ($G\alpha_{Olf}$, ACIII, CNG channels) are expressed in ciliated sensory neurons, while ORA (V1R), OlfC (V2R), FPR and the associated signal transduction proteins ($G\alpha_i/G\alpha_0$, PLC, TRPC channels) are found in microvillar sensory neurons (Buck and Axel 1991; Menco et al. 1992; Menco et al. 2001; Hansen, Anderson, and Finger 2004). An association of TRPC channels and PLC with microvilli and of CNG channels with cilia is found in several types of sensory cells and across animal taxa (see Chapters 3 and 4 in Schlosser 2021), suggesting that the localization of these channels to microvilli and cilia, respectively, may be an evolutionarily ancient trait of vertebrate chemoreceptor cells (Fain, Hardie, and Laughlin 2010; Johnson and Leroux 2010; Lange 2011).

In accordance with this, cartilaginous fishes, which have only microvillar receptors, have almost lost OR, TAAR, and ORA (V1R) receptors (with only a handful of genes remaining in the genome) but have a greatly expanded repertoire of OlfC (V2R) (Grus and Zhang 2009; Hussain, Saraiva, and Korsching 2009; Niimura 2009; Marra et al. 2019; Sharma et al. 2019). Since all receptor protein families except for the OR family are vertebrate specific, it is tempting to speculate that the ORA (V1R) or OlfC (V2R) expressing microvillar chemosensory neurons derived from the olfactory placode may have evolved from ciliary chemosensory neurons (which possibly also carried microvilli) by elaboration of TRPC signaling linked to microvilli.

There are, however, some notable exceptions to this association of different receptors and signal transduction pathways with cilia or microvilli. Cartilaginous fishes, in spite of possessing only microvillar olfactory receptor cells, retain a few genes encoding OR and TAAR receptor proteins (see above), while lampreys, which have only ciliary receptors, have ORA (V1R) genes (Eisthen 1992; Laframboise et al. 2007; Libants et al. 2009; Grus and Zhang 2009). Unless there are rare ciliary or microvillar cell types in cartilaginous fishes and lampreys, respectively, that have not yet been discovered, this suggests that OR/TAAR receptors may be localized to microvilli and ORA receptors to cilia in these groups.

Furthermore, in mammals, two types of ciliated sensory neurons in the main olfactory epithelium have been identified that express TRPC2 in addition to CNGA2 (Omura and Mombaerts 2014). While type A cells of unknown function co-express a full canonical OR transduction cascade including $G\alpha_{olf}$, ACIII and sometimes even OR receptors, type B cells, which serve as sensors of low oxygen levels, express Guanylylcyclase1b2 suggesting a divergent signal transduction mechanism (Omura and Mombaerts 2014; Omura and Mombaerts 2015; Saraiva et al. 2015; Bleymehl et al. 2016). Another type of ciliated sensory neuron in the main olfactory epithelium co-expresses TRPM5 with elements of the canonical OR signaling cascade such as ACIII and CNGA2 and has been shown to respond to pheromones (Lin et al. 2008; López et al. 2014). Conversely, the CNGA4 modulatory subunit of CNG channels is expressed in some microvillar cells of the vomeronasal epithelium but does not appear to participate in canonical CNG signaling (Berghard, Buck, and Liman 1996; Broillet and Firestein 1997). Taken together this suggests that new subtypes of ciliary or microvillar receptor cells with modified signal transduction mechanisms have evolved in mammals and probably other vertebrates by recruiting TRP or CNG channels, respectively, or retaining them from a multifunctional precursor.

The mapping of different functions to receptor cells expressing particular receptor proteins is also complex. Typically, OR and TAAR expressing cells respond to general odorants and ORA (V1R), OlfC (V2R), and FPR to species-specific signals like pheromones or kairomones (Touhara and Vosshall 2009; Kaupp 2010; Dalton and Lomvardas 2015). However, in teleosts, some ORA (V1R) and OlfC (V2R) expressing sensory neurons, which are located in the single olfactory epithelium of these fishes, were shown to be involved in general odorant (amino acid) detection (Speca et al. 1999; Luu et al. 2004; Hansen and Zielinski 2005; Alioto and Ngai 2006; Behrens et al. 2014). Cartilaginous fishes, which have lost almost all *OR* genes, but express an expanded repertoire of *OlfC* (*V2R*) genes together with $G\alpha_0$ (no $G\alpha_{olf}$) in their single olfactory epithelium, most likely rely predominantly on OlfC (V2R) type receptors for general odorant detection (Ferrando et al. 2009; Ferrando and Gallus 2013). Conversely, species-specific evolutionary rates for some *OR* or *TAAR* gene clades suggest that they possibly play a role in the detection of species-specific signals like pheromones in teleosts (e.g. Hashiguchi, Furuta, and Nishida 2008).

Nevertheless, recent findings in teleosts indicate that their microvillar receptor cells mostly express ORA (V1R) and OlfC (V2R) and/or their associated signal transduction proteins (e.g. $G\alpha_i$ or $G\alpha_0$, TRPC2), whereas ciliary receptor cells mostly express OR and TAAR receptors and/or their associated signal transduction proteins (e.g. $G\alpha_{olf}$, CNGA2) (Hansen et al. 2003; Hansen, Anderson, and Finger 2004; Hansen and Zielinski 2005; Sato, Miyasaka, and Yoshihara 2005; Saraiva and Korsching 2007; Hussain, Saraiva, and Korsching 2009; Oka and Korsching 2011; Oka, Saraiva, and Korsching 2012; Biechl et al. 2017). Furthermore, central projections resembling the accessory olfactory pathways of tetrapods have been shown to originate from microvillar and crypt cells in teleosts (Sato, Miyasaka, and Yoshihara 2005; Biechl et al. 2017). Taken together this lends support to the idea that some functional separation between a general odorant detecting main olfactory system using ciliary receptor cells expressing OR/TAAR and a system involved in intra- or interspecific communication utilizing microvillar receptor cells expressing ORA/OlfC originated already in osteichthyans (or possibly even gnathostomes or vertebrates). Subsequently, anatomical separation of the receptor cells mediating these different functions into distinct main olfactory and vomeronasal epithelia has evolved in the last ancestor of lungfishes and tetrapods (Gonzalez et al., 2010; Silva and Antunes 2017). As a caveat, while the main olfactory epithelium in mammals contains predominantly OR and TAAR expressing ciliary sensory neurons, in amphibians it contains a mixture of microvillar and ciliary receptor cells, some of which express V1R, V2R, and TRPC2 (Hansen et al. 1998; Oikawa et al. 1998; Hagino-Yamagishi et al. 2004; Date-Ito et al. 2008; Syed et al. 2013; Sansone et al. 2014). This suggests that complete anatomical separation of both functional subsystems may not even be achieved in all tetrapods.

6.5.1.5 Other Cell Types Derived from the Olfactory Placode

The receptor cells of the main olfactory and vomeronasal epithelia are surrounded by a special type of supporting cell known as sustentacular cell, which typically carry microvilli, although ciliated supporting cells have been found in some fishes (Figs. 1.10, 6.9) (Hansen and Zielinski 2005). These are glial-like cells, which help in the maintenance of ion balance, degradation of odorants and toxic environmental compounds (xenobiotics), and phagocytosis of damaged cells (Yu et al. 2005; Sokpor et al. 2018). Olfactory and vomeronasal epithelia also produce another non-neuronal cell type, the cells of the exocrine Bowman's gland, which secrete mucus and other compounds including xenobiotic detoxifying enzymes (Yu et al. 2005; Solbu and Holen 2012).

In addition, the main olfactory epithelium also gives rise to a number of neurosecretory cells. Several types of these cells have been described as microvillous cells without an axon in the main olfactory epithelium of mammals, which probably modulate the activity of adjacent cells by secretion of neuropeptides or other neurotransmitters. First, there are cells that express the signal transduction molecule PLCβ2 but not the TRP channel TRPM5 and differentiate into cells secreting neuropeptide Y (NPY) (Elsaesser et al. 2005; Montani et al. 2006; Weng, Vinjamuri,

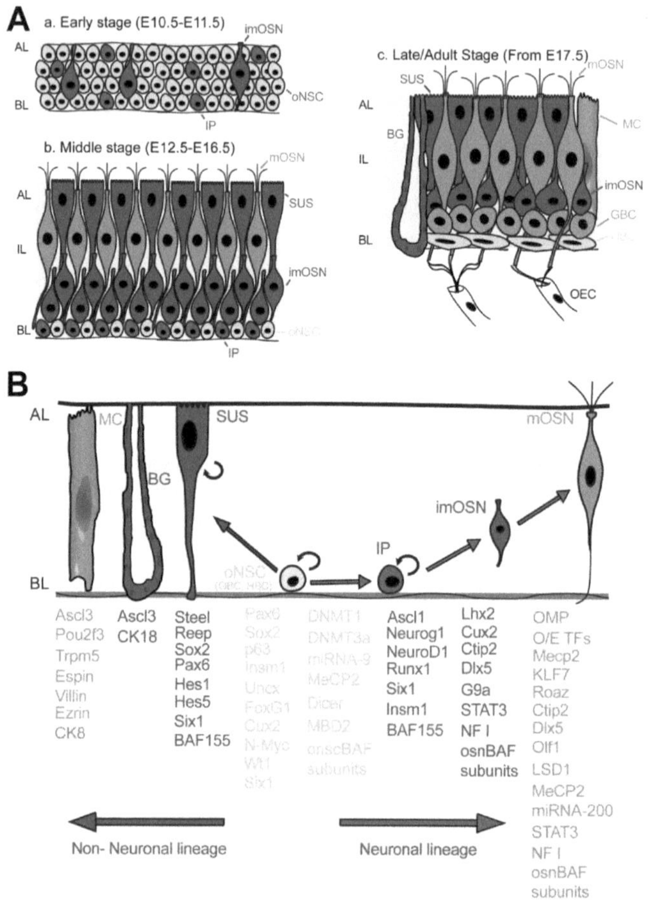

FIGURE 6.9 Development of the olfactory epithelium in the mouse. (**A**) During early embryonic stages (E10.5–E11.5), the olfactory epithelium consists mainly of olfactory neural stem cell (oNSCs), some of which form intermediate progenitors (IPs). A few early differentiating, but still immature olfactory sensory neurons (imOSN; dark, spindle shaped) are also present at this stage. By mid-embryonic stages (E12.5–E16.5), three layers of the olfactory epithelium can be distinguished. The apical layer (AL) harbors sustentacular (SUS) cells; the intermediary layer (IL) contains mature (mOSN) and immature (imOSN) olfactory sensory neurons (OSNs); the basal layer (BL) contains oNSCs and IPs. At late-embryonic or adult stages, round globose basal cells (GBCs) are distinguishable on top of flattened horizontal basal cells (HBCs) in the basal layer. In the middle layer, imOSNs mature into mOSNs. The apical layer mainly contains the cells bodies of SUS cells and microvillar cells (MCs). Bowman's glands (BGs) are typically located deep in the olfactory epithelium. OEC, olfactory ensheathing cells. (**B**) Transcription factors required for cell specification in the olfactory epithelium. Progenitors (oNSCs) require a large set of transcription factors (indicated below) for proliferation and survival. These progenitors may either be specified in a stepwise fashion as olfactory sensory neurons (right of oNSCs) or adopt a non-neural fate (left of oNSCs) mediated by transcription factors indicated below. (Reprinted with permission from Sokpor et al. 2018, after Bachmann et al. 2016).

and Ovitt 2016). Second, there are two morphologically distinct classes of cells that express TRPM5 and differentiate into cholinergic cells (Lin et al. 2008, Hansen and Finger 2008; Ogura et al. 2011; Fu et al. 2018).

Third, and most importantly, there are cells that migrate out of the border region between olfactory and respiratory epithelium and express GnRH or other neuropeptides (reviewed in Somoza et al. 2002; Wray 2010, Forni and Wray 2015). Gnathostomes produce three types of GnRH, each encoded by a different gene, which probably arose by duplications of one GnRH gene in the ancestor of vertebrates (Decatur et al. 2013). The olfactory placode only produces GnRH1 and GnRH3 cells (the latter only found in teleosts), while GnRH2 cells are neural tube derived (reviewed in Roch, Busby, and Sherwood 2011; Forni and Wray 2015). The GnRH1 cells migrate into the forebrain, where they colonize the septo-preoptic nuclei of the hypothalamus that regulate the release of gonadotropic hormones (LH, FSH) from the anterior pituitary (Fig. 1.11B). Some GnRH1 cells also contribute to ganglia of the terminal nerve, while GnRH3 cells contribute to these ganglia in teleosts (reviewed in Muske and Moore 1988; Demski 1993; Von Bartheld 2004). The GnRH cells of the terminal nerve have neuromodulatory effects on the olfactory receptor cells and other neurons and neurosecretory cells (Abe and Oka 2000; Eisthen et al. 2000; Park and Eisthen 2003). In lampreys, only orthologs to gnathostome GnRH2 (lamprey GnRH II) and GnRH3 (lamprey GnRH I and III) have been found, suggesting that GnRH1 may have been lost (Decatur et al. 2013).

Finally, we need to mention the cells of the respiratory epithelium, which surrounds the olfactory epithelium (Fig. 6.7D, E). Most of these cells are characterized by multiple motile cilia employed in moving the mucus of the nose, which is important for clearance of toxins and the prevention of infections (reviewed in Brooks and Wallingford 2014; Spassky and Meunier 2017). The multiciliary cells of the placode-derived respiratory epithelium resemble the multiciliary cells found in the adjacent endodermal airway epithelia and in other epithelia of the body, for example in brain ependyma, where they promote circulation of cerebrospinal fluid; in oviducts, where they propel the ovum; or in the epidermis of amphibian larvae. In each of these cells, the multiple basal bodies, to which the cilia are attached develop in a peculiar process that does not require a mother centriole as template, different from mono-ciliated cells (reviewed in Brooks and Wallingford 2014; Spassky and Meunier 2017). In addition, the respiratory epithelium contains mucus producing goblet-cells resembling those of the digestive tract (McCauley and Guasch 2015).

Apart from the various neuronal and non-neuronal differentiated cells, the olfactory and vomeronasal epithelia also contain several populations of progenitor cells, which will be discussed in the following section.

6.5.2 Patterning of the Olfactory Placode and Development of Different Cell Types

Experiments in chick and mouse embryos have shown that once the olfactory placode is specified, it is subdivided into the two adjacent territories of the olfactory and the respiratory epithelium by opposing BMP and FGF signals from adjacent ectoderm and other tissues (Maier et al. 2010). While BMP signaling promotes the formation of the respiratory epithelium, FGF and retinoic acid signaling is required for the development of the olfactory epithelium (Kawauchi et al. 2005; Maier et al. 2010; Paschaki et al. 2013). Signals from the neural-crest derived mesenchyme surrounding the invaginating olfactory pit are also required for the proper differentiation of both the olfactory and vomeronasal epithelium (Mantia et al. 2000; Bhasin et al. 2003; La Rawson et al. 2010).

Like other placode-derived epithelia, the olfactory and vomeronasal epithelia initially form a single-layer pseudostratified epithelium but become multilayered during later embryonic development (Fig. 6.9A). Olfactory neural stem and progenitor cells reside in a basal layer, while the cell bodies of immature and differentiated olfactory sensory neurons occupy the intermediate layer and the cell bodies of sustentacular cells and the microvillar neurosecretory cells the apical layer. Apart from a multipotent population of stem-cell like progenitors, there appear to be two types of progenitors with more restricted developmental potential, which give rise to the sensory neurons and the non-neural cells of the olfactory epithelium, respectively (DeHamer et al. 1994; Goldstein et al. 1998; Chen, Fang, and Schwob 2004). The vomeronasal epithelium is organized slightly differently with progenitor cells located basally as well as at the margins of the epithelium and two different populations of sensory neurons occupying the apical (V1R and $G\alpha_i$ expressing cells) and basal (V2R and $G\alpha_0$ expressing cells) part of the epithelium (Dulac and Axel 1995; Halpern, Shapiro, and Jia 1995; Berghard and Buck 1996; Herrada and Dulac 1997; de la Rosa-Prieto et al. 2010). However, the differentiation of the various cell types in the olfactory and vomeronasal epithelia is regulated by similar transcription factors (Fig. 6.9B).

Sox2, Six1, and Hes1/5 are expressed throughout the embryonic olfactory epithelium, with higher Sox2 levels medial than lateral, probably reflecting a graded distribution of different types of progenitors (Cau et al. 2000; Kawauchi et al. 2005, Ikeda et al. 2007; Chen, Kim, and Xu 2009; Tucker et al. 2010). These transcription factors together with Pax6, FoxG1, Meis1, and p63, which are similarly expressed, play important roles in the maintenance of olfactory and vomeronasal progenitors (Collinson et al. 2003; Donner, Episkopou, and Maas 2007; Ikeda et al. 2007; Duggan et al. 2008; Chen, Kim, and Xu 2009; Kawauchi et al. 2009 Tucker et al. 2010; Fletcher et al. 2011; Packard et al. 2011). Whereas stem cell-like progenitors are enriched laterally, progenitor cells that can undergo only a limited number of divisions (so-called transit amplifying cells) are enriched medially (Tucker et al. 2010). High levels of Sox2 have been shown to promote Ascl1 expression, leading to upregulation of Ascl1 in the medial progenitors (Tucker et al. 2010).

Ascl1 was shown to be essential in initiating differentiation of both olfactory and vomeronasal sensory neurons in the mouse (Guillemot and Joyner 1993; Cau et al. 1997; Cau, Casarosa, and Guillemot 2002; Murray et al. 2003). However, an early population of vomeronasal sensory neurons was only mildly affected in Ascl1 mutants suggesting that other, unknown transcription factors may be able to substitute for Ascl1 in these cells (Cau, Casarosa, and Guillemot 2002). Downstream of Ascl1, Neurog1, and NeuroD1 are required for differentiation of olfactory and vomeronasal sensory neurons (Cau, Casarosa, and Guillemot 2002; Cau et al. 1997). Differentiating neurons repress Ascl1 in these cells and Neurog1 expression in adjacent progenitors of the olfactory epithelium by Notch-mediated lateral inhibition via upregulation of Hes1 and Hes5 (Cau et al. 2000).

In the adult, the olfactory epithelium continues to produce olfactory sensory neurons as part of normal turnover in all vertebrates. Its capacity to generate sensory neurons in the adult also enables it to regenerate after tissue damage, and in mammals it is the only placode to do so (reviewed in Beites et al. 2005; Murdoch and Roskams 2007; Maier et al. 2014; Schwob et al. 2017; Sokpor et al. 2018). Two types of stem/progenitor cells persist in the adult mammalian olfactory epithelium (Fig. 6.9A). The flattened horizontal basal cells (HBCs), which express Sox2 and Pax6, are relatively quiescent stem cell like progenitors residing next to the basal lamina (Holbrook, Szumowski, and Schwob 1995; Carter, MacDonald, and Roskams 2004; Leung, Coulombe, and Reed 2007; Iwai et al. 2008; Guo et al. 2010; Suzuki and Osumi 2015). They normally divide at a very slow rate, but serve as a repository for new cells during normal turnover and injury-induced regeneration. The round globose basal cells (GBCs) instead are rapidly dividing progenitor cells located right next to HBCs (Caggiano, Kauer, and Hunter 1994; DeHamer et al. 1994; Schwob et al. 1994; Goldstein et al. 1998; Huard et al. 1998; Chen, Fang, and Schwob 2004; Beites et al. 2005). They undergo a limited number of cell divisions and then differentiate either into olfactory sensory cells or into one of the non-neural cells of the olfactory epithelium such as sustentacular, Bowman's gland, or microvillar neurosecretory cells (Fig. 6.9B). Similar to the embryo, the neurogenic GBC population in the adult is characterized by the expression of Ascl1 and Neurog1, which initiate neuronal differentiation (Gordon et al. 1995; Schwob et al. 2017; Sokpor et al. 2018).

6.5.3 Transcription Factors Involved in Specifying Identity of Olfactory Sensory Neurons and Other Cell Types Derived from the Olfactory Placode

Since the bHLH transcription factors that initiate sensory differentiation in the olfactory placode (Ascl1, Neurog1, NeuroD1) are also expressed in other placodes, additional transcription factors with a more restricted expression must be involved in the specification of olfactory and vomeronasal sensory neurons. POU4f1, which plays a central role for the differentiation of somatosensory neurons, does not appear to be essential for the differentiation of neurons derived from the olfactory placode, where it is only weakly and transiently expressed (Hutcheson and Vetter 2001). However, several other transcription factors including Six1, Six4, Lhx2, DMRT4/5, FoxG1, Sp8, Runx1, the COE-class transcription factors EBF2/3, KLF7, and Atoh7 (in *Xenopus*) are expressed in differentiating olfactory and vomeronasal sensory neurons and have been implicated in regulating various aspects of their differentiation or maturation program (Burns and Vetter 2002; Bell et al. 2003; Hirota and Mombaerts 2004; Kolterud et al. 2004; Wang et al. 2004; Huang et al. 2005; Laub et al. 2005; Theriault et al. 2005; Hirota, Omura, and Mombaerts 2007; Chen, Kim, and Xu 2009; Berghard et al. 2012; Heron et al. 2013; Parlier et al. 2013).

However, the functions of many of these transcription factors for olfactory neurogenesis are still poorly understood and only Lhx2 and DMRT4/5 proteins become specifically confined to the olfactory placode (see Chapter 4), making them the most promising candidates for transcription factors specifying the identity of olfactory and/or vomeronasal sensory cells. DMRT4 and DMRT5 are indeed required for olfactory neurogenesis upstream of proneural bHLH transcription factors and EBF2/3 (Huang et al. 2005; Parlier et al. 2013). Lhx2 is subsequently required for the differentiation of olfactory sensory neurons and cooperatively binds with EBF2/3 on *OR* enhancers (Zhang et al. 2016; Monahan et al. 2017).

It is currently unresolved, how many different types of sensory neurons can be distinguished in the olfactory epithelium. Even though expression of different ORs on different olfactory sensory neurons is essential for proper function of the olfactory system, selection of different odorants depends largely on a random process as discussed above and does not require different CoRNs of transcription factors. Recent single-cell sequencing studies indeed shows tight clustering of expression profiles for olfactory sensory cells suggesting that the majority of sensory neurons represent a single cell type (Hanchate et al. 2015; Saraiva et al. 2015; Tan et al. 2015; Scholz et al. 2016; Fletcher et al. 2017). However, a recent study in zebrafish revealed differences in transcription factor profiles between subpopulations of olfactory sensory neurons with different projection patterns (Dang et al. 2018). This suggests that there may be a coarse subdivision into regional subtypes of olfactory receptor neurons, which may also underlie the zonality of *OR* gene expression in the mammalian olfactory epithelium (Miyamichi et al. 2005). Moreover, the identification of scattered TRPC2 or TRPM5 expressing ciliary neurons in the main olfactory epithelium (see above) indicates that there are likely to be a number of additional rare cell types, and a recent single cell sequencing study has confirmed that TRPC2 neurons are also characterized by the expression of a unique profile of transcription factors including Emx1 (Saraiva et al. 2015).

Apart from the differential expression of transcriptions factors Fezf1 and Fezf2, which are required for the identity of olfactory sensory neurons and the survival of vomeronasal sensory neurons, respectively (Eckler et al. 2011), there are some quantitative differences in transcription factor expression between olfactory and vomeronasal epithelia (Shimizu and Hibi 2009; Chang and Parrilla 2016; Parrilla et al. 2016). However, overall the expression profile of transcription factors in olfactory and vomeronasal sensory cells is very similar. This has been recently confirmed in single-cell sequencing studies (Tietjen et al. 2003; Tietjen, Rihel, and Dulac 2005). Furthermore, as discussed above, molecular phylogenetic studies suggest that among all receptor protein families involved in vertebrate olfaction, only the OR family predates vertebrates, whereas all other families (TAAR, ORA/V1R, OlfC/V2R, FPR) originated by vertebrate specific expansions. Taken together, this suggests that olfactory and vomeronasal cell types may be sister cell types that evolved from a common precursor within vertebrates and acquired their morphological (cilia vs. microvilli) and molecular differences (receptor and signal transduction proteins) relatively recently. Although we do not know, how this common precursor cell looked like, the co-occurrence of ciliary and microvillar features and TRPC and CNG receptors in some olfactory cells (see above) makes it tempting to speculate that this precursor may have carried both cilia and microvilli and expressed their associated signal transduction pathways (cilia: CNG, cyclic nucleotide mediated signaling; microvilli: TRPC channels, PLC mediated signaling). These may then have been differentially lost in many vomeronasal and olfactory cells, respectively. It must be emphasized here that this scenario implies that the ciliary and microvillous olfactory cells of vertebrates are unlikely to be homologous to ciliary and microvillous chemosensory cells of invertebrates, respectively, but rather represent a vertebrate-specific diversification of a single cell type.

We currently know very little about the transcription factors regulating differentiation of the neurosecretory cells derived from the olfactory placode, but recent evidence suggests that these are substantially different from those of sensory cells. The microvillous PLCβ2/NeuropeptideY-positive cells of the olfactory epithelium are characterized by persistent expression of Ascl3, although this transcription factor does not appear to be essential for their differentiation (Weng et al. 2016; Fletcher et al. 2017). In contrast, the transcription factor POU2f3 plays an essential role for differentiation of the microvillous cholinergic TRPM5-positive cells (Yamaguchi et al. 2014; Fletcher et al. 2017). The gene expression profile of these TRPM5 expressing cells in the olfactory epithelium closely resembles chemosensory cells in the airway epithelium (innervated by the trigeminal nerve); type II (sweet/bitter/umami) taste receptor cells of the tongue (innervated by the facial, glossopharyngeal and vagal nerves); and the endodermally derived tuft cells of the gut (Tizzano et al. 2010; O'Leary, Schneider, and Locksley 2019). All of these cells express a common battery of differentiation markers including T1R (sweet, umami)/T2R (bitter)-type taste receptors, TRPM5 and the enzyme choline acetyltransferase (ChAT), and require transcription factor POU2f3 for their differentiation suggesting that they represent a single neurosecretory cell type, the so-called tuft cell (O'Leary, Schneider, and Locksley 2019).

Although the enhancers controlling GnRH expression in differentiated cells are well studied and many upstream transcription factors have been identified (Lee, Lee, and Chow 2008), neither Ascl1, nor Neurog1, Neurog2, or NeuroD1 is expressed in precursors of mammalian GnRH1 expressing neurons and it is currently unknown, which transcription factors regulate their early specification or that of other migratory neurosecretory cells derived from the olfactory placode (Kramer and Wray 2000; Wray 2010; Forni and Wray 2015). While Islet1/2 expression is confined to the teleost GnRH3 neurons in the olfactory epithelium, the role of these transcription factors in specification of GnRH3 neurons remains to be experimentally confirmed (Aguillon et al. 2018).

In contrast to the sensory neurons and neurosecretory cells derived from the olfactory placode, its sustentacular cells continue to express progenitor-type transcription factors such as Pax6 and Sox2 and retain the capacity to divide, even though their mitotic activity is low (Davis and Reed 1996; Murray et al. 2003; Guo et al. 2010). Recent single-cell sequencing data suggest that in contrast to the neuronal and neurosecretory cells of the olfactory epithelium, sustentacular cells can arise directly from HBCs without passing through an intermediate GBC progenitor state (Fletcher et al. 2017).

The final cell types to be discussed here are two cell types of the respiratory epithelium, multiciliary cells, and goblet cells. The multiciliary cells have recently been recognized to share a CoRN with the multiciliary cells of the airway epithelium and other epithelia (Fig. 6.7D, E) comprising transcription factors GEMC1 and Multicilin (MCIDAS) in association with E2F4/5 as well as their downstream effectors FoxJ1 and Rfx2/3 (reviewed in Brooks and Wallingford 2014; Meunier and Azimzadeh 2016; Spassky and Meunier 2017). The mucus secreting goblet cell closely resembles the goblet cells of the digestive tract and appears to be specified like the latter by a combination of Atoh1 (in mammals) or Ascl1 (in zebrafish), SPDEF, and KLF4 (Yang et al. 2001; Gupta et al. 2011; Flasse et al. 2013; Roach et al. 2013; McCauley and Guasch 2015; Chen et al. 2009).

In summary, this survey of cell types derived from the olfactory placode shows that in spite of the staggering diversity of olfactory and vomeronasal receptor proteins and a striking morphological diversity of olfactory and vomeronasal receptor cells, most sensory cells derived from this placode are specified by the same or closely related CoRN involving Ascl1, probably Lhx2 and DMRT4/5, and possibly additional transcription factors. In contrast, the olfactory placode gives rise to multiple types of neurosecretory cells, each of which maybe specified by a radically different CoRN. One of these neurosecretory cell types, a cholinergic microvillar cell that requires POU2f3 for specification, is not uniquely derived from the olfactory placode but is also generated from other epithelia. Similarly, the multiciliary (specified by GEMC1, Multicilin, E2F4/5) and goblet cells (Atoh1, SPDEF, KLF4) of the olfactory respiratory epithelium also arise from other epithelia. Taken together this indicates that olfactory and vomeronasal receptor cells are sister cell types that originated from a common ancestral cell type in stem vertebrates. In contrast, other cell types derived from the olfactory placode probably have a different evolutionary history, a topic that will be explored further in the second volume (Schlosser 2021).

7 Differentiation of Photoreceptors

After reviewing the differentiation of neurons and sensory cells from individual placodes in the last chapter, I will sketch in the present chapter how photoreceptors and other cells type differentiate from the vertebrate retina. Photoreceptors share many of the general mechanisms of sensory and neuronal differentiation discussed in Chapter 5 with these placode-derived sensory cells and neurons. As explained in Chapter 1, the photoreceptors found in the retina of the vertebrate eye or in the pineal body do not develop from cranial placodes but from the neural tube. However, I will try to show in Volume 2 (Schlosser 2021) that vertebrate photoreceptors are evolutionarily closely related to other sensory cell types, such as mechanoreceptors, that are placode derived. To make this argument, I will briefly introduce vertebrate photoreceptors here and summarize how their specification and differentiation is transcriptionally regulated. I may be excused in drawing heavily on previous reviews in the following paragraphs since it is impossible to do justice to the vast literature on this topic in the short space available. I will not cover the early development of the retina here, which has been extensively reviewed elsewhere (Livesey and Cepko 2001; Hatakeyama and Kageyama 2004; Brzezinski and Reh 2015; Cepko 2015; Reh 2018).

7.1 PHOTORECEPTORS IN THE VERTEBRATE RETINA AND CNS

While cells in several vertebrate tissues (including skin and pigment cells) are sensitive to light, specialized photoreceptors are only found in structures derived from the neural tube (reviewed in Vigh et al. 2002; Ekström and Meissl 2003; Lamb, Collin, and Pugh 2007; Graw 2010; Masland 2012; Lamb 2013; Cepko 2014; Porter 2016). These include, most prominently, the neural retinae of the paired eyes, which develop as lateral outpocketings from the diencephalon (part of the forebrain) (Fig. 1.5B). In addition, there are photoreceptors in the pineal and parapineal organs which develop as unpaired dorsal outgrowths of the diencephalon. Finally, there is a special class of cerebrospinal fluid contacting-cells in the hypothalamus and septum of the diencephalon, which serve as deep-brain photoreceptors. Only the rod- and cone-type photoreceptors in the retina are involved in image formation and, thus, serve a visual function. All other types of photoreceptors (including other cell types in the retina) serve as detectors of general light levels (irradiance). These irradiance detectors play an important role in adjusting the circadian clock, thereby synchronizing daily rhythms of behavioral activity with environmental light levels (reviewed in Vigh et al. 2002; Ekström and Meissl 2003; Falcon et al. 2009; Davies, Hankins, and Foster 2010). In the following, I will provide an overview over retinal photoreceptors first, followed by a brief summary of other photoreceptors.

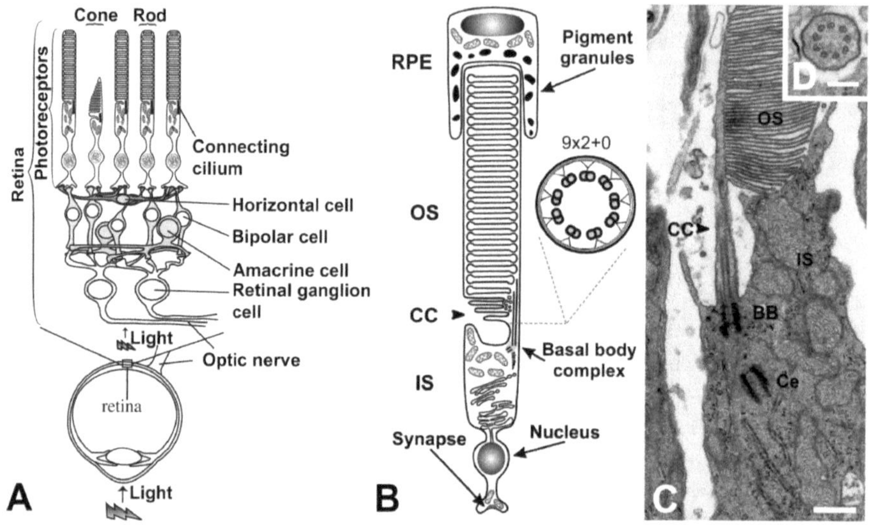

FIGURE 7.1 Photoreceptors in the mammalian retina. (**A**) Diagram of the mammalian retina with photoreceptors, bipolar cells, horizontal cells, amacrine cells, and ganglion cells. (**B**) Diagram of a vertebrate rod photoreceptor. The light-sensitive outer segment (OS) is a modified primary cilium, which is linked via the connecting cilium (CC) to the metabolically active inner segment (IS). RPE: retinal pigment epithelium. (**C**) Transmission electron micrograph of a connecting cilium, which connects OS and IS. The basal body complex comprising the basal body (BB) and its centriole (Ce) is evident (scale bar: 500 nm). (**D**) A cross-section of the CC reveals the characteristic ($9 \times 2 + 0$) microtubule array (scale bar: 200 nm). ([A] Reprinted with modification with permission from Shichida and Matsuyama, 2009; [B,C,D] Reprinted with permission from Falk et al. 2015.)

Rods and cones occupy the outermost layer of the retina (Fig. 7.1). Rods are much more light-sensitive than cones and can respond to single photons (reviewed in Lamb et al. 2007; Lamb 2009, 2013; Shichida and Matsuyama 2009). However, while there is only one type of rod responding to intermediate wavelengths, there are several types of cones (four types in ancestral gnathostomes) tuned to different ranges of wavelengths. Therefore, rods are used for vision in dim light (e.g. at night time) without allowing discrimination of different colors, whereas cones enable color vision in bright light (e.g. daytime). Both rods and cones use a modified cilium for light perception and are, therefore, known as ciliary photoreceptors (reviewed in Lamb et al. 2007; Lamb 2013; Falk et al. 2015) (Fig. 7.1B–D). The basal body of the cilium is located in the inner segment of the photoreceptors, which also contains mitochondria and other cell organelles. On the apical side of the inner segment arises a short connecting cilium with an axoneme composed of $9 \times 2 + 0$ microtubules (nine microtubule doublets in the periphery of the cilium; no microtubules in the center), which connects the basal body to the outer segment of rods and cones. The outer segments are modified distal cilia, which contain the membrane bound photopigments and proteins of the phototransduction cascade. These proteins are concentrated in a series of stacked discs, which form by infolding of the plasma membrane (Steinberg,

Fisher, and Anderson 1980; Volland et al. 2015; Schietroma et al. 2017). While these membrane folds remain connected to the surface in cones, they become separated from it in rods, forming membrane discs. On its basal side, the inner segment is attached to the main cell body containing the nucleus and forming ribbon synapses (similar to those in hair cells; see Chapter 6) with the adjacent bipolar cells.

The photopigments of rods and cones are G protein-coupled receptor (GPCR) proteins of the ciliary opsin (c-opsin) family (Feuda et al. 2012; Ramirez et al. 2016). These opsins (e.g. the rhodopsin of rods) form a covalent bond with retinal, a vitamin A derivative, which is able to undergo a conformational change from the kinked isomer 11-cis retinal to the straight form all-trans retinal in response to light (Fig. 7.2A, C, D). This change leads to a conformational change in the opsin which then triggers the activation of the G-protein transducin. The alpha-subunit $G\alpha_t$ of transducin, which belongs to the inhibitory $G\alpha_i$ family, then activates the enzyme phosphodiesterase (PDE), which hydrolyzes cGMP. The resulting decrease in intracellular cGMP concentration leads to the closure of cation channels of the cyclic nucleotide-gated (CNG) family and consequently to a hyperpolarization (inhibition) of the photoreceptor (Fig. 7.2A) (reviewed in Lamb et al. 2007; Lamb 2009, 2013; Shichida and Matsuyama 2009; Arshavsky and Burns 2012). Consequently, and somewhat counterintuitively, ciliary photoreceptors are inhibited by increasing light levels and activated when light levels decrease. Phosphorylation of the photoactivated opsin and binding of the protein arrestin subsequently leads to the deactivation of opsin (Fig. 7.2B). All trans-retinal is then released, reduced to vitamin A, and taken up by cells in the retinal pigmented epithelium, where 11-cis retinal is reconstituted in a light-independent mechanism and then transported back to the neural retina (reviewed in Lamb 2009, 2013). Hydrolysis of GTP to GDP then reconstitutes inactivated PDE, while elevation of cGMP levels by guanylylcyclases opens CNG channels again.

Comparative studies of the various cone and rod opsins have concluded that these ciliary opsins are derived from a common ancestral opsin, which first duplicated in the earliest vertebrates to give rise to a short-wave sensitive (SWS) and long-wave sensitive (LWS) opsin. Subsequently, the SWS opsin duplicated twice to give rise to SWS1, SWS2 and two rhodopsin (Rh)-like opsins, Rh1 (rhodopsin) and Rh2 (Okano et al. 1992; reviewed in Davies, Collin, and Hunt 2012; Lamb 2013; Musser and Arendt 2017). Some genes encoding proteins of the phototransduction cascade (e.g. the transducin alpha subunit) are located on the same chromosome near the *opsin* genes and were co-duplicated with them facilitating the co-adaptation of different members of the phototransduction cascade during cone diversification (Larhammar, Nordstrom, and Larsson 2009; Lagman et al. 2013). Rh1 opsin has become specialized for vision in dim light due to its ability to detect single photons when rods evolved out of cones (reviewed in Shichida and Matsuyama 2009; Lamb 2009, 2013; Davies et al. 2012). In lampreys, two types of photoreceptors (long and short) can be distinguished, both of which have cone-like morphology. However, their short photoreceptors express an opsin (RhA), which appears to be the ortholog of Rh1 in gnathostomes, and are capable of single-photon detection (Collin et al. 2003; Pisani et al. 2006; Collin 2009; Lamb 2013; Asteriti, Grillner, and Cangiano 2015; Morshedian and Fain 2015, 2017). This suggests that rods originated already in the last common ancestor of vertebrates.

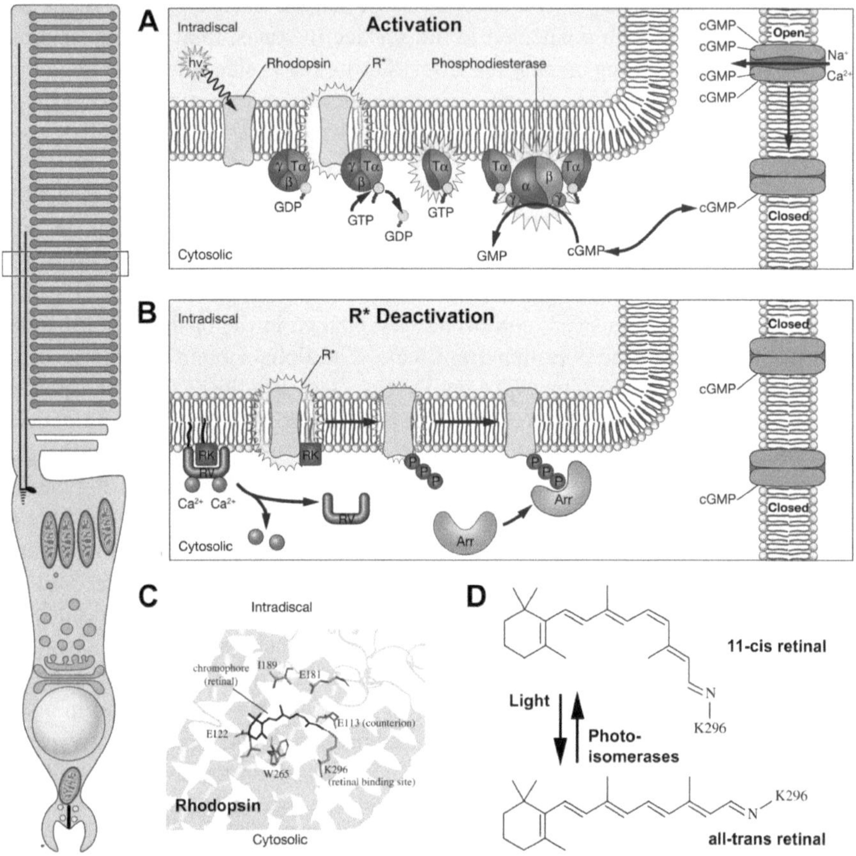

FIGURE 7.2 Phototransduction in a vertebrate ciliary photoreceptor (rod). Phototransduction takes place in the membraneous discs of the outer segment (boxed area). (**A**) Phototransduction begins when a photon (hν) is absorbed by the chromophore retinal of rhodopsin resulting in active rhodopsin (R*). R* then interacts with the heterotrimeric G-protein transducin and promotes the exchange of GTP for GDP on the α-subunit (Tα). Tα in turn disinhibits phosphodiesterase (PDE), thereby allowing the hydrolysis of cGMP and the closure of cyclic nucleotide gated (CNG) channels. This blocks the flow of Na^+ and $Ca2^+$ into the photoreceptor, thereby hyperpolarizing the membrane potential. (**B**) Deactivation of R* is required to quench the phototransduction cascade. R* deactivation is initiated by its phosphorylation by rhodopsin kinase (RK). RK is inhibited by recoverin (Rv), an inhibition that is released when $Ca2^+$ concentration falls after CNG channel closure. The R* catalytic activity is further quenched by the binding of visual arrestin (Arr) to phosphorylated R*. (**C**) Binding pocket of rhodopsin (shaded in grey) for the chromophore retinal. Some key amino acids including lysine 296 (K296), which covalently binds to retinal, are indicated. (**D**) Photoisomerization of retinal from 11-cis to all-trans retinal. Photoisomerases are able to reconstitute 11-cis-retinal. (Rod diagram and [A,B] reprinted with modification with permission from Chen and Sampath 2013; [C] Reprinted with modification with permission from Shichida and Matsuyama 2009.)

The five opsins that were present in ancestral vertebrates (LWS, SWS1, SWS2, Rh1, and Rh2) are maintained in some extant vertebrate groups (e.g. lampreys, lungfishes, lizards, and birds) but are further expanded in some vertebrate groups and lost in others (reviewed in Davies et al. 2012; Musser and Arendt 2017). For example, teleosts have expanded their opsin repertoire, while placental mammals have lost the SWS2 and Rh2 opsins and only retain the cone pigments LWS, SWS1, and the rod pigment Rh1. It has been proposed that mammals acquired their reduced, dichromatic color vision (with only two types of cones) during early mammalian evolution, when dinosaurs ruled the earth in broad daylight and mammals were evolving as predominantly nocturnal animals to escape competition (Walls 1942; Davies et al. 2012; Borges et al. 2018). LWS opsins later duplicated in Old World monkeys and diverged in their spectral sensitivity, endowing these secondarily diurnal animals with trichromatic color vision (Nathans, Thomas, and Hogness 1986; Davies et al. 2012).

In contrast to the ciliary photoreceptors of the vertebrate retina, many invertebrate photoreceptors (e.g. the photoreceptors of the insect eyes) are of the so-called rhabdomeric type, which have their photopigments concentrated in expanded microvilli rather than in modified cilia (see Chapter 4 in Volume 2; Schlosser 2021). The distinction between ciliary and rhabdomeric photoreceptors has first been introduced by Richard M. Eakin on purely morphological grounds (Eakin 1963, 1979). However, it has since become clear that ciliary and rhabdomeric photoreceptors also use opsins that belong to two different families (c-opsins vs. r-opsins), use different G-protein alpha subunits ($G\alpha_i$ or $G\alpha_0$ vs. $G\alpha_q$), rely on different signal transduction mechanisms (involving PDE vs. PLC), and depend on different ion channels (CNG vs. TRP) (Yokoyama 2000; Arendt and Wittbrodt 2001; Arendt 2003; Plachetzki, Fong, and Oakley 2010). Whereas ciliary photoreceptors hyperpolarize in response to light, rhabdomeric photoreceptors become depolarized. Ciliary and rhabdomeric photoreceptors also employ different mechanisms to reconstitute the photopigment. In contrast to c-opsins, the activated form of r-opsins (with all-trans retinal) can revert back to the inactivated form (with 11-cis retinal) by absorption of a second (lower energy) photon either by the activated opsin itself or by a distinct photoisomerase in a light-dependent process (Lamb 2009, 2013).

Although, Eakin initially proposed that ciliary photoreceptors are confined to cnidarians and deuterostomes and rhabdomeric photoreceptors to protostomes (Eakin 1963), there is now substantial evidence that both types of photoreceptors occur in most bilaterians (see Chapter 4 in Volume 2; Schlosser 2021). In vertebrates, melanopsin, a member of the r-opsin family has been shown to be expressed in a subset of retinal ganglion cells (RGCs) that are intrinsically photosensitive and serve as irradiance detectors (Hattar et al. 2002; Arendt 2003; Fu et al. 2005). Phototransduction in these cells appears to rely on $G\alpha_q$, PLC, and TRP channels (Panda et al. 2005; Isoldi et al. 2005; Contin, Verra, and Guido 2006; Graham et al. 2008) suggesting that photosensitive RGCs are modified rhabdomeric photoreceptors. Based on size, physiology, localization, and connectivity, several different subtypes of intrinsically photosensitive RGCs plus many other RGCs have been identified (Masland 2012; Sanes and Masland 2015). Other types of retinal cells (Fig. 7.1A) are probably evolutionarily derived from either retinal ganglion cells (amacrine and horizontal cells) or rod/cone-type photoreceptors (bipolar cells) as will be discussed in the next section.

Although the co-expression of r-opsin with many "rhabdomeric" signal transduction members supports the identification of RGCs with rhabdomeric photoreceptors, some of these melanopsin expressing cells also co-express members of the c-opsin family (in particular va-opsin) (Davies et al. 2010). Conversely, melanopsin in zebrafish has been shown to be expressed in all retinal cells including the presumably "ciliary"-type rods and cones (Davies et al. 2011). This suggests that r-opsins and c-opsins were occasionally coopted into photoreceptors of the other type (ciliary and rhabdomeric, respectively) similar to the cooption of TRP and CNG channels observed in some olfactory cell types as discussed in Chapter 6.

In addition to the retina, both c-opsin and r-opsin expressing photoreceptors are also found as part of the pineal and parapineal organs or as deep-brain photoreceptors in other parts of the brain (reviewed in Vigh et al. 2002; Ekström and Meissl 2003; Falcon et al. 2009). The pineal organ, which is present in most vertebrates, is usually photoreceptive (and even gives rise to a third, so-called frontal eye in frogs), but has lost its light sensitivity in mammals. The parapineal organ has been lost in many vertebrate groups, but in some groups (e.g. lizards) it forms a third, so-called parietal eye on the dorsal side of the head. Most deep-brain photoreceptors or photoreceptors of the pineal and parapineal organs are of the ciliary type, express c-opsins, and synapse via ribbon synapses onto pineal projection neurons. When stimulated, many pineal photoreceptors also release melatonin and other neurohormones which serve to synchronize daily rhythms of physiological and behavioral activity. The pinealocytes of mammals are thought to have evolved from these photoreceptors, which have lost their photosensitivity (relying on input from the retina instead), and their neural synapses and act predominantly as melatonin secreting neurosecretory cells.

While most deep-brain and pineal photoreceptors appear to be ciliary photoreceptors, the r-opsin melanopsin was found to be expressed in some of these pineal and deep-brain receptors in teleosts and birds (Bailey and Cassone 2005; Chaurasia et al. 2005; Holthues et al. 2005; Matos-Cruz et al. 2011; Fernandes et al. 2012; Eilertsen et al. 2014). However, it remains to be clarified, whether these melanopsin expressing cells represent modified ciliary photoreceptors that have recruited melanopsin as an additional photopigment, or rather are more closely related to rhabdomeric photoreceptors.

7.2 TRANSCRIPTION FACTORS INVOLVED IN SPECIFYING THE IDENTITY OF VERTEBRATE PHOTORECEPTORS

Similar to other neuronal or sensory cells, proneural transcription factors of the basic helix-loop-helix family are essential for initiating neuronal differentiation in all retinal cell types (see Chapter 5). While Atoh7 (also known as Math5) plays this role for RGCs, Ascl1 is required for the differentiation of all other retinal cell types including photoreceptors (reviewed in Swaroop, Kim, and Forrest 2010; Brzezinski and Reh 2015; Cepko 2015). Together with another bHLH transcription factor, Neurog1, Ascl1 is also essential for the development of photoreceptors and neurons in the pineal gland (Cau and Wilson 2003). Another bHLH transcription factor, Olig2, is expressed in all retinal cell types except retinal ganglion cells and Müller glia

(Hafler et al. 2012). This suggests that it may have some role in lineage specification, which is however poorly understood.

bHLH transcription factors work together with many other transcription factors to generate the various cell types of the retina in a stereotypical temporal sequence and ensure their correct positioning to retinal layers and wiring with other cell types (reviewed in Livesey and Cepko 2001; Agathocleous and Harris 2009). Many of these transcription factors have multiple functions in retinal patterning and cell type specification. The Pax6 and Rx (or Rax) transcription factors, in particular, initially regulate the formation and proliferation of progenitors for all retinal cell types, but later are maintained only in subsets of differentiating retinal cell types and down-regulated in photoreceptors (Belecky-Adams et al. 1997; Perron et al. 1998; Ashery-Padan and Gruss 2001; Bailey et al. 2004; Muranishi, Terada, and Furukawa 2012; Shaham et al. 2012). Pax6 is subsequently required for specification of RGCs and horizontal cells (Marquardt et al. 2001; Oron-Karni et al. 2008). In contrast, Rx/Rax family members are required for the specification of photoreceptors and bipolar cells either upstream of or in parallel to Otx2 and directly regulate transcription of *c-opsin* genes (Kimura et al. 2000; Chen and Cepko 2002; Wang et al. 2004; Pan et al. 2006; Wu, Perron, and Hollemann 2009; Nelson, Park, and Stenkamp 2009; Pan et al. 2010; Muranishi et al. 2011). While Rx/Rax factors are also expressed in the pineal body, they appear not to be required for specification of neurosecretory pinealocytes in mammals, but instead modulate circadian gene expression (Casarosa et al. 1997; Rohde et al. 2011, 2017).

7.2.1 Transcription Factors Specifying Identities of Ciliary Photoreceptors (rods and cones)

Members of the Otx family of homeodomain transcription factors play particularly important roles for the specification of rod/cone-type photoreceptors in the vertebrate retina (reviewed in Hennig, Peng, and Chen 2008; Swaroop et al. 2010; Brzezinski and Reh 2015; Cepko 2015; Viets, Eldred, and Johnston 2016). The vertebrate Otx family evolved from a single Otd transcription factor in invertebrates and diversified into three major subfamilies (Otx1, Otx2, and Otx5) in gnathostomes (Plouhinec et al. 2003; Ranade et al. 2008). Crx in mammals evolved as a divergent member of the Otx5 family, and is not closely related to teleost "Crx", which originated by a teleost-specific duplication of Otx5 (Plouhinec et al. 2003). While Otx2 has an earlier broader role as an anterior patterning factor in all germ layers (see Chapter 4), it later is specifically required for the differentiation of a few specific cell types. In all gnathostomes, Otx2 serves as upstream regulator of both photoreceptor and bipolar cell development (Bovolenta et al. 1997; Martinez-Morales et al. 2001; Nishida et al. 2003; Viczian et al. 2003; Omori et al. 2011; Emerson et al. 2013). Mutual antagonism between two direct Otx2 targets, Vsx2 and PRDM, appear then to be crucial for the decision between these two fates with Vsx2 promoting bipolar fate and repressing photoreceptors and PRDM promoting photoreceptors and repressing bipolar fate (reviewed in Brzezinski and Reh 2015). Because bipolar cells share their developmental lineage and Otx2 dependence with rods and cones, express many of the same effector proteins and also form ribbon synapses, it has been proposed that bipolar cells evolved from the same ciliary photoreceptor cell type in ancestral vertebrates than rods and cones, but subsequently lost their photosensitivity (Arendt 2003; Lamb et al. 2007; Lamb 2009).

Downstream of Otx2, Crx or other Otx5-related transcription factors are core regulators of terminal differentiation in all rods and cones (Fig. 7.3) (Chen et al. 1997; Furukawa, Morrow, and Cepko 1997; Furukawa et al. 1999; Sauka-Spengler et al. 2001; Viczian et al. 2003; Shen and Raymond 2004; Qian et al. 2005; Hennig et al. 2008).

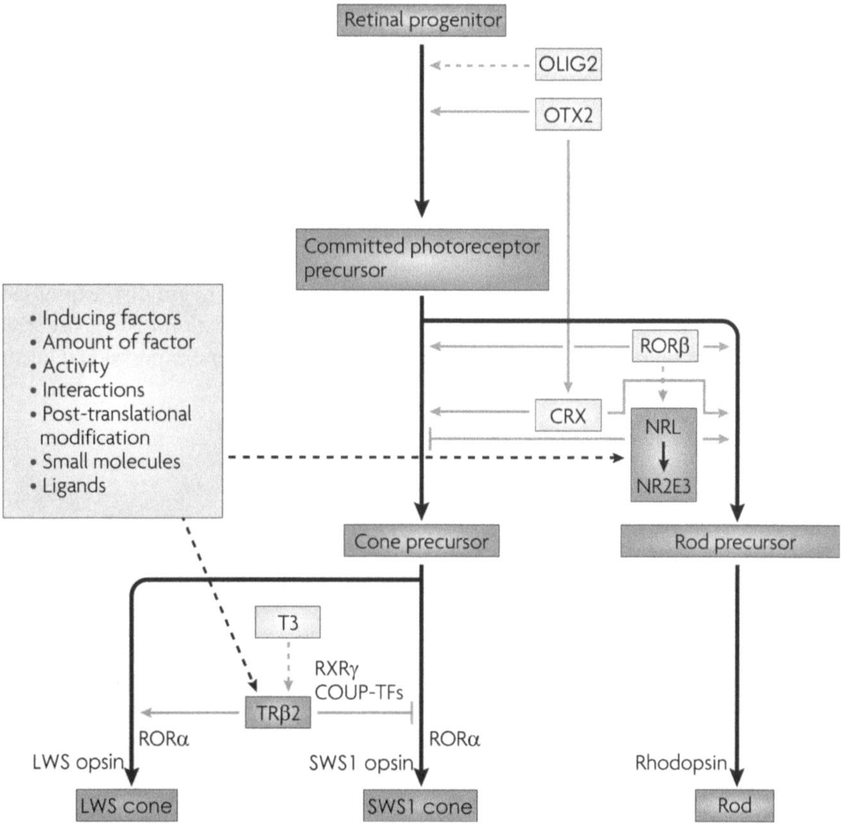

FIGURE 7.3 Transcription factors involved in the specification of photoreceptors. Progenitors expressing Otx2 and Olig2 give rise to all retinal cell types except for retinal ganglion cells and Müller glia cells. Crx (which is activated by Otx2) promotes a photoreceptor fate. Expression of NRL and its downstream target photoreceptor-specific nuclear receptor Nr2e3 determine whether a precursor becomes a rod or a cone, and the thyroid hormone receptor β2 (TRβ2) and its ligand triiodothyronine (T3) promote LWS cone versus SWS1 cone identity. Whether a photoreceptor expresses Nrl or TRβ2 depends on a range of factors (largely unknown) as indicated. Other factors involved in fate decision including retinoid X receptor γ (RXRγ), retinoic acid receptors RORβ and RORα, and COUP transcription factors (TF) are indicated. See text for details. (Reprinted with modification with permission from Swaroop et al. 2010.)

In Crx mutant mice, the outer segments of rods and cones do not develop and photoreceptors degenerate (Furukawa et al. 1997, 1999). Ectopic expression of Crx, in turn, can transform iris-derived cells into photoreceptors, probably due to the ability of Crx to directly activate many photoreceptor-specific genes (Blackshaw et al. 2001; Haruta et al. 2001; Akagi et al. 2004; Peng and Chen 2005; Hennig et al. 2008). Together with Otx2, Otx5/Crx is also required for the specification of photoreceptors and proper circadian gene expression in the pineal body (Furukawa et al. 1999; Gamse et al. 2002; Nishida et al. 2003; Rovsing et al. 2011). Taken together, this suggests that Crx is at the center of the core regulatory network (CoRN) for the specification of ciliary photoreceptors in retina and pineal body (Fig. 7.3).

Together with two other transcription factors, neural retina leucine zipper protein (Nrl) and nuclear receptor subfamily 2 group E member 3 (Nr2e3), which are themselves Crx targets, Crx promotes specification of rods, while inhibiting cone formation (Akhmedov et al. 2000; Haider, Naggert, and Nishina 2001; Mears et al. 2001; Corbo and Cepko 2005; Chen, Rattner, and Nathans 2005; Peng and Chen 2005; Cheng et al. 2006; Oh et al. 2007; Hennig et al. 2008). Crx forms a protein complex and cooperates with Nr2e3 and Nrl to promote expression of rod-specific genes but repress expression of cone-specific genes (Chen et al. 1997; Mitton et al. 2000; Cheng et al. 2004; Chen, Rattner, and Nathans 2005; Peng and Chen 2005, 2007). In the absence of Nrl and Nr2e3, Crx instead synergizes with the retinoic acid receptor-related orphan receptor Rorβ2 to promote cone differentiation (Fig. 7.3) (Srinivas et al. 2006).

Interactions with additional transcription factors are probably involved in specifying the different types of cones found in vertebrates (Fig. 7.3). For example, the thyroid hormone receptor TRβ and the retinoid X receptor RXRγ, which are expressed at elevated levels in LWS cones, were shown to promote LWS cones and represses SWS1 cones (Ng et al. 2001; Yanagi, Takezawa, and Kato 2002; Roberts et al. 2006; Enright et al. 2015; Peng et al. 2019). In contrast, Tbx2, which is expressed at high level in SWS1 cones in teleosts, birds and mammals, has been shown to be essential for SWS1 cone specification in zebrafish (Alvarez-Delfin et al. 2009; Enright et al. 2015; Peng et al. 2019). Little is currently known about the transcription factors involved in the specification of SWS2 and Rh2 cones (reviewed in Musser and Arendt 2017).

Recent RNA-sequencing of fluorescently labeled retinal lineages or single cells in birds and mammals confirm these transcriptional codes for rod (Nrl, Nr2e3), LWS (TRβ, RXRγ), and SWS1 (Tbx2) cones and suggest that these are well-individualized photoreceptor subtypes (Siegert et al. 2012; Macosko et al. 2015; Shekhar et al. 2016; Peng et al. 2019; Collin et al. 2019). However, the situation is different for the M- and L-cones of Old World monkeys. These respond to light of medium or long wavelength, respectively, and together with the SWS1 cone responsive to short wavelengths endow these primates with trichromatic vision. A recent single-cell RNA-sequencing study of the retina of the macaque, an Old World monkey, revealed that expression of one or the other of the Old World monkey-specific duplicate of the *LWS-opsin* gene was the only molecular difference between M- and L-cones (Peng et al. 2019). This indicates that M- and L-cones are specified by the same CoRN and, thus, do not represent two different cell types. Instead, LWS cones in Old World monkeys have probably evolved a molecular mechanism ensuring a stochastic choice of expression between the two duplicate *LWS opsin* genes (Wang et al. 1999), similar to the OR choice in olfactory receptor neurons that I discussed above.

7.2.2 Transcription Factors Specifying Identity of Retinal Ganglion Cells

Whereas the CoRN underlying rod and cone differentiation is relatively well understood, less is known about specification of retinal ganglion cells. Together with Pax6, Atoh7 (also known as Math5 in the mouse) is required for RGC specification and can promote the ectopic formation of RGCs after overexpression (Brown et al. 1998, 2001; Kay et al. 2001; Liu, Mo, and Xiang 2001; Wang et al. 2001). However, not all Atoh7 expressing cells differentiate into RGCs indicating that additional transcription factors are required for RGC specification (Yang et al. 2003; Poggi et al. 2005). Indeed, two transcription factors, POU4f2 (= Brn3b) and Islet1, have been shown to be essential downstream of Atoh7 for RGC specification, and POU4f2 expression is restricted to RGCs in the retina (Xiang et al. 1995; Gan et al. 1999; Wang et al. 2000; Mu et al. 2004, 2008; Pan et al. 2008). Another target gene of Atoh7, Gfi1, is also confined to RGCs in the retina but its function for RGC specification or differentiation has not been analyzed (Yang et al. 2003; Wallis et al. 2003; Dufourcq et al. 2004).

The CoRN for specification of RGCs resembles those of rhabdomeric photoreceptors, which also rely on Atonal-, POU4-, and Gfi1-related transcription factors, suggesting that RGCs are modified rhabdomeric photoreceptors (Arendt 2003; Lamb et al. 2007) (see Chapter 4 in Volume 2; Schlosser 2021). The CoRN of RGCs also shows striking similarity to the CoRNs regulating specification of hair cells (Atoh1, POU4f3, Gfi1) and somatosensory neurons (POU4f1, Islet1, Tlx1/3), including some of the same transcription factors (Islet1 and possibly Gfi1) or paralogs arising from vertebrate-specific gene duplications (Atoh7, POU4f2). These similarities point to evolutionary relationships between these various sensory cells and neurons that will be discussed in Volume 2 (Schlosser 2021). Additional transcription factors that are specifically expressed in RGCs and may potentially contribute to RGC specification have been identified in recent RNA-seq and single-cell RNA-Seq studies and include other POU4 (POU4f1, POU4f3) and Islet (Islet2) paralogs, Sox4, Sox11/12, and EBF1 (Siegert et al. 2012; Langer et al. 2018; Rheaume et al. 2018). Other transcription factors that are expressed only in subsets of RGCs allow to distinguish up to 40 RGC subtypes in mice based on their combinatorial expression. However, apart from the melanopsin expressing intrinsically photosensitive RGCs, these do not appear to be evolutionarily conserved between the mouse and macaque retina (Rheaume et al. 2018; Peng et al. 2019).

7.3 OTHER RETINAL CELL TYPES AND THEIR POTENTIAL EVOLUTIONARY HISTORY

Based on the shared dependence on maintained Pax6 expression and some other similarities, it has been proposed that the amacrine and horizontal cells of the retina may be evolutionarily derived, like RGCs, from a rhabdomeric photoreceptor in the ancestral vertebrate (Arendt 2003; Lamb et al. 2007; Lamb 2009). Although this proposal remains largely speculative at present, a recent single-cell RNA-seq study showed that the transcriptomes of amacrine and horizontal cells cluster with RGCs as would be predicted under this proposal (Macosko et al. 2015). Taken together with the proposed derivation of bipolar cells from ciliary photoreceptors (see above), this supports a scenario, according to which the vertebrate retina evolved from a simpler structure composed of only two different cell types, rhabdomeric and ciliary photoreceptors with ciliary photoreceptors contacting rhabdomeric photoreceptors, which provide the connection to the brain (Lamb 2009). Photoreceptive organs that maintain this simple structure can still be found in the bilayered retinae of hagfishes and larval lampreys, in which ciliary photoreceptors make direct synaptic contacts with retinal ganglion cells and in some pineal or parapineal organs (Holmberg 1971; Fernholm and Holmberg 1975; Holmberg, Ohman, and Dreyfert 1977; De Miguel, Rodicio, and Anadón 1989; Rubinson and Cain 1989; Locket and Jørgensen 1998; Suzuki and Grillner 2018).

Lamb has proposed that such a bilayered retina may have evolved from a light sensitive organ in the brain composed of only rhabdomeric photoreceptors, when ancestral chordates invaded the deeper ocean (Lamb 2009). Since rhabdomeric photoreceptors depend on light to reconstitute the ground state of opsin with its attached chromophore 11-cis-retinal, they would have become less efficient in environments of low light levels. In contrast, reconstitution of the ground state is not light dependent in ciliary photoreceptors (Fig. 7.2B). In low-light environments, it may, therefore have become advantageous for ciliary photoreceptors located nearby the ancient light sensitive organ to form synaptic contacts with its rhabdomeric photoreceptors, thereby substituting for their role in photoreception, while taking over their connections to the brain's visual centers. Although largely speculative, this is an interesting scenario for the evolution of the vertebrate retina from a situation, where ciliary and rhabdomeric photoreceptors co-exist as found in amphioxus. I will, therefore, discuss this scenario in more detail in Chapter 4 of Volume 2 (Schlosser 2021) after introducing the photoreceptors of other chordates.

In summary, this survey of vertebrate photoreceptors and related cell types suggests that they fall into two broad classes suggesting an evolutionary origin from two distinct types of photoreceptors in the vertebrate ancestor. The rods and cones of the retina as well as most deep-brain photoreceptors or photoreceptors in the pineal and parapineal bodies belong to the lineage of ciliary photoreceptors, which rely on c-opsin, $G\alpha_t$ or $G\alpha_i$, PDE and CNG channels for phototransduction and depend on a CoRN involving persistent Rx/Rax, Otx2 and Otx5/Crx. Bipolar cells may be evolutionarily derived from such ciliary photoreceptors, but have lost photosensitivity. Intrinsically photosensitive RGCs, on the other hand, are modified rhabdomeric photoreceptors, which rely on r-opsin (melanopsin), $G\alpha_q$, PLC and TRP channels and depend on a CoRN involving persistent Pax6, Atoh7, POU4f2, and Islet1 expression. Other RGCs, amacrine and horizontal cells may also be evolutionarily derived from rhabdomeric photoreceptors, but have lost photosensitivity.

8 Differentiation of Cell Types from Non-Neurogenic Placodes

After reviewing the differentiation of neurons and sensory cells from neurogenic placodes in Chapter 6 and of photoreceptors in the last chapter, I will turn in this chapter to the cell types derived from the lens and adenohypophyseal placodes. These two placodes are the only placodes in vertebrates that do not give rise to any neurons or sensory cells and are, thus, known as non-neurogenic placodes. Like in the last two chapters, I will first briefly introduce the cell types originating from these two placodes and will then review how their specification and differentiation is transcriptionally regulated.

8.1 LENS PLACODE

8.1.1 Differentiation of Lens Fiber Cells

The lens placode gives rise to a single differentiated cell type, the elongated lens fiber cell, while progenitors of these fiber cells persist as lens epithelial cells on its exterior side (facing the cornea) and contribute to lens growth (reviewed in Cvekl and Ashery-Padan 2014; Cvekl and Zhang 2017). After invagination of the lens placode and formation of the lens vesicles, the fibers on the internal face of the lens vesicle (facing the retina) first elongate until they reach the base of the lens epithelial cells, completely filling up the cavity of the vesicle (Fig. 8.1A–F). These are the so-called primary lens fibers. The epithelial cells continue to divide but elongate and differentiate into secondary lens fiber cells when they reach the equator of the lens (Fig. 8.1G). Differentiation of both primary and sensory lens fibers begins with elongation of the cells employing lens-specific intermediary filament proteins (Bfsp1/Bfsp2) (Rao and Maddala 2006; Fudge et al. 2011). Lens fiber cells then transcribe large amounts of crystallins, water–soluble proteins that render the lens transparent at high concentrations. They also express various gap junction connexins and membrane bound transport proteins to balance water, ion, and nutrient levels across the lens (Mathias, White, and Gong 2010). Finally, lens fiber cells develop an organelle free zone by losing their nuclei and other organelles (mitochondria, endoplasmatic reticulum, Golgi apparatus) in a process resembling the early stages of apoptosis (Bassnett and Mataic 1997; Bassnett 2009; Wride 2011). As a consequence, lens fiber cells are freed from light scattering particles and have homogeneous refractive properties.

The crystallins are a group of diverse proteins that have similar refractive properties at high concentrations, but belong to unrelated protein families (reviewed in

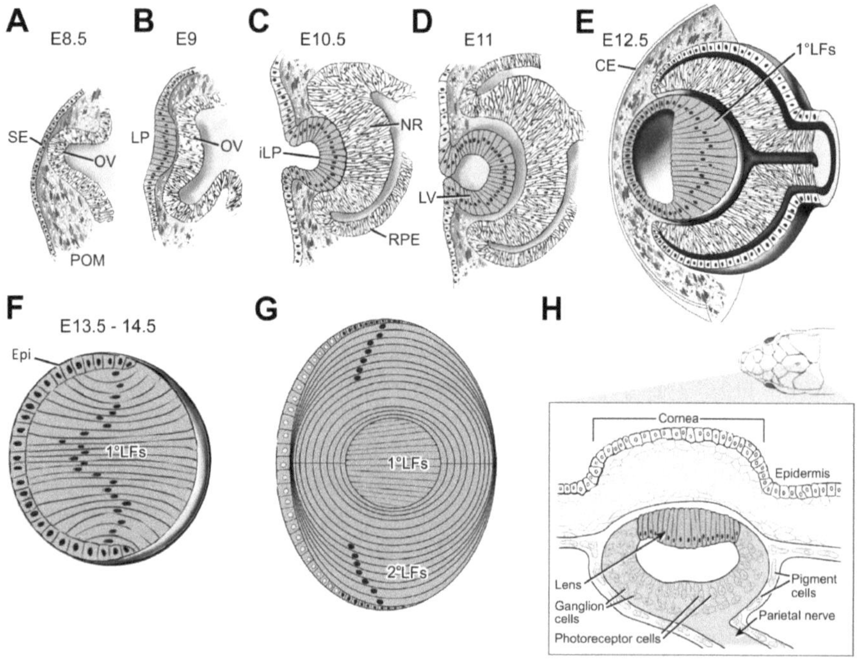

FIGURE 8.1 Lens development and lens fiber differentiation. (**A–F**) Development of the mouse lens from embryonic day (E) 8.5 to E14.5. After thickening of the lens placode (E9) and invagination of the lens vesicle (E10.5), primary lens fibers begin to differentiate at E12.5. Primary lens fiber elongation is completed and secondary lens fibers begin to form at E13.5. (**G**) Mature mouse lens. Note that the initial compartment formed by the primary lens, now forms the central lens nucleus devoid of intracellular organelles. (**H**) The parietal eye of lizards. The lens is formed by the outward-facing layer of this eye, which develops as a dorsal outpocketing of the diencephalon. The lens is composed of elongated nucleated cells and is covered by a transparent layer resembling the cornea. CE, corneal epithelium; Epi, lens epithelium; iLP, invaginating lens placode; LC, lens capsule; 1°LFs, primary lens fibers; 2°LFs, secondary lens fibers; LP, lens placode; NR, neuroretina; OV, optic vesicle; POM, periocular mesenchyme; PLE, prospective lens ectoderm; RPE, retinal pigmented epithelium; SE, surface ectoderm. Reprinted with permission from Cvekl and Zhang 2017. ([A–G] adapted from Griep and Zhang 2004 and Cvekl and Ashery-Padan 2014; [H] based on Solessio and Engbretson 1993.)

Cvekl and Piatigorsky 1996; Piatigorsky 1998). Many crystallins are expressed at lower concentrations elsewhere in the body, where they perform completely different functions. This has been called "gene sharing" (Piatigorsky et al. 1988) and it suggests that proteins from different families have independently acquired a secondary role as crystallins in the lens because they make the lens transparent at high concentrations. For example, the α-crystallins, which are encoded by the *αA* and *αB* genes, belong to the family of small heat shock proteins and still serve as stress-inducible proteins (αB) or chaperones, which prevent the misfolding and aggregation of proteins (αA), in several other tissues (Klemenz et al. 1991; Horwitz 1992;

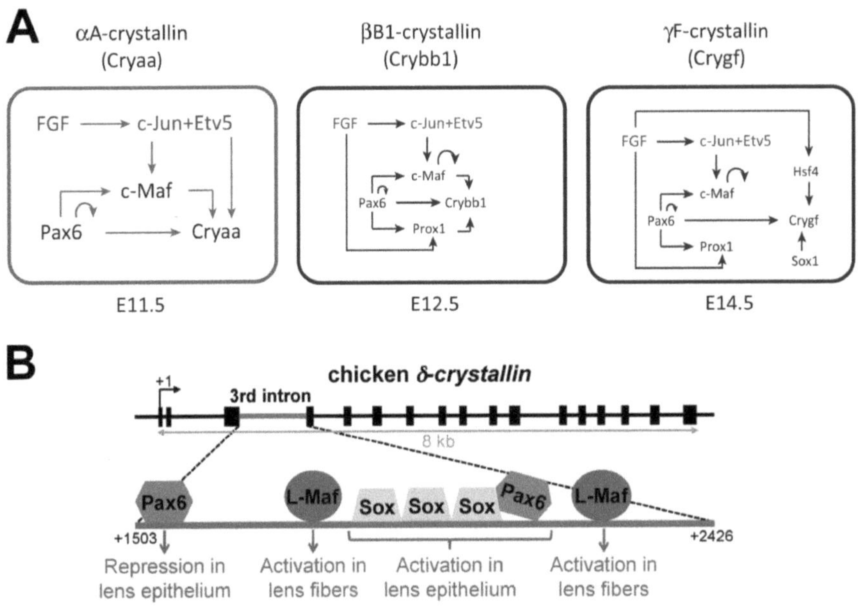

FIGURE 8.2 Transcriptional regulation of crystallin expression in the lens. (**A**) Gene regulatory networks (GRNs) underlying the regulation of various crystallins during lens cell differentiation at mouse embryonic day (E) E11.5, E12.5, and E14.5. The core regulatory network consists of Pax6, c-Maf and the crystallins. Pax6 activates both *c-Maf* and their *crystallin* target genes in a so-called feed-forward loop. Pax6 also positively autoregulates its own expression. Additional signaling molecules (grey) and transcription factors (black) contribute to transcriptional activation of the different crystallins. (**B**) Cis-regulation of the chicken δ-*crystallin* gene. The lens enhancer of this gene is located in the third intron. This enhancer is activated in lens epithelial cells by Pax6 binding conjointly with Sox2/3 to the central binding sites. L-Maf-binding to the flanking sites strongly activate this enhancer in lens fiber cells, whereas Pax6-binding on the 5′-end (to the left) attenuates enhancer activity. ([A] Reprinted with permission from Cvekl and Zhang 2017; [B] Reprinted with permission from Ogino et al. 2012.)

Aoyama et al. 1993; de Jong, Leunissen, and Voorter 1993; Sax and Piatigorsky 1994). The δ- and ε-crystallins of birds, in contrast, correspond to arginine succinate lyase and lactate dehydrogenase, respectively, which maintain a metabolic enzymatic function in other tissues (Wistow and Piatigorsky 1987; Wistow, Mulders, and de Jong 1987). Lens-specific enhancers of genes encoding the various crystallins usually have multiple binding sites for Pax6 or other transcription factors expressed in the developing lens (Fig. 8.2) (reviewed in Cvekl and Piatigorsky 1996). This suggests that acquisition of such binding sites allowed these proteins to acquire a novel expression domain and function during evolution of the vertebrate lens. One group of crystallins comprising the α- and βγ-crystallins are found ubiquitously in vertebrates indicating that they acquired their function as crystallins in the ancestral vertebrate lineage. The remaining crystallins are taxon-specific and are found only in subgroups of vertebrates, for example δ-crystallins in reptiles and birds and ε-crystallins in crocodiles and some birds, suggesting that they acquired their new function as crystallins only in the ancestors of these taxa.

In spite of their distinct function and morphology, lens fiber cells share a number of characteristics with neurons. They employ mechanisms for polarized intracellular vesicle transport similar to neurons, form dendrite-like protrusions and express a large number of neuronal genes including genes encoding synaptic vesicle proteins and neurotransmitter receptors for glutamate and GABA (reviewed in Frederikse, Kasinathan, and Kleiman 2012; Frederikse and Kasinathan 2017). Recent RNA-seq data in mice indicate that lens fiber cells also express various proteins characteristic for ciliary photoreceptors at relatively high levels including the opsins of rods (rhodopsin) and LWS cones (Khan et al. 2015). Conversely, photoreceptors are known to express α-, β-, and γ-crystallins, which are specifically associated with the membranes of their outer segments and may play a role in the renewal of outer segment membranes (Deretic et al. 1994; Xi, Bai, and Andley 2003; Andley 2007; Organisciak et al. 2011). More generally, they have been shown to act as chaperones with a neuroprotective effect on photoreceptors in response to potentially damaging strong light (Rao et al. 2008, 2012; Organisciak et al. 2011).

8.1.2 Transcription Factors Involved in Specification of Lens Fiber Cell Identity

Whereas Six1, Six4, and Eya are expressed in the part of the pre-placodal ectoderm destined to become lens, they are subsequently downregulated in the lens placode and are not required for lens fiber specification (Pieper et al. 2011; Ogino et al. 2012). In contrast, several other transcription factors including Pax6 and FoxE3 that are initially expressed broadly in the extended anterior placodal area (see Chapter 4) ultimately become confined to the lens placode and subsequently are specifically required for lens development (reviewed in Ogino et al. 2012; Cvekl and Ashery-Padan 2014; Cvekl and Zhang 2017). FoxE3 together with Pitx3 continues to be expressed in lens epithelial cell, maintaining their progenitor status, and repressing differentiation (Grimm et al. 1998; Kenyon, Moody, and Jamrich 1999; Blixt et al. 2000; Brownell, Dirksen, and Jamrich 2000; Ho et al. 2009; Medina-Martinez, Shah, and Jamrich 2009). Pax6, in contrast, together with its protein binding partners from the SoxB1 family activates other transcription factors, including Maf proteins (c-Maf, MafA) and Prox1 (Fig. 8.2A) (Reza, Ogino, and Yasuda 2002; Xie and Cvekl 2011; Sun et al. 2015). These transcription factors synergize with Pax6 itself and HSF4 in directly activating a battery of lens terminal differentiation genes including the *crystallin* genes, thus forming a core regulatory network (CoRN) for specifying the identity of lens fiber cells (Fig. 8.2A, B) (Kawauchi et al. 1999; Wigle et al. 1999; Ring et al. 2000; Reza et al. 2002; Cui et al. 2004; Fujimoto et al. 2004; reviewed in Ogino et al. 2012; Cvekl and Ashery-Padan 2014; Cvekl and Zhang 2017).

The large Maf proteins, members of the basic leucine zipper (bZIP) family of transcription factors, play a particularly important role for lens fiber specification (reviewed in Reza and Yasuda 2004; Reza, Urano et al. 2007). Due to the two rounds of genome duplication in ancestral vertebrates, four paralogs are found in vertebrates

– MafA (or L-Maf), MafB, c-Maf, and Nrl – compared to only one in invertebrates (Coolen et al. 2005). Misexpression of MafA, MafB, or c-Maf is able to activate expression of crystallins and multiple other lens differentiation markers in the retina and non-neural ectoderm of chick and frog embryos, but falls short of converting ectodermal cells into fully differentiated lens fiber cells (Ogino and Yasuda 1998; Ishibashi and Yasuda 2001; Reza, Nishi et al. 2007; Reza, Urano et al. 2007). This suggests that large Maf transcription factors are important members of the lens fiber CoRN but need to interact with additional transcription factors (e.g. Prox1, HSF4, and Pax6) for full lens fiber specification.

Various combinations of large Maf proteins are expressed in the developing lens of different vertebrate species with at least one of these being essential for lens differentiation. This indicates that large Maf proteins have partly redundant functions with different paralogs adopting central roles for lens specification in different vertebrates. For example, loss of function studies have shown that c-Maf is required for lens differentiation in the mouse (Kawauchi et al. 1999; Ring et al. 2000), while MafA is indispensable in the chick embryo (Reza et al. 2002). The fourth member of the large Maf proteins, Nrl plays a decisive role in the specification of rod photoreceptors as discussed in Chapter 7. In addition, it is also expressed in the lens of zebrafish and *Xenopus* together with other large Maf proteins; it remains yet to be determined, which of these are most important for lens specification in these species (Ishibashi and Yasuda 2001; Coolen et al. 2005).

In summary, the four large Mafs arose by vertebrate-specific gene duplication events, appear to maintain partly redundant functions during lens development and play central roles in the CoRN specifying cell identity of lens fiber cells (MafA, c-Maf, and possibly MafB and Nrl as well) and a subset of ciliary photoreceptors in the retina (Nrl). Interestingly, other transcription factors with essential roles in specifying retinal photoreceptors such as Otx5/Crx are also expressed in the lens although no function for lens development has yet been identified (Sauka-Spengler et al. 2001; Khan et al. 2015). Taken together with the many neuronal genes expressed in lens fiber cells (including rhodopsin and LWS opsin) and the role of crystallins as chaperones in photoreceptor cells, discussed above, this invites speculation that lens fiber cells may have evolved as a highly specialized neuronal cell type, possibly from ciliary photoreceptors.

One possible scenario would be that photoreceptors initially expressed crystallins for their neuroprotective effect. Accumulation of high levels of crystallins may then have been selected for in a subset of photoreceptors located in the outward-facing portion of the retina because it rendered them transparent and increased the amount of light reaching into the depth of the eyecup. These cells may have become genetically distinct from other photoreceptors by expression of particular Maf paralogs (MafA, c-Maf) and thus were able to evolve along an independent evolutionary trajectory. The new refractive properties of these cells may, as a side effect, have led to some sharpening of light projection onto the deep retina. This may have initiated selection for better image forming properties resulting in the evolution of lens fiber cells which completely lost their original function as photoreceptors. Initially such lens fiber cells would have formed the outward-facing portion of the retina as is still the case for the parietal eyes of some lizards today (Fig. 8.1H) (McDevitt 1972).

Finally, the CoRN for lens fiber formation may have been activated in the non-neural ectoderm adjacent to the eye cup, which already co-expressed some of the regional transcription factors (e.g. Six3/6, Pax6, Otx2) serving as upstream regulators of this network in the retina, followed by a loss of this CoRN from the retina. This proposed evolutionary origin of the lens from the retina and, thus, from neural ectoderm may help to explain, why the vertebrate retina has the capacity to transdifferentiate into lens tissue and that the iris, which develops from the retina, can regenerate a lens in newts and some teleosts (Wolff 1895; Okada et al. 1975; Del Rio-Tsonis and Tsonis 2003; Iida, Ishii, and Kondoh 2017).

8.2 ADENOHYPOPHYSEAL PLACODE

8.2.1 Differentiation of Adenohypophyseal Cells

The neurosecretory (endocrine) cells of the adenohypophysis produce a variety of hormones, which belong to three different families: peptide hormones, dimeric glycoprotein hormones, and four-helix cytokine-like proteins (Figs. 1.11B and 8.3) (reviewed in Campbell, Satoh, and Degnan 2004; Kawauchi and Sower 2006; Norris and Carr 2020). All hormones of the adenohypophysis are proteins or peptides. They are all produced in the rough endoplasmatic reticulum, followed by further processing in the Golgi apparatus and packaging into secretory vesicles (Burgoyne and Morgan 2003; Norris and Carr 2020). The *peptide hormones* are produced by the proteolytic processing of larger precursor proteins (prohormones) (Fig. 8.4). Two important peptide hormones produced in the anterior pituitary are adrenocorticotropic hormone (ACTH) and the alpha melanocyte stimulating hormone (α-MSH) (Fig. 8.3). Both are generated from the same precursor protein proopiomelanocortin (POMC) which also produces β-MSH, γ-MSH, and the opioid β-endorphin. Due to tissue-specific differences in POMC processing, cells will either produce ACTH or α-MSH which is generated by further cleavage of ACTH (Fig. 8.4) (Bicknell 2008). In lampreys, the *POMC* gene has been duplicated into *POM* and *POC* genes with subsequent differential loss of some peptide encoding regions from each copy (Takahashi et al. 2005; Kawauchi and Sower 2006; Dores and Baron 2011).

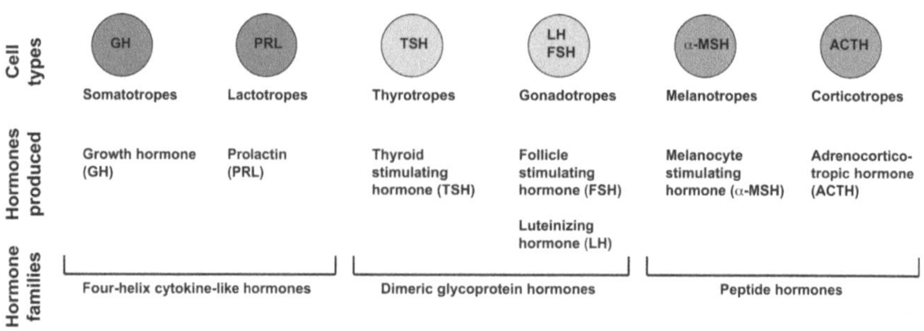

FIGURE 8.3 Neurosecretory cell types and hormone classes in the anterior pituitary.

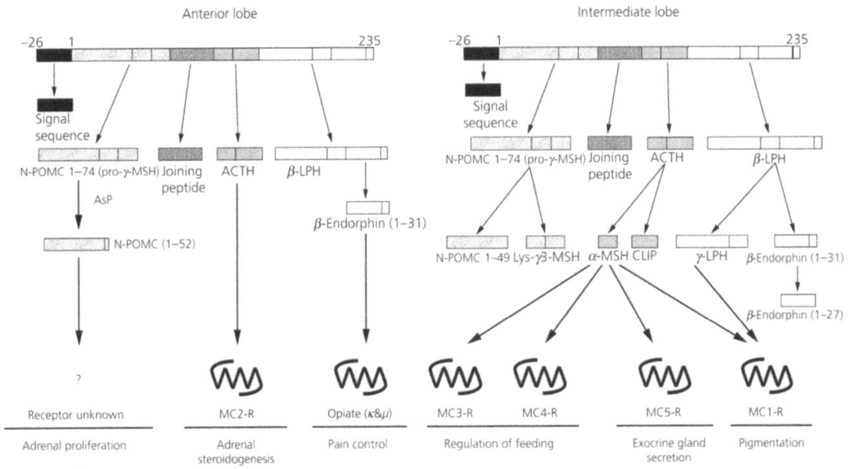

FIGURE 8.4 Processing of pro-opiomelanocortin (POMC) and function of POMC-derived peptides. POMC is processed differently in the anterior and intermediate lobes of the pituitary by prohormone convertases resulting in different peptides (e.g. ACTH in the anterior lobe, but α-MSH in the intermediate lobe). POMC-derived peptides act mainly through a family of five G-protein coupled receptors known as the melanocortin receptors (MCX-R where X is a number from 1 to 5). β-endorphin acts via the κ and μ opiate receptors. ACTH, adrenocorticotrophin; CLIP, corticotrophin-like intermediate peptide; LPH, lipotrophin; MSH, melanocyte-stimulating hormone. (Reprinted with permission from Bicknell 2008.)

The *four-helix cytokine-like hormones*, secreted by the anterior pituitary include growth hormone (GH), prolactin (PRL), and, in some actinopterygian fishes, somatolactin (SL) (Fig. 8.3). In lampreys and hagfishes only GH but not PRL or SL have been identified suggesting that PRL and SL may have originated only in gnathostomes due to duplication of the ancestral *GH* gene (Kawauchi and Sower 2006; Nozaki 2008). A new analysis, however, suggests that GH, PRL, and SL may already have been present in stem vertebrates, with PRL and SL being secondarily lost in lampreys (Ocampo Daza and Larhammar 2018).

The *dimeric glycoprotein hormones* secreted by the adenohypophysis consist of a common alpha subunit (known as GPH-α or α-GSU) and specific beta subunits, which differ for each of the three hormones thyroid stimulating hormone (TSH), luteinizing hormone (LH), and follicle stimulating hormone (FSH) (Fig. 8.3). The TSH contains TSH-β, while the two gonadotropic (gonad stimulating) hormones, LH, and FSH, contain LH-β and FSH-β, respectively. In lampreys and hagfishes, only one representative of the adenohypophyseal glycoprotein beta subunits (GPH-β) is present, which when combined with an alpha subunit probably acts via two different receptors both as a thyroid stimulating and as a gonadotropic hormone (Sower et al. 2006; Sower, Freamat, and Kavanaugh 2009; Uchida et al. 2010; Nozaki 2013). The TSH, LH, and FSH beta subunits then probably evolved by duplication of GPH-β in gnathostomes. Because GPH-α has been identified in hagfishes and gnathostomes, it most likely originated in the ancestors of all vertebrates (Uchida et al. 2010, 2013). However, lampreys appear to have secondarily lost GPH-α again and produce a unique glycoprotein hormone combining the alpha subunit GPA2 of another glycoprotein hormone, thyrostimulin, with GPH-β. This hormone (GPA2/GPH-β) is co-expressed with thyrostimulin (GPA2/GPB5) in the same adenohypophyseal cells suggesting that both hormones have overlapping functions (Sower et al. 2015; Marquis et al. 2017; Hausken et al. 2018).

None of the vertebrate adenohypophyseal hormones were found in the genomes of amphioxus, *Ciona* or other invertebrates (Dehal et al. 2002; Holland et al. 2008; Putnam et al. 2008). This suggests that the hormones of the anterior pituitary originated in the vertebrate lineage from other members of the same hormone families, which can be traced back to the earliest metazoans (Campbell et al. 2004; Jékely 2013; Mirabeau and Joly 2013; Roch and Sherwood 2014). This will be discussed in more detail in Chapter 5 of Volume 2 (Schlosser 2021).

Many cells in the anterior pituitary are specialized for the production of only one (or sometimes two in the case of LH/FSH) of these hormones and are classified accordingly (Fig. 8.3). Similar neurosecretory cells producing the same hormones have also been found in the brain and other tissues outside of the anterior pituitary (Murphy and Harvey 2001; So, Kwok, and Ge 2005; Bicknell 2008). In the adenohypophysis of gnathostomes at least six different neurosecretory cell types can be distinguished (four in lampreys, Marquis et al. 2017). The neuropeptide hormones are secreted by corticotropes (ACTH) and melanotropes (α-MSH), the glycoprotein hormones by gonadotropes (LH and/or FSH) and thyrotropes (TSH), and the cytokine-like hormones by somatotropes (GH) and lactotropes (PRL) (reviewed in Scully and Rosenfeld 2002; Ooi, Tawadros, and Escalona 2004; Rizzoti and Lovell-Badge 2005; Zhu, Gleiberman, and Rosenfeld 2007; Kelberman et al. 2009; Pogoda and Hammerschmidt 2009; Davis et al. 2013; Rizzoti 2015). Contrary to traditional assumptions, two recent single-cell RNA-Seq studies of the mouse pituitary failed to identify a well-defined population of dedicated thyrotropes; TSH expressing cells often co-expressed other hormones, mostly GH and/or PRL, and tended to cluster with other endocrine cell types (Cheung, George et al. 2018; Ho et al. 2020). One of these studies also identified a major adenohypophyseal cell cluster, which produces multiple hormones in various combinations (Ho et al. 2020). These findings suggest close developmental and probably evolutionary relationships between the different hormone producing cells of the adenohypophysis (see Chapter 5 of Schlosser 2021). They may also indicate that thyrotropes may comprise those cells that fail to be specified to any other cell type and thus maintain an incompletely differentiated state as the default fate of the adenohypophysis. However, this needs to be confirmed in further studies.

The various hormone producing cell types of the anterior pituitary are packed with secretory vesicles like endocrine cells in other glands. Although they lack dendrites, axons, and synapses, they display many other neuron-like properties and are therefore classified as so-called neurosecretory cells (Scharrer and Scharrer 1945; Scharrer 1987; reviewed in De Camilli and Jahn 1990; Stojilkovic 2006; Norris and Carr 2020). Like neurons, adenohypophyseal cells are excitable cells. They have a resting membrane potential and various voltage gated ion channels endowing them with the ability to form action potentials in response to stimulation by the releasing hormones secreted by the hypothalamus (Stojilkovic 2006). These action potentials control calcium-influx through voltage-gated calcium-channels, which drives the release of secretory vesicles resembling the events at the synapse of a neuron (Stojilkovic 2006). Secretory vesicles then fuse with the cell membrane using an exocytosis machinery similar to the one used for synaptic vesicles, employing proteins of the SNARE complex (Snap-25, Syntaxin1, Synaptobrevin) together with Rab3 and Synaptophysin (De Camilli and Jahn 1990; Jacobson and Meister 1996; Salinas, Quintanar, and Reig 1999; Matsuno et al. 2003, 2011). Gonadotropes and thyrotropes further resemble glutamatergic neurons in expressing the vesicular glutamate transporter Vglut2 (Hrabovszky and Liposits 2008).

Moreover, endocrine cells of the adenohypophysis secrete their hormones in a synchronized oscillatory manner. This is known as pulsatile secretion and is thought to help avoid desensitization of responding tissues, which has been observed after release of hormones at a constant rate (e.g. Belchetz et al. 1978; Brabant, Prank, and Schofl 1992). While pulsatile secretion of adenohypophyseal cells is partly driven by similar oscillations in the secretion of releasing hormones (e.g. GnRH) from hypothalamic neurons, the anterior pituitary is also able to produce such pulses in the absence of hypothalamic input (reviewed in Le Tissier, Fiordelisio Coll, and Mollard 2018).

Both, feedback from target tissues (e.g. regulation of LH/FSH secretion by hormones secreted in the gonads) and communication between the endocrine cells of the anterior pituitary play a role for generating these oscillations (Le Tissier et al. 2018). Endocrine cells of the same type (e.g. gonadotropes) have been shown to be in physical contact and form adherens junctions with each other, establishing extensive three-dimensional networks of interconnected cells within the anterior pituitary (Bonnefont et al. 2005; Budry et al. 2011). Cells within the networks are often linked by gap junctions and may also communicate by paracrine signals allowing coordination of calcium influx and secretory activity among the cells (Bonnefont et al. 2005; Denef 2008; Sánchez-Cardenas et al. 2010; Hodson et al. 2012). In some networks (e.g. lactotropes), pacemaker cells have been identified, which can locally synchronize the activity of interconnected cells (Hodson et al. 2012). The clustering and arrangement of cells within these networks can change in response to new functional demands leading for example, to sexually dimorphic reorganization of the GH and PRL networks in puberty (Bonnefont et al. 2005; Sánchez-Cardenas et al. 2010; Hodson et al. 2012). Taken together with their other neuron-like properties mentioned above and their neuron-like specification and differentiation discussed in the next section, the coordinated activity of endocrine cells in such homotypic endocrine networks lends further support to their neuron-like, neurosecretory identity.

In addition to the neurosecretory cells, the anterior pituitary also contains several types of non-endocrine folliculostellate cells and a very small population of Sox2-positive stem-cells, which can give rise to all endocrine cell types as well as folliculostellate cells in the adult (Fauquier et al. 2008; Andoniadou et al. 2013; Rizzoti 2015; Ho et al. 2020). Folliculostellate cells are star shaped and follicle-forming cells that have long processes and are connected via gap junctions to form a three-dimensional network (Rinehart and Farquhar 1953; Inoue et al. 1999; Shirasawa et al. 2004; Devnath and Inoue 2008). Apart from structural and metabolic support functions and scavenging roles, these interconnected cells allow the rapid propagation of Ca^{2+} currents throughout the pituitary, suggesting important but poorly understood roles in long-distance coordination of its cellular activities (Fauquier et al. 2001; Devnath and Inoue 2008).

The distribution of different types of neurosecretory cells in the adenohypophysis is not homogeneous and different cell types are enriched in different regions. For example, gonadotropes form in its most ventral part, while melanotropes form in its dorsal-most part, which is in contact with the neurohypophysis (and comprises the so-called intermediate lobe of the pituitary in mammals). This regional patterning of the anterior pituitary is established in response to several signaling molecules released by adjacent tissues such as the ventral diencephalon, which is in contact with the dorsal adenohypophysis (FGFs, BMP4, surrounded by Shh), the mesenchyme on its ventral side (BMP2), and the underlying oral ectoderm (Shh). The resulting signaling gradients established within the developing adenohypophysis, for example with high FGF concentrations dorsally and high BMP2 concentrations ventrally, have been shown to affect the spatial arrangement of various cell types within the pituitary (Ericson et al. 1998; Treier et al. 1998). A more detailed discussion of these patterning mechanisms is beyond the scope of this book and the reader is referred to previous reviews (e.g. Scully and Rosenfeld 2002; Zhu et al. 2007; Kelberman et al. 2009; Davis et al. 2013; Rizzoti 2015).

8.2.2 Transcription Factors Involved in Specification of Anterior Pituitary Cell Identities

The specification of different adenohypophyseal cell types is regulated by multiple transcription factors acting sequentially and in a combinatorial manner (Fig. 8.5A). These transcription factors have partly redundant functions and many of them play pleiotropic roles by regulating different cellular processes during early and late stages of pituitary development or by activating different gene batteries in different cell lineages in a context-dependent fashion. In addition, there are species-specific differences. Taken together, this results in a rather complex network of regulatory interactions, which we are only beginning to understand. I will, therefore, limit myself here to some general observations based on functional studies in mouse and zebrafish embryos (for detailed reviews see Scully and Rosenfeld 2002; Savage et al. 2003; Rizzoti and Lovell-Badge 2005; Zhu et al. 2007; Kelberman et al. 2009; Pogoda and Hammerschmidt 2009; Davis et al. 2013; Rizzoti 2015).

Like in other placodes, Six1 and/or Eya are also involved in regulating cell specification and differentiation of some cell types derived from the adenohypophyseal placodes. Whereas no pituitary defects were reported in Eya1 or Six1 single mutants in the mouse, the pituitary was strongly reduced in size in Eya1/Six1 double mutants (Li et al. 2003). In zebrafish, Eya1 but not Six1 was shown to be required for the differentiation of several adenohypophyseal cell types (gonadotropes, corticotropes, melanotropes) (Nica et al. 2006). Whether Six4 or other Six family transcription

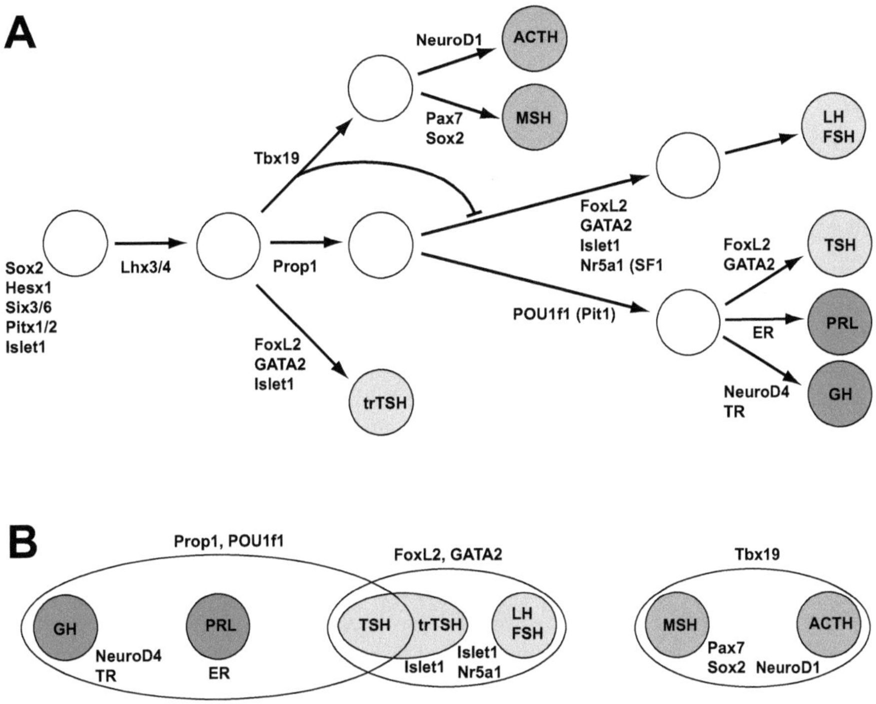

FIGURE 8.5 Transcription factors involved in the specification of neurosecretory cells in the anterior pituitary. (**A**) Role of transcription factors in cell lineage decisions in the mammalian pituitary. trTSH, transitory thyrotropes. Hormones belonging to the same families (see Fig. 8.3) are shown in the same shades of gray. (**B**) Generally hormones belonging to the same family are specified by the same transcription factors (Prop1, POU1f11: four helix cytokine like hormones GH and PRL; FoxL2, GATA2: dimeric glycoprotein hormones TSH, FSH, LH; Tbx19: peptide hormones MSH and ACTH). An exception to this rule is a subpopulation of TSH producing cells that are also dependent on Prop1 and POU1f1 (Pit1). (Based on Zhu et al. 2007, Kelberman et al. 2009, and Rizzoti 2015.)

factors may be able to compensate for the role of Six1 in the pituitary (or other Eya family members for Eya1) has not been addressed. This suggests an important role of Eya and Six family members in the specification of the neurosecretory cells of the adenohypophysis, which require however further investigation.

Similar to the neurogenic placodes, transcription factors of the SoxB1 family (Sox2 and/or Sox3) are expressed in the adenohypophyseal placode from early, pre-placodal stages onward and contribute to maintaining a proliferative progenitor state together with other transcription factors such as Six3/6 (Kelberman et al. 2006; Kelberman et al. 2008). Sox2 expression is then downregulated when endocrine cell types differentiate, but has been shown to persist in a small population of cells that can give rise to all adenohypophyseal cell types in culture, indicating that they represent pituitary stem cells (Fauquier et al. 2008; Rizzoti, Akiyama, and Lovell-Badge 2013). This suggests that during development of neurosecretory cells in the adenohypophysis,

SoxB1 factors play equivalent roles in promoting a progenitor state and repressing differentiation than during development of neurons and sensory cells from other placodes (see Chapter 5). Another important regulator of early adenohypophyseal development, Hesx1, acts as a repressor. It limits progenitor proliferation and prevents precocious differentiation of all adenohypophyseal cell types, which can only proceed after Hesx1 is downregulated (Dasen et al. 2001; Gaston-Massuet et al. 2008).

A third group of transcription factors play dual roles in the maintenance of adenohypophyseal progenitors and the subsequent differentiation of endocrine cell types. This group includes the Pitx transcription factors Pitx1, Pitx2, and Pitx3 and the LIM-domain transcription factors Islet1 (=Isl1), Lhx3, and Lhx4. Expression of at least some members of these Pitx (e.g. Pitx1/2 in mouse, Pitx3 in frogs and zebrafish) and LIM-domain factors (Lhx3/4) becomes restricted to the adenohypophyseal placode in all vertebrates. Knockdown of Pitx1/2, Lhx3/4, or Islet1 in the mouse results in smaller pituitaries indicating that these transcription factors are required for progenitor maintenance in a partly redundant fashion (Sheng et al. 1997; Takuma et al. 1998; Raetzman, Ward, and Camper 2002; Charles et al. 2005; Zhao et al. 2006; Ellsworth, Butts, and Camper 2008). In addition, differentiation of all neurosecretory cell types is compromised in mutants of Pitx1, Pitx2 or Lhx3 (Sheng et al. 1997; Szeto et al. 1999; Ellsworth et al. 2008). Islet1 instead is maintained only in a transitory population of thyrotropes (rostral-tip thyrotropes) and in gonadotropes, but whether it is involved in the specification of these cell types has not been investigated (Ericson et al. 1998; Schang et al. 2013). Both Pitx1 and Lhx3 have been shown to promote expression of hormone encoding genes and/or lineage-specific transcription factors in most adenohypophyseal cell types by direct binding to their cis-regulatory region in cooperation with other transcription factors (e.g. POU1f1; see further below in this section) (Bach et al. 1995; Sloop et al. 1999; Szeto et al. 1996; Tremblay, Lanctot, and Drouin 1998; Tremblay et al. 1999; Tremblay and Drouin 1999; Poulin, Turgeon, and Drouin 1997; Lamolet et al. 2001; West et al. 2004).

In zebrafish, the bHLH transcription factor Ascl1, which plays a central role in specifying the sensory neurons from the olfactory placode (see Chapter 6), has also been shown to be required for the differentiation of all neurosecretory cell types in the adenohypophysis (Pogoda et al. 2006). In mammals, Ascl1 is required for the differentiation of some endodermally-derived neurosecretory cells, but loss of Ascl1 does not appear to compromise differentiation of endocrine cells in the anterior pituitary (Guillemot and Joyner 1993; Borges et al. 1997; Lanigan, DeRaad, and Russo 1998; Huber et al. 2002). However, somato-, cortico-, and gonado tropes were recently shown to be drastically reduced in triple mutants of Ascl1, Atoh3 (=NeuroD4), and NeuroD1 in mice (Ando et al. 2018). In addition, these triple mutant mice upregulated Neurog2 expression, which may possibly account for the differentiation of the remaining endocrine cell types, if Neurog2 is able to compensate for the loss of the other three bHLH factors in these cell types (Ando et al. 2018). Overall, these findings suggest that proneural bHLH factors are essential, but act in a partly redundant fashion during specification of neurosecretory cells in the anterior pituitary.

Taken together, the studies reviewed so far demonstrate that Pitx1/2, Lhx3/4 are important members of the CoRN regulating the specification of neurosecretory cells in the adenohypophysis, while Ascl and/or other proneural bHLH transcription factors may be required, like in neurons, for the initiation of differentiation in all of its endocrine cell lines. A multitude of additional transcription factors is then required in defining more specific subtypes of neurosecretory cells (Fig. 8.5A). First, Tbx19 (=Tpit) is required for the melanotrope and corticotrope lineages, with NeuroD1 co-expression promoting corticotrope fate, while the co-expression of Pax7 and Sox2 promotes melanotrope fate (Pulichino, Vallette-Kasic, Tsai et al. 2003; Pulichino, Vallette-Kasic, Couture et al. 2003; Lavoie et al. 2008; Budry et al. 2012; Goldsmith, Lovell-Badge, and Rizzoti 2016). Second, a combination of FoxL2, GATA2, Islet1, and the steroid receptor Nr5a1 (=SF1) are essential for gonadotrope differentiation (Zhao et al. 2001; Charles et al. 2006; Ellsworth et al. 2006; Fortin et al. 2009, 2014; Schang et al. 2013). Third, Prop1 initiates POU1f1 (=Pit1) expression (Nasonkin et al. 2004; Olson et al. 2006), which is required for most thyrotropes as well as for lactotropes and somatotropes (Camper et al. 1990; Li et al. 1990; Dasen et al. 1999; Nica et al. 2004). Within the POU1f1 lineage, then, FoxL2 and GATA2 (which promote gonadotropes in the absence of POU1f1) promote thyrotrope development (Charles et al. 2006; Ellsworth et al. 2006), while NeuroD4 and the thyroid hormone receptor (TR) is essential for somatotrope and the estrogen receptor (ER) for lactotrope differentiation (Day et al. 1990; Simmons et al. 1990; Nowakowski and Maurer 1994; Zhu et al. 2006).

In addition, there is a small and somewhat enigmatic subpopulation of thyrotropes that is independent of POU1f1 but expresses GATA2, Islet1, and FoxL2 (Fig. 8.5A). These cells transiently occupy the rostral tip of the embryonic anterior pituitary but disappear at birth (Lin et al. 1994; Ericson et al. 1998; Ellsworth et al. 2006; Kelberman et al. 2009). The adoption of specific neurosecretory fates also requires the inhibition of alternative fates by various mechanisms such as the cross-repressive interactions between Tbx19 and Nr5a1 (Fig. 8.5A) and the upregulation of Notch2 by Prop1, which inhibits Tbx19 and thus interferes with formation of melano- and cortico tropes (Pulichino, Vallette-Kasic, Tsai et al. 2003; Raetzman et al. 2004; Goldberg, Aujla, and Raetzman 2011; Cheung, Le Tissier et al. 2018).

The cis-regulatory regions of genes encoding the hormones and other proteins specific for the different endocrine cell types of the adenohypophysis are relatively well characterized (reviewed in Savage et al. 2003; Zhu et al. 2007). Most of these genes are regulated by a complex combination of general adenohypophyseal transcription factors (e.g. Pitx1, Lhx3) with more lineage specific ones (e.g. GATA2, SF1, NeuroD4), often involving direct and cooperative protein-protein interactions between different transcription factors. For example, joint GATA2-POU1f11 binding to adjacent binding sites on the promoter of the *TSH-β* gene is required for its activation (Dasen et al. 1999).

This combinatorial regulation of cell type identity suggests that all endocrine cells of the anterior pituitary can be considered subtypes of a single neurosecretory cell type with overlapping CoRNs that share some but not all transcription factors with each other (see Chapter 2 and fig. 2.14 in Schlosser 2021). It is, therefore, likely that all adenohypophyseal endocrine cells have ultimately evolved from a single neurosecretory cell type that possibly co-expressed multiple hormones similar to some of the "thyrotropes" and multifunctional cells identified in recent single-cell sequencing studies.

The hierarchy of lineage decisions in the pituitary may offer some clues on how this ancestral multifunctional neurosecretory cell type diversified further. Generally, cells producing related hormones tend to depend on the same transcription factors for specification (Fig. 8.5B). Melanotropes and corticotropes, which produce POMC-derived peptide hormones, both require Tbx19 (T-pit). Thyrotropes and gonadotropes, which produce dimeric glycoprotein hormones, both require FoxL2 and GATA2. And somatotropes and lactotropes, which produce four-helix cytokine-like hormones, require POU1f1 (Pit1) in the absence of GATA2. Although POU1f1 is also required for most thyrotropes, there is also a transient subpopulation of thyrotropes developing early at the rostral tip, which is POU1f1 independent. This raises the possibility, that this transient population of thyrotropes represents the evolutionarily primitive condition, while POU1f-1 dependence of the remaining thyrotropes may be evolutionarily derived.

Taken together with evidence that lampreys have only one cell type producing glycoprotein hormones (a combined thyro/gonadotrope) and one cell type producing cytokine like hormones (a combined lacto/somatotrope) and the observation that melanotropes and corticotropes express the same pro-hormone POMC and only differ in its processing, this suggests that three specialized subtypes of adenohypophyseal endocrine cells were probably present at the base of vertebrates. Each of these subtypes was probably dedicated to the production of one of the three hormone classes. During vertebrate evolution these three subtypes diversified further into the six or more endocrine cell types found in extant gnathostomes. This scenario leaves open whether the origin of the three subtypes from a single multifunctional neurosecretory cell type happened in the stem lineage of vertebrates or can be traced back to ancestral chordates or even more distantly related ancestors. Comparisons of vertebrates with other chordates will be required to resolve this (and argue for an origin in stem vertebrates) as will be discussed in Chapter 5 of Volume 2 (Schlosser 2021). The many similarities between the neurosecretory cells of the pituitary and neurons highlighted above also suggest that the adenohypophyseal neurosecretory cells are evolutionarily related to neurons and/or sensory cells. Again, a broader and evolutionary perspective is necessary to shine light on this issue. After completion of my survey of the sensory and neurosecretory cells of the vertebrate head, I will now take this broader perspective and consider the evolutionary origin of these cell types and of vertebrate cranial placodes in volume 2 (Schlosser 2021).

References

Abdel Samad, O., Y. Liu, F. C. Yang, I. Kramer, S. Arber, and Q. Ma. 2010. Characterization of two Runx1-dependent nociceptor differentiation programs necessary for inflammatory versus neuropathic pain. *Mol Pain* 6:45.

Abdelhak, S., V. Kalatzis, R. Heilig, S. Compain, D. Samson, C. Vincent, D. Weil, C. Cruaud, I. Sahly, M. Leibovici, M. Bitner-Glindzicz, M. Francis, D. Lacombe, J. Vigneron, R. Charachon, K. Boven, P. Bedbeder, N. Van Regemorter, J. Weissenbach, and C. Petit. 1997. A human homologue of the *Drosophila* eyes absent gene underlies branchio-oto-renal (BOR) syndrome and identifies a novel gene family. *Nat Genet* 15:157–164.

Abdolazimi, Y., Z. Stojanova, and N. Segil. 2016. Selection of cell fate in the organ of Corti involves the integration of Hes/Hey signaling at the Atoh1 promoter. *Development* 143:841–850.

Abe, H., and Y. Oka. 2000. Modulation of pacemaker activity by salmon gonadotropin-releasing hormone (sGnRH) in terminal nerve (TN)-GnRH neurons. *J Neurophysiol* 83:3196–3200.

Abelló, G., S. Khatri, F. Giráldez, and B. Alsina. 2007. Early regionalization of the otic placode and its regulation by the Notch signaling pathway. *Mech Dev* 124:631–645.

Abelló, G., S. Khatri, M. Radosevic, P. J. Scotting, F. Giráldez, and B. Alsina. 2010. Independent regulation of Sox3 and Lmx1b by FGF and BMP signaling influences the neurogenic and non-neurogenic domains in the chick otic placode. *Dev Biol* 339:166–178.

Abu-Elmagd, M., Y. Ishii, M. Cheung, M. Rex, D. Le Rouedec, and P. J. Scotting. 2001. cSox3 expression and neurogenesis in the epibranchial placodes. *Dev Biol* 237:258–269.

Acampora, D., S. Mazan, Y. Lallemand, V. Avantaggiato, M. Maury, A. Simeone, and P. Brulet. 1995. Forebrain and midbrain regions are deleted in Otx2-/- mutants due to a defective anterior neuroectoderm specification during gastrulation. *Development* 121:3279–3290.

Ache, B. W., and J. M. Young. 2005. Olfaction: diverse species, conserved principles. *Neuron* 48:417–430.

Adachi, K., and H. R. Schöler. 2012. Directing reprogramming to pluripotency by transcription factors. *Curr Opin Genet Dev* 22:416–422.

Adam, J., A. Myat, I. Leroux, M. Eddison, D. Henrique, D. Ishhorowicz, and J. Lewis. 1998. Cell fate choices and the expression of Notch, Delta and Serrate homologues in the chick inner ear: parallels with *Drosophila* sense-organ development. *Development* 125:4645–4654.

Agathocleous, M., and W. A. Harris. 2009. From progenitors to differentiated cells in the vertebrate retina. *Annu Rev Cell Dev Biol* 25:45–69.

Agathocleous, M., I. Iordanova, M. I. Willardsen, X. Y. Xue, M. L. Vetter, W. A. Harris, and K. B. Moore. 2009. A directional Wnt/beta-catenin-Sox2-proneural pathway regulates the transition from proliferation to differentiation in the *Xenopus* retina. *Development* 136:3289–3299.

Aghaallaei, N., B. Bajoghli, and T. Czerny. 2007. Distinct roles of Fgf8, Foxi1, Dlx3b and Pax8/2 during otic vesicle induction and maintenance in medaka. *Dev Biol* 307:408–420.

Agius, E., M. Oelgeschläger, O. Wessely, C. Kemp, and E. M. DeRobertis. 2000. Endodermal Nodal-related signals and mesoderm induction in *Xenopus*. *Development* 127:1173–1183.

Aguillon, R., J. Batut, A. Subramanian, R. Madelaine, P. Dufourcq, T. F. Schilling, and P. Blader. 2018. Cell-type heterogeneity in the early zebrafish olfactory epithelium is generated from progenitors within preplacodal ectoderm. *Elife* 7:e32041.

Aguillon, R., P. Blader, and J. Batut. 2016. Patterning, morphogenesis, and neurogenesis of zebrafish cranial sensory placodes. *Methods Cell Biol* 134:33–67.

Aguirre, A., M. E. Rubio, and V. Gallo. 2010. Notch and EGFR pathway interaction regulates neural stem cell number and self-renewal. *Nature* 467:323–327.

Aguirre, C. E., S. Murgan, A. E. Carrasco, and S. L. López. 2013. An intact brachyury function is necessary to prevent spurious axial development in *Xenopus laevis*. *PLoS One* 8:e54777.

Ahmad, N., M. Aslam, D. Muenster, M. Horsch, M. A. Khan, P. Carlsson, J. Beckers, and J. Graw. 2013. Pitx3 directly regulates Foxe3 during early lens development. *Int J Dev Biol* 57:741–751.

Ahmed, M., E. Y. Wong, J. Sun, J. Xu, F. Wang, and P. X. Xu. 2012. Eya1-Six1 interaction is sufficient to induce hair cell fate in the cochlea by activating atoh1 expression in cooperation with Sox2. *Dev Cell* 22:377–390.

Ahmed, M., J. Xu, and P. X. Xu. 2012. EYA1 and SIX1 drive the neuronal developmental program in cooperation with the SWI/SNF chromatin-remodeling complex and SOX2 in the mammalian inner ear. *Development* 139:1965–1977.

Ahrens, K., and G. Schlosser. 2005. Tissues and signals involved in the induction of placodal Six1 expression in *Xenopus laevis*. *Dev Biol* 288:40–59.

Akagi, T., M. Mandai, S. Ooto, Y. Hirami, F. Osakada, R. Kageyama, N. Yoshimura, and M. Takahashi. 2004. Otx2 homeobox gene induces photoreceptor-specific phenotypes in cells derived from adult iris and ciliary tissue. *Invest Ophthalmol Vis Sci* 45:4570–4575.

Akhmedov, N. B., N. I. Piriev, B. Chang, A. L. Rapoport, N. L. Hawes, P. M. Nishina, S. Nusinowitz, J. R. Heckenlively, T. H. Roderick, C. A. Kozak, M. Danciger, M. T. Davisson, and D. B. Farber. 2000. A deletion in a photoreceptor-specific nuclear receptor mRNA causes retinal degeneration in the rd7 mouse. *Proc Natl Acad Sci USA* 97:5551–5556.

Al Anber, A., and B. L. Martin. 2019. Transformation of a neural activation and patterning model. *EMBO Rep* 20:e48060.

Ali, F., C. Hindley, G. McDowell, R. Deibler, A. Jones, M. Kirschner, F. Guillemot, and A. Philpott. 2011. Cell cycle-regulated multi-site phosphorylation of Neurogenin 2 coordinates cell cycling with differentiation during neurogenesis. *Development* 138:4267–4277.

Ali, F. R., K. Cheng, P. Kirwan, S. Metcalfe, F. J. Livesey, R. A. Barker, and A. Philpott. 2014. The phosphorylation status of Ascl1 is a key determinant of neuronal differentiation and maturation in vivo and in vitro. *Development* 141:2216–2224.

Alioto, T. S., and J. Ngai. 2006. The repertoire of olfactory C family G protein-coupled receptors in zebrafish: candidate chemosensory receptors for amino acids. *BMC Genomics* 7:309.

Alkobtawi, M., and A. H. Monsoro-Burq. 2020. The neural crest, a vertebrate invention. In *Evolving neural crest cells*, edited by B. F. Eames, D. Meulemans Medeiros and I. Adameyko, 5–66. Boca Raton: CRC Press.

Allan, D. W., and S. Thor. 2015. Transcriptional selectors, masters, and combinatorial codes: regulatory principles of neural subtype specification. *Wiley Interdiscip Rev Dev Biol* 4:505–528.

Allis, E. P. 1897. The cranial muscles and cranial and first spinal nerves in *Amia calva*. *J Morphol* 12:487–769.

Alsina, B., G. Abelló, E. Ulloa, D. Henrique, C. Pujades, and F. Giráldez. 2004. FGF signaling is required for determination of otic neuroblasts in the chick embryo. *Dev Biol* 267:119–134.

Alsina, B., F. Giráldez, and C. Pujades. 2009. Patterning and cell fate in ear development. *Int J Dev Biol* 53:1503–1513.

Alsina, B., and T. T. Whitfield. 2017. Sculpting the labyrinth: morphogenesis of the developing inner ear. *Semin Cell Dev Biol* 65:47–59.

Alsina, B. 2020. Mechanisms of cell specification and differentiation in vertebrate cranial sensory systems. *Curr Opin Cell Biol* 67:79–85.

Altmann, C. R., R. L. Chow, R. A. Lang, and A. Hemmati-Brivanlou. 1997. Lens induction by Pax-6 in *Xenopus laevis. Dev Biol* 185:119–123.

Altmann, C. R., and A. Hemmati-Brivanlou. 2001. Neural patterning in the vertebrate embryo. *Int Rev Cytol* 203:447–482.

Alvarez-Delfin, K., A. C. Morris, C. D. Snelson, J. T. Gamse, T. Gupta, F. L. Marlow, M. C. Mullins, H. A. Burgess, M. Granato, and J. M. Fadool. 2009. Tbx2b is required for ultraviolet photoreceptor cell specification during zebrafish retinal development. *Proc Natl Acad Sci USA* 106:2023–2028.

Alvarez-Rodriguez, R., and S. Pons. 2009. Expression of the proneural gene encoding Mash1 suppresses MYCN mitotic activity. *J Cell Sci* 122:595–599.

Alvarez, Y., M. T. Alonso, V. Vendrell, L. C. Zelarayan, P. Chamero, T. Theil, M. R. Bosl, S. Kato, M. Maconochie, D. Riethmacher, and T. Schimmang. 2003. Requirements for FGF3 and FGF10 during inner ear formation. *Development* 130:6329–6338.

Amador-Arjona, A., F. Cimadamore, C. T. Huang, R. Wright, S. Lewis, F. H. Gage, and A. V. Terskikh. 2015. SOX2 primes the epigenetic landscape in neural precursors enabling proper gene activation during hippocampal neurogenesis. *Proc Natl Acad Sci USA* 112:E1936–E1945.

Andermann, P., J. Ungos, and D. W. Raible. 2002. Neurogenin1 defines zebrafish cranial sensory ganglia precursors. *Dev Biol* 251:45–58.

Andersson, E. K., D. K. Irvin, J. Ahlsio, and M. Parmar. 2007. Ngn2 and Nurr1 act in synergy to induce midbrain dopaminergic neurons from expanded neural stem and progenitor cells. *Exp Cell Res* 313:1172–1180.

Andley, U. P. 2007. Crystallins in the eye: function and pathology. *Prog Retin Eye Res* 26:78–98.

Ando, M., M. Goto, M. Hojo, A. Kita, M. Kitagawa, T. Ohtsuka, R. Kageyama, and S. Miyamoto. 2018. The proneural bHLH genes Mash1, Math3 and NeuroD are required for pituitary development. *J Mol Endocrinol* 61:127–138.

Ando, Z., S. Sato, K. Ikeda, and K. Kawakami. 2005. Slc12a2 is a direct target of two closely related homeobox proteins, Six1 and Six4. *FEBS J* 272:3026–3041.

Andoniadou, C. L., D. Matsushima, S. N. Mousavy Gharavy, M. Signore, A. I. Mackintosh, M. Schaeffer, C. Gaston-Massuet, P. Mollard, T. S. Jacques, P. Le Tissier, M. T. Dattani, L. H. Pevny, and J. P. Martinez-Barbera. 2013. Sox2(+) stem/progenitor cells in the adult mouse pituitary support organ homeostasis and have tumor-inducing potential. *Cell Stem Cell* 13:433–445.

Aoki, Y., N. Saint-Germain, M. Gyda, E. Magner-Fink, Y. H. Lee, C. Credidio, and J. P. Saint-Jeannet. 2003. Sox10 regulates the development of neural crest-derived melanocytes in *Xenopus. Dev Biol* 259:19–33.

Aoyama, A., E. Frohli, R. Schafer, and R. Klemenz. 1993. Alpha B-crystallin expression in mouse NIH 3T3 fibroblasts: glucocorticoid responsiveness and involvement in thermal protection. *Mol Cell Biol* 13:1824–1835.

Appler, J. M., C. C. Lu, N. R. Druckenbrod, W. M. Yu, E. J. Koundakjian, and L. V. Goodrich. 2013. Gata3 is a critical regulator of cochlear wiring. *J Neurosci* 33:3679–3691.

Arendt, D. 2003. Evolution of eyes and photoreceptor cell types. *Int J Dev Biol* 47:563–571.

Arendt, D., J. M. Musser, C. V. Baker, A. Bergman, C. Cepko, D. H. Erwin, M. Pavlicev, G. Schlosser, S. Widder, M. D. Laubichler, and G. P. Wagner. 2016. The origin and evolution of cell types. *Nat Rev Genet* 17:744–757.

Arendt, D., and J. Wittbrodt. 2001. Reconstructing the eyes of Urbilateria. *Philos Trans R Soc Lond B Biol Sci* 356:1545–1563.

Arshavsky, V. Y., and M. E. Burns. 2012. Photoreceptor signaling: supporting vision across a wide range of light intensities. *J Biol Chem* 287:1620–1626.

Ashery-Padan, R., and P. Gruss. 2001. Pax6 lights-up the way for eye development. *Curr Opin Cell Biol* 13:706–714.

Asteriti, S., S. Grillner, and L. Cangiano. 2015. A Cambrian origin for vertebrate rods. *Elife* 4:e07166.

Ataliotis, P., S. Ivins, T. J. Mohun, and P. J. Scambler. 2005. XTbx1 is a transcriptional activator involved in head and pharyngeal arch development in *Xenopus laevis*. *Dev Dyn* 232:979–991.

Atkinson, P. J., E. Huarcaya Najarro, Z. N. Sayyid, and A. G. Cheng. 2015. Sensory hair cell development and regeneration: similarities and differences. *Development* 142:1561–1571.

Aybar, M. J., M. A. Nieto, and R. Mayor. 2003. Snail precedes slug in the genetic cascade required for the specification and migration of the *Xenopus* neural crest. *Development* 130:483–494.

Ayer-LeLièvre, C. S., and N. M. Le Douarin. 1982. The early development of cranial sensory ganglia and the potentialities of their component cells studied in quail-chick chimeras. *Dev Biol* 94:291–310.

Ayers, H. 1892. Vertebrate cephalogenesis. *J Morphol* 6:1–360.

Aydin, B., A. Kakumanu, M. Rossillo, M. Moreno-Estelles, G. Garipler, N. Ringstad, N. Flames, S. Mahony, and E. O. Mazzoni. 2019. Proneural factors Ascl1 and Neurog2 contribute to neuronal subtype identities by establishing distinct chromatin landscapes. *Nat Neurosci* 22:897–908.

Bach, I., S. J. Rhodes, R. V. Pearse, 2nd, T. Heinzel, B. Gloss, K. M. Scully, P. E. Sawchenko, and M. G. Rosenfeld. 1995. P-Lim, a LIM homeodomain factor, is expressed during pituitary organ and cell commitment and synergizes with Pit-1. *Proc Natl Acad Sci USA* 92:2720–2724.

Bachmann, C., H. Nguyen, J. Rosenbusch, L. Pham, T. Rabe, M. Patwa, G. Sokpor, R. H. Seong, R. Ashery-Padan, A. Mansouri, A. Stoykova, J. F. Staiger, and T. Tuoc. 2016. mSWI/SNF (BAF)complexes are indispensable for the neurogenesis and development of embryonic olfactory epithelium. *PLoS Genet* 12:e1006274.

Bae, C. J., B. Y. Park, Y. H. Lee, J. W. Tobias, C. S. Hong, and J. P. Saint-Jeannet. 2014. Identification of Pax3 and Zic1 targets in the developing neural crest. *Dev Biol* 386:473–483.

Bailey, A. P., S. Bhattacharyya, M. Bronner-Fraser, and A. Streit. 2006. Lens specification is the ground state of all sensory placodes, from which FGF promotes olfactory identity. *Dev Cell* 11:505–517.

Bailey, A. P., and A. Streit. 2006. Sensory organs: making and breaking the pre-placodal region. *Curr Top Dev Biol* 72:167–204.

Bailey, M. J., and V. M. Cassone. 2005. Melanopsin expression in the chick retina and pineal gland. *Brain Res Mol Brain Res* 134:345–348.

Bailey, T. J., H. El-Hodiri, L. Zhang, R. Shah, P. H. Mathers, and M. Jamrich. 2004. Regulation of vertebrate eye development by Rx genes. *Int J Dev Biol* 48:761–770.

Bajoghli, B., N. Aghaallaei, G. Jung, and T. Czerny. 2009. Induction of otic structures by canonical Wnt signalling in medaka. *Dev Genes Evol* 219:391–398.

Baker, C. V. H., and M. Bronner-Fraser. 2001. Vertebrate cranial placodes. I. Embryonic induction. *Dev Biol* 232:1–61.

Baker, C. V. H., and M. S. Modrell. 2018. Insights into electroreceptor development and evolution from molecular comparisons with hair cells. *Integr Comp Biol* 58:329–340.

Baker, C. V., M. S. Modrell, and J. A. Gillis. 2013. The evolution and development of vertebrate lateral line electroreceptors. *J Exp Biol* 216:2515–2522.

Baker, C. V., P. O'Neill, and R. B. McCole. 2008. Lateral line, otic and epibranchial placodes: developmental and evolutionary links? *J Exp Zoolog B Mol Dev Evol* 310:370–383.

Baker, C. V., M. R. Stark, and M. Bronner-Fraser. 2002. Pax3-expressing trigeminal placode cells can localize to trunk neural crest sites but are committed to a cutaneous sensory neuron fate. *Dev Biol* 249:219–236.

Baker, C. V., M. R. Stark, C. Marcelle, and M. Bronner-Fraser. 1999. Competence, specification and induction of Pax-3 in the trigeminal placode. *Development* 126:147–156.

Baker, J. C., R. S. P. Beddington, and R. M. Harland. 1999. Wnt signaling in *Xenopus* embryos inhibits Bmp4 expression and activates neural development. *Gene Develop* 13:3149–3159.

Baker, N. E., and N. L. Brown. 2018. All in the family: proneural bHLH genes and neuronal diversity. *Development* 145:dev159426.

Balemans, W., and W. Van Hul. 2002. Extracellular regulation of BMP signaling in vertebrates: a cocktail of modulators. *Dev Biol* 250:231–250.

Balinsky, B. I. 1970. *An introduction to embryology.* 3rd ed. Philadelphia: Saunders.

Balmer, C. W., and A. S. LaMantia. 2005. Noses and neurons: induction, morphogenesis, and neuronal differentiation in the peripheral olfactory pathway. *Dev Dyn* 234:464–481.

Bang, A. G., N. Papalopulu, M. D. Goulding, and C. Kintner. 1999. Expression of Pax-3 in the lateral neural plate is dependent on a Wnt-mediated signal from posterior nonaxial mesoderm. *Dev Biol* 212:366–380.

Bani-Yaghoub, M., R. G. Tremblay, J. X. Lei, D. Zhang, B. Zurakowski, J. K. Sandhu, B. Smith, M. Ribecco-Lutkiewicz, J. Kennedy, P. R. Walker, and M. Sikorska. 2006. Role of Sox2 in the development of the mouse neocortex. *Dev Biol* 295:52–66.

Barald, K. F., and M. W. Kelley. 2004. From placode to polarization: new tunes in inner ear development. *Development* 131:4119–4130.

Barembaum, M., and M. Bronner-Fraser. 2010. Pax2 and Pea3 synergize to activate a novel regulatory enhancer for spalt4 in the developing ear. *Dev Biol* 340:222–231.

Barlow, L. A. 2015. Progress and renewal in gustation: new insights into taste bud development. *Development* 142:3620–3629.

Barlow, L. A., and R. G. Northcutt. 1995. Embryonic origin of amphibian taste buds. *Dev Biol* 169:273–285.

Barnea, G., S. O'Donnell, F. Mancia, X. Sun, A. Nemes, M. Mendelsohn, and R. Axel. 2004. Odorant receptors on axon termini in the brain. *Science* 304:1468.

Barraud, P., A. A. Seferiadis, L. D. Tyson, M. F. Zwart, H. L. Szabo-Rogers, C. Ruhrberg, K. J. Liu, and C. V. Baker. 2010. Neural crest origin of olfactory ensheathing glia. *Proc Natl Acad Sci USA* 107:21040–21045.

Barresi, M. J., and S. F. Gilbert. 2019. *Developmental biology.* 12th ed. New York: Oxford University Press.

Bassnett, S. 2009. On the mechanism of organelle degradation in the vertebrate lens. *Exp Eye Res* 88:133–139.

Bassnett, S., and D. Mataic. 1997. Chromatin degradation in differentiating fiber cells of the eye lens. *J Cell Biol* 137:37–49.

Baxi, K. N., K. M. Dorries, and H. L. Eisthen. 2006. Is the vomeronasal system really specialized for detecting pheromones? *Trends Neurosci* 29:1–7.

Bayramov, A. V., N. Y. Martynova, F. M. Eroshkin, G. V. Ermakova, and A. G. Zaraisky. 2004. The homeodomain-containing transcription factor X-nkx-5.1 inhibits expression of the homeobox gene Xanf-1 during the *Xenopus laevis* forebrain development. *Mech Dev* 121:1425–1441.

Begbie, J., M. Ballivet, and A. Graham. 2002. Early steps in the production of sensory neurons by the neurogenic placodes. *Mol Cell Neurosci* 21:502–511.

Begbie, J., J. F. Brunet, J. L. R. Rubenstein, and A. Graham. 1999. Induction of the epibranchial placodes. *Development* 126:895–902.

Begbie, J., and A. Graham. 2001a. The ectodermal placodes: a dysfunctional family. *Philos Trans R Soc Lond B Biol Sci* 356:1655–1660.

Begbie, J., and A. Graham. 2001b. Integration between the epibranchial placodes and the hindbrain. *Science* 295:595–598.

Behrens, M., O. Frank, H. Rawel, G. Ahuja, C. Potting, T. Hofmann, W. Meyerhof, and S. Korsching. 2014. ORA1, a zebrafish olfactory receptor ancestral to all mammalian V1R genes, recognizes 4-hydroxyphenylacetic acid, a putative reproductive pheromone. *J Biol Chem* 289:19778–19788.

Beites, C. L., S. Kawauchi, C. E. Crocker, and A. L. Calof. 2005. Identification and molecular regulation of neural stem cells in the olfactory epithelium. *Exp Cell Res* 306:309–316.

Belchetz, P. E., T. M. Plant, Y. Nakai, E. J. Keogh, and E. Knobil. 1978. Hypophysial responses to continuous and intermittent delivery of hypopthalamic gonadotropin-releasing hormone. *Science* 202:631–633.

Belecky-Adams, T., S. Tomarev, H. S. Li, L. Ploder, R. R. McInnes, O. Sundin, and R. Adler. 1997. Pax-6, Prox 1, and Chx10 homeobox gene expression correlates with phenotypic fate of retinal precursor cells. *Invest Ophthalmol Vis Sci* 38:1293–1303.

Bell, D., A. Streit, I. Gorospe, I. Varela-Nieto, B. Alsina, and F. Giráldez. 2008. Spatial and temporal segregation of auditory and vestibular neurons in the otic placode. *Dev Biol* 322:109–120.

Bell, S. M., C. M. Schreiner, R. R. Waclaw, K. Campbell, S. S. Potter, and W. J. Scott. 2003. Sp8 is crucial for limb outgrowth and neuropore closure. *Proc Natl Acad Sci USA* 100:12195–12200.

Bellefroid, E. J., C. Bourguignon, T. Hollemann, Q. Ma, D. J. Anderson, C. Kintner, and T. Pieler. 1996. X-MyT1, a *Xenopus* C2HC-type zinc finger protein with a regulatory function in neuronal differentiation. *Cell* 87:1191–1202.

Bellefroid, E. J., A. Kobbe, P. Gruss, T. Pieler, J. B. Gurdon, and N. Papalopulu. 1998. Xiro3 encodes a *Xenopus* homolog of the *Drosophila* Iroquois genes and functions in neural specification. *EMBO J* 17:191–203.

Bellmeyer, A., J. Krase, J. Lindgren, and C. LaBonne. 2003. The protooncogene c-myc is an essential regulator of neural crest formation in *Xenopus*. *Dev Cell* 4:827–839.

Bellono, N. W., D. B. Leitch, and D. Julius. 2017. Molecular basis of ancestral vertebrate electroreception. *Nature* 543:391–396.

Belluscio, L., G. Koentges, R. Axel, and C. Dulac. 1999. A map of pheromone receptor activation in the mammalian brain. *Cell* 97:209–220.

Bennett, J. T., K. Joubin, S. Cheng, P. Aanstad, R. Herwig, M. Clark, H. Lehrach, and A. F. Schier. 2007. Nodal signaling activates differentiation genes during zebrafish gastrulation. *Dev Biol* 304:525–540.

Berghard, A., and L. B. Buck. 1996. Sensory transduction in vomeronasal neurons: evidence for G alpha o, G alpha i2, and adenylyl cyclase II as major components of a pheromone signaling cascade. *J Neurosci* 16:909–918.

Berghard, A., L. B. Buck, and E. R. Liman. 1996. Evidence for distinct signaling mechanisms in two mammalian olfactory sense organs. *Proc Natl Acad Sci USA* 93:2365–2369.

Berghard, A., A. C. Hagglund, S. Bohm, and L. Carlsson. 2012. Lhx2-dependent specification of olfactory sensory neurons is required for successful integration of olfactory, vomeronasal, and GnRH neurons. *FASEB J* 26:3464–3472.

Bergsland, M., D. Ramskold, C. Zaouter, S. Klum, R. Sandberg, and J. Muhr. 2011. Sequentially acting Sox transcription factors in neural lineage development. *Genes Dev* 25:2453–2464.

Bermingham, N. A., B. A. Hassan, S. D. Price, M. A. Vollrath, N. Ben Arie, R. A. Eatock, H. J. Bellen, A. Lysakowski, and H. Y. Zoghbi. 1999. Math1: an essential gene for the generation of inner ear hair cells. *Science* 284:1837–1841.

Berninger, B., M. R. Costa, U. Koch, T. Schroeder, B. Sutor, B. Grothe, and M. Götz. 2007. Functional properties of neurons derived from in vitro reprogrammed postnatal astroglia. *J Neurosci* 27:8654–8664.

Bertrand, N., D. S. Castro, and F. Guillemot. 2002. Proneural genes and the specification of neural cell types. *Nat Rev Neurosci* 3:517–530.

Bessarab, D. A., S. W. Chong, and V. Korzh. 2004. Expression of zebrafish six1 during sensory organ development and myogenesis. *Dev Dyn* 230:781–786.

Betancur, P., M. Bronner-Fraser, and T. Sauka-Spengler. 2010. Assembling neural crest regulatory circuits into a gene regulatory network. *Annu Rev Cell Dev Biol* 26:581–603.

Betancur, P., T. Sauka-Spengler, and M. Bronner. 2011. A Sox10 enhancer element common to the otic placode and neural crest is activated by tissue-specific paralogs. *Development* 138:3689–3698.

Beurg, M., X. Tan, and R. Fettiplace. 2013. A prestin motor in chicken auditory hair cells: active force generation in a nonmammalian species. *Neuron* 79:69–81.

Bhasin, N., T. M. Maynard, P. A. Gallagher, and A. S. La Mantia. 2003. Mesenchymal/epithelial regulation of retinoic acid signaling in the olfactory placode. *Dev Biol* 261:82–98.

Bhat, N., H. J. Kwon, and B. B. Riley. 2012. A gene network that coordinates preplacodal competence and neural crest specification in zebrafish. *Dev Biol* 373:107–117.

Bhat, N., and B. B. Riley. 2011. Integrin-alpha5 coordinates assembly of posterior cranial placodes in zebrafish and enhances Fgf-dependent regulation of otic/epibranchial cells. *PLoS One* 6:e27778.

Bhattacharyya, S., A. P. Bailey, M. Bronner-Fraser, and A. Streit. 2004. Segregation of lens and olfactory precursors from a common territory: cell sorting and reciprocity of Dlx5 and Pax6 expression. *Dev Biol* 271:403–414.

Bhattacharyya, S., and M. Bronner-Fraser. 2008. Competence, specification and commitment to an olfactory placode fate. *Development* 135:4165–4177.

Bicknell, A. B. 2008. The tissue-specific processing of pro-opiomelanocortin. *J Neuroendocrinol* 20:692–699.

Biechl, D., K. Tietje, S. Ryu, B. Grothe, G. Gerlach, and M. F. Wullimann. 2017. Identification of accessory olfactory system and medial amygdala in the zebrafish. *Sci Rep* 7:44295.

Birol, O., T. Ohyama, R. K. Edlund, K. Drakou, P. Georgiades, and A. K. Groves. 2016. The mouse Foxi3 transcription factor is necessary for the development of posterior placodes. *Dev Biol* 409:139–151.

Bissonnette, J. P., and D. M. Fekete. 1996. Standard atlas of the gross anatomy of the developing inner ear of the chicken. *J Comp Neurol* 368:620–630.

Blackshaw, S., R. E. Fraioli, T. Furukawa, and C. L. Cepko. 2001. Comprehensive analysis of photoreceptor gene expression and the identification of candidate retinal disease genes. *Cell* 107:579–589.

Blaxter, J. H. S. 1987. Structure and development of the lateral line. *Biol Rev* 62:471–514.

Bleymehl, K., A. Pérez-Gómez, M. Omura, A. Moreno-Pérez, D. Macias, Z. Bai, R. S. Johnson, T. Leinders-Zufall, F. Zufall, and P. Mombaerts. 2016. A sensor for low environmental oxygen in the mouse main olfactory epithelium. *Neuron* 92:1196–1203.

Blixt, A., M. Mahlapuu, M. Aitola, M. Pelto-Huikko, S. Enerback, and P. Carlsson. 2000. A forkhead gene, FoxE3, is essential for lens epithelial proliferation and closure of the lens vesicle. *Genes Dev* 14:245–254.

Bodznick, D., and J. C. Montgomery. 2005. The physiology of low-frequency electrosensory systems In *Electroreception*, edited by T. H. Bullock, C. D. Hopkins, A. N. Popper and R. R. Fay, 132–153. New York: Springer.

Bok, J., W. Chang, and D. K. Wu. 2007. Patterning and morphogenesis of the vertebrate inner ear. *Int J Dev Biol* 51:521–533.

Bok, J., S. Raft, K. A. Kong, S. K. Koo, U. C. Drager, and D. K. Wu. 2011. Transient retinoic acid signaling confers anterior-posterior polarity to the inner ear. *Proc Natl Acad Sci USA* 108:161–166.

Bonanomi, D., and S. L. Pfaff. 2010. Motor axon pathfinding. *Cold Spring Harb Perspect Biol* 2:a001735.

Bonev, B., P. Stanley, and N. Papalopulu. 2012. MicroRNA-9 Modulates Hes1 ultradian oscillations by forming a double-negative feedback loop. *Cell Rep* 2:10–18.

Bonnefont, X., A. Lacampagne, A. Sánchez-Hormigo, E. Fino, A. Creff, M. N. Mathieu, S. Smallwood, D. Carmignac, P. Fontanaud, P. Travo, G. Alonso, N. Courtois-Coutry, S. M. Pincus, I. C. Robinson, and P. Mollard. 2005. Revealing the large-scale network organization of growth hormone-secreting cells. *Proc Natl Acad Sci USA* 102:16880–16885.

Borday, C., K. Parain, H. Thi Tran, K. Vleminckx, M. Perron, and A. H. Monsoro-Burq. 2018. An atlas of Wnt activity during embryogenesis in *Xenopus tropicalis*. *PLoS One* 13:e0193606.

Borges, M., R. I. Linnoila, H. J. van de Velde, H. Chen, B. D. Nelkin, M. Mabry, S. B. Baylin, and D. W. Ball. 1997. An achaete-scute homologue essential for neuroendocrine differentiation in the lung. *Nature* 386:852–855.

Borges, R., W. E. Johnson, S. J. O'Brien, C. Gomes, C. P. Heesy, and A. Antunes. 2018. Adaptive genomic evolution of opsins reveals that early mammals flourished in nocturnal environments. *BMC Genomics* 19:121.

Bosman, E. A., E. Quint, H. Fuchs, de Angelis Hrabe, and K. P. Steel. 2009. Catweasel mice: a novel role for Six1 in sensory patch development and a model for branchio-oto-renal syndrome. *Dev Biol* 328:285–296.

Bosse, A., A. Zulch, M. B. Becker, M. Torres, J. L. Gómez-Skarmeta, J. Modolell, and P. Gruss. 1997. Identification of the vertebrate Iroquois homeobox gene family with overlapping expression during early development of the nervous system. *Mech Dev* 69:169–181.

Bouchard, M., D. de Caprona, M. Busslinger, P. Xu, and B. Fritzsch. 2010. Pax2 and Pax8 cooperate in mouse inner ear morphogenesis and innervation. *BMC Dev Biol* 10:89.

Bourguignon, C., J. Li, and N. Papalopulu. 1998. XBF-1, a winged helix transcription factor with dual activity, has a role in positioning neurogenesis in *Xenopus* competent ectoderm *Development* 125:4889–4900.

Bovolenta, P., A. Mallamaci, P. Briata, G. Corte, and E. Boncinelli. 1997. Implication of otx2 in epithelium determination and neural retina differentiation. *J Neurosci* 17:4243–4252.

Brabant, G., K. Prank, and C. Schofl. 1992. Pulsatile patterns in hormone secretion. *Trends Endocrinol Metab* 3:183–190.

Braun, C. B. 1996. The sensory biology of the living jawless fishes: a phylogenetic assessment. *Brain Behav Evol* 48:262–276.

Braun, C. B., and R. G. Northcutt. 1997. The lateral line system of hagfishes (Craniata: Myxinoidea). *Acta Zool* 78:247–268.

Braun, M. M., A. Etheridge, A. Bernard, C. P. Robertson, and H. Roelink. 2003. Wnt signaling is required at distinct stages of development for the induction of the posterior forebrain. *Development* 130:5579–5587.

Bray, S. J. 2006. Notch signalling: a simple pathway becomes complex. *Nat Rev Mol Cell Biol* 7:678–689.

Breau, M. A., and S. Schneider-Maunoury. 2014. Mechanisms of cranial placode assembly. *Int J Dev Biol* 58:9–19.

Breau, M. A., and S. Schneider-Maunoury. 2015. Cranial placodes: models for exploring the multi-facets of cell adhesion in epithelial rearrangement, collective migration and neuronal movements. *Dev Biol* 401:25–36.

Breer, H., J. Fleischer, and J. Strotmann. 2006. The sense of smell: multiple olfactory subsystems. *Cell Mol Life Sci* 63:1465–1475.

Breuskin, I., M. Bodson, N. Thelen, M. Thiry, L. Borgs, L. Nguyen, P. P. Lefebvre, and B. Malgrange. 2009. Sox10 promotes the survival of cochlear progenitors during the establishment of the organ of Corti. *Dev Biol* 335:327–339.

Breuskin, I., M. Bodson, N. Thelen, M. Thiry, L. Borgs, L. Nguyen, C. Stolt, M. Wegner, P. P. Lefebvre, and B. Malgrange. 2010. Glial but not neuronal development in the cochleo-vestibular ganglion requires Sox10. *J Neurochem* 114:1827–1839.

Bricaud, O., and A. Collazo. 2006. The transcription factor six1 inhibits neuronal and promotes hair cell fate in the developing zebrafish (*Danio rerio*) inner ear. *J Neurosci* 26:10438–10451.

Brigande, J. V., and S. Heller. 2009. Quo vadis, hair cell regeneration? *Nat Neurosci* 12:679–685.

Briggs, J. A., C. Weinreb, D. E. Wagner, S. Megason, L. Peshkin, M. W. Kirschner, and A. M. Klein. 2018. The dynamics of gene expression in vertebrate embryogenesis at single-cell resolution. *Science* 360:eaar5780.

Brignull, H. R., D. W. Raible, and J. S. Stone. 2009. Feathers and fins: non-mammalian models for hair cell regeneration. *Brain Res* 1277:12–23.

Broccoli, V., E. Boncinelli, and W. Wurst. 1999. The caudal limit of Otx2 expression positions the isthmic organizer. *Nature* 401:164–168.

Broillet, M. C., and S. Firestein. 1997. Beta subunits of the olfactory cyclic nucleotide-gated channel form a nitric oxide activated Ca^{2}+ channel. *Neuron* 18:951–958.

Brooks, E. R., and J. B. Wallingford. 2014. Multiciliated cells. *Curr Biol* 24:R973–R982.

Brown, N. L., S. Kanekar, M. L. Vetter, P. K. Tucker, D. L. Gemza, and T. Glaser. 1998. Math5 encodes a murine basic helix-loop-helix transcription factor expressed during early stages of retinal neurogenesis. *Development* 125:4821–4833.

Brown, N. L., S. Patel, J. Brzezinski, and T. Glaser. 2001. Math5 is required for retinal ganglion cell and optic nerve formation. *Development* 128:2497–2508.

Brown, S. T., J. Wang, and A. K. Groves. 2005. Dlx gene expression during chick inner ear development. *J Comp Neurol* 483:48–65.

Brownell, I., M. Dirksen, and M. Jamrich. 2000. Forkhead Foxe3 maps to the dysgenetic lens locus and is critical in lens development and differentiation. *Genesis* 27:81–93.

Brownell, W. E., C. R. Bader, D. Bertrand, and Y. de Ribaupierre. 1985. Evoked mechanical responses of isolated cochlear outer hair cells. *Science* 227:194–196.

Brugmann, S. A., P. D. Pandur, K. L. Kenyon, F. Pignoni, and S. A. Moody. 2004. Six1 promotes a placodal fate within the lateral neurogenic ectoderm by functioning as both a transcriptional activator and repressor. *Development* 131:5871–5881.

Brulet, R., T. Matsuda, L. Zhang, C. Miranda, M. Giacca, B. K. Kaspar, K. Nakashima, and J. Hsieh. 2017. NEUROD1 instructs neuronal conversion in non-reactive astrocytes. *Stem Cell Reports* 8:1506–1515.

Brunet, J. F., and A. Pattyn. 2002. Phox2 genes – from patterning to connectivity. *Curr Opin Genet Dev* 12:435–440.

Brunjes, P. C., and L. L. Frazier. 1986. Maturation and plasticity in the olfactory system of vertebrates. *Brain Res* 396:1–45.

Brzezinski, J. A., and T. A. Reh. 2015. Photoreceptor cell fate specification in vertebrates. *Development* 142:3263–3273.

Buck, L., and R. Axel. 1991. A novel multigene family may encode odorant receptors: a molecular basis for odor recognition. *Cell* 65:175–187.

Buck, L. B. 2004. Olfactory receptors and odor coding in mammals. *Nutr Rev* 62:S184–S188.

Budry, L., A. Balsalobre, Y. Gauthier, K. Khetchoumian, A. L'Honoré, S. Vallette, T. Brue, D. Figarella-Branger, B. Meij, and J. Drouin. 2012. The selector gene Pax7 dictates alternate pituitary cell fates through its pioneer action on chromatin remodeling. *Genes Dev* 26:2299–2310.

Budry, L., C. Lafont, T. El Yandouzi, N. Chauvet, G. Conejero, J. Drouin, and P. Mollard. 2011. Related pituitary cell lineages develop into interdigitated 3D cell networks. *Proc Natl Acad Sci USA* 108:12515–12520.

Bufe, B., T. Schumann, and F. Zufall. 2012. Formyl peptide receptors from immune and vomeronasal system exhibit distinct agonist properties. *J Biol Chem* 287:33644–33655.

Buitrago-Delgado, E., K. Nordin, A. Rao, L. Geary, and C. LaBonne. 2015. Shared regulatory programs suggest retention of blastula-stage potential in neural crest cells. *Science* 348:1332–1335.

Buitrago-Delgado, E., E. N. Schock, K. Nordin, and C. LaBonne. 2018. A transition from SoxB1 to SoxE transcription factors is essential for progression from pluripotent blastula cells to neural crest cells. *Dev Biol* 444:50–61.

Bulankina, A. V., and T. Moser. 2012. Neural circuit development in the mammalian cochlea. *Physiology* 27:100–112.

Bullock, T. H., D. A. Bodznick, and R. G. Northcutt. 1983. The phylogenetic distribution of electroreception: evidence for convergent evolution of a primitive vertebrate sense modality. *Brain Res Rev* 6:25–46.

Bullock, T. H., and W. Heiligenberg. 1986. *Electroreception*. New York: Wiley.

Burgoyne, R. D., and A. Morgan. 2003. Secretory granule exocytosis. *Physiol Rev* 83:581–632.

Burkhardt, P., and S. G. Sprecher. 2017. Evolutionary origin of synapses and neurons – bridging the gap. *Bioessays* 39: 1700024.

Burns, C. J., and M. L. Vetter. 2002. Xath5 regulates neurogenesis in the *Xenopus* olfactory placode *Dev Dyn* 225:536–543.

Burns, J. C., M. C. Kelly, M. Hoa, R. J. Morell, and M. W. Kelley. 2015. Single-cell RNA-Seq resolves cellular complexity in sensory organs from the neonatal inner ear. *Nat Commun* 6:8557.

Burton, Q., L. K. Cole, M. Mulheisen, W. Chang, and D. K. Wu. 2004. The role of Pax2 in mouse inner ear development. *Dev Biol* 272:161–175.

Butler, A. B., and W. Hodos. 2005. *Comparative vertebrate neuroanatomy*. 2nd ed. New York: Wiley-Liss.

Bylund, M., E. Andersson, B. G. Novitch, and J. Muhr. 2003. Vertebrate neurogenesis is counteracted by Sox1-3 activity. *Nat Neurosci* 6:1162–1168.

Byrd, N. A., and E. N. Meyers. 2005. Loss of Gbx2 results in neural crest cell patterning and pharyngeal arch artery defects in the mouse embryo. *Dev Biol* 284:233–245.

Caggiano, M., J. S. Kauer, and D. D. Hunter. 1994. Globose basal cells are neuronal progenitors in the olfactory epithelium: a lineage analysis using a replication-incompetent retrovirus. *Neuron* 13:339–352.

Cai, T., and A. K. Groves. 2015. The role of atonal factors in mechanosensory cell specification and function. *Mol Neurobiol* 52:1315–1329.

Cai, T., M. L. Seymour, H. Zhang, F. A. Pereira, and A. K. Groves. 2013. Conditional deletion of Atoh1 reveals distinct critical periods for survival and function of hair cells in the organ of Corti. *J Neurosci* 33:10110–10122.

Campbell, R. K., N. Satoh, and B. M. Degnan. 2004. Piecing together evolution of the vertebrate endocrine system. *Trends Genet* 20:359–366.

Camper, S. A., T. L. Saunders, R. W. Katz, and R. H. Reeves. 1990. The Pit-1 transcription factor gene is a candidate for the murine Snell dwarf mutation. *Genomics* 8:586–890.

Canning, C. A., L. Lee, S. X. Luo, A. Graham, and C. M. Jones. 2008. Neural tube derived Wnt signals cooperate with FGF signaling in the formation and differentiation of the trigeminal placodes. *Neural Develop* 3:35.

Cao, Y., D. Siegel, and W. Knöchel. 2006. *Xenopus* POU factors of subclass V inhibit activin/nodal signaling during gastrulation. *Mech Dev* 123:614–625.

Carl, M., F. Loosli, and J. Wittbrodt. 2002. Six3 inactivation reveals its essential role for the formation and patterning of the vertebrate eye. *Development* 129:4057–4063.

Carney, T. J., K. A. Dutton, E. Greenhill, M. Delfino-Machin, P. Dufourcq, P. Blader, and R. N. Kelsh. 2006. A direct role for Sox10 in specification of neural crest-derived sensory neurons. *Development* 133:4619–4630.

Carpenter, E. 1937. The head pattern in *Amblystoma* studied by vital staining and transplantation methods. *J exp Zool* 75:103–129.

Carron, C., and D. L. Shi. 2016. Specification of anteroposterior axis by combinatorial signaling during *Xenopus* development. *Wiley Interdiscip Rev Dev Biol* 5:150–168.

Carter, L. A., J. L. MacDonald, and A. J. Roskams. 2004. Olfactory horizontal basal cells demonstrate a conserved multipotent progenitor phenotype. *J Neurosci* 24:5670–5683.

Casarosa, S., M. Andreazzoli, A. Simeone, and G. Barsacchi. 1997. Xrx1, a novel *Xenopus* homeobox gene expressed during eye and pineal gland development. *Mech Dev* 61:187–198.

Casarosa, S., C. Fode, and F. Guillemot. 1999. Mash1 regulates neurogenesis in the ventral telencephalon. *Development* 126:525–534.

Castro, D. S., B. Martynoga, C. Parras, V. Ramesh, E. Pacary, C. Johnston, D. Drechsel, M. Lebel-Potter, L. G. Garcia, C. Hunt, D. Dolle, A. Bithell, L. Ettwiller, N. Buckley, and F. Guillemot. 2011. A novel function of the proneural factor Ascl1 in progenitor proliferation identified by genome-wide characterization of its targets. *Genes Dev* 25:930–945.

Cau, E., S. Casarosa, and F. Guillemot. 2002. Mash1 and Ngn1 control distinct steps of determination and differentiation in the olfactory sensory neuron lineage. *Development* 129:1871–1880.

Cau, E., G. Gradwohl, S. Casarosa, R. Kageyama, and F. Guillemot. 2000. Hes genes regulate sequential stages of neurogenesis in the olfactory epithelium. *Development* 127:2323–2332.

Cau, E., G. Gradwohl, C. Fode, and F. Guillemot. 1997. Mash1 activates a cascade of bhlh regulators in olfactory neuron progenitors. *Development* 124:1611–1621.

Cau, E., and S. W. Wilson. 2003. Ash1a and Neurogenin1 function downstream of Floating head to regulate epiphysial neurogenesis. *Development* 130:2455–2466.

Cavallaro, M., J. Mariani, C. Lancini, E. Latorre, R. Caccia, F. Gullo, M. Valotta, S. DeBiasi, L. Spinardi, A. Ronchi, E. Wanke, S. Brunelli, R. Favaro, S. Ottolenghi, and S. K. Nicolis. 2008. Impaired generation of mature neurons by neural stem cells from hypomorphic Sox2 mutants. *Development* 135:541–557.

Cepko, C. 2014. Intrinsically different retinal progenitor cells produce specific types of progeny. *Nat Rev Neurosci* 15:615–627.

Cepko, C. L. 2015. The determination of rod and cone photoreceptor fate. *Annu Rev Vis Sci* 1:211–234.

Chae, J. H., G. H. Stein, and J. E. Lee. 2004. NeuroD: the predicted and the surprising. *Mol Cells* 18:271–288.

Chalfie, M. 2009. Neurosensory mechanotransduction. *Nat Rev Mol Cell Biol* 10:44–52.

Chanda, S., C. E. Ang, J. Davila, C. Pak, M. Mall, Q. Y. Lee, H. Ahlenius, S. W. Jung, T. C. Südhof, and M. Wernig. 2014. Generation of induced neuronal cells by the single reprogramming factor ASCL1. *Stem Cell Rep* 3:282–296.

Chang, I., and M. Parrilla. 2016. Expression patterns of homeobox genes in the mouse vomeronasal organ at postnatal stages. *Gene Expr Patterns* 21:69–80.

Chang, Y. K., Y. Srivastava, C. Hu, A. Joyce, X. Yang, Z. Zuo, J. J. Havranek, G. D. Stormo, and R. Jauch. 2017. Quantitative profiling of selective Sox/POU pairing on hundreds of sequences in parallel by Coop-seq. *Nucleic Acids Res* 45:832–845.

Chapman, S. C., A. L. Sawitzke, D. S. Campbell, and G. C. Schoenwolf. 2005. A three-dimensional atlas of pituitary gland development in the zebrafish. *J Comp Neurol* 487:428–440.

Charles, M. A., T. L. Saunders, W. M. Wood, K. Owens, A. F. Parlow, S. A. Camper, E. C. Ridgway, and D. F. Gordon. 2006. Pituitary-specific Gata2 knockout: effects on gonadotrope and thyrotrope function. *Mol Endocrinol* 20:1366–1377.

Charles, M. A., H. Suh, T. A. Hjalt, J. Drouin, S. A. Camper, and P. J. Gage. 2005. PITX genes are required for cell survival and Lhx3 activation. *Mol Endocrinol* 19:1893–1903.

Chaurasia, S. S., M. D. Rollag, G. Jiang, W. P. Hayes, R. Haque, A. Natesan, M. Zatz, G. Tosini, C. Liu, H. W. Korf, P. M. Iuvone, and I. Provencio. 2005. Molecular cloning, localization and circadian expression of chicken melanopsin (Opn4): differential regulation of expression in pineal and retinal cell types. *J Neurochem* 92:158–170.

Chen, B., E. H. Kim, and P. X. Xu. 2009. Initiation of olfactory placode development and neurogenesis is blocked in mice lacking both Six1 and Six4. *Dev Biol* 326:75–85.

Chen, C. L., D. C. Broom, Y. Liu, J. C. de Nooij, Z. Li, C. Cen, O. A. Samad, T. M. Jessell, C. J. Woolf, and Q. Ma. 2006. Runx1 determines nociceptive sensory neuron phenotype and is required for thermal and neuropathic pain. *Neuron* 49:365–377.

Chen, C. M., and C. L. Cepko. 2002. The chicken RaxL gene plays a role in the initiation of photoreceptor differentiation. *Development* 129:5363–5375.

Chen, G., T. R. Korfhagen, Y. Xu, J. Kitzmiller, S. E. Wert, Y. Maeda, A. Gregorieff, H. Clevers, and J. A. Whitsett. 2009. SPDEF is required for mouse pulmonary goblet cell differentiation and regulates a network of genes associated with mucus production. *J Clin Invest* 119:2914–2924.

Chen, H., A. Thiagalingam, H. Chopra, M. W. Borges, J. N. Feder, B. D. Nelkin, S. B. Baylin, and D. W. Ball. 1997. Conservation of the *Drosophila* lateral inhibition pathway in human lung cancer: a hairy-related protein (HES-1) directly represses achaete-scute homolog-1 expression. *Proc Natl Acad Sci USA* 94:5355–5360.

Chen, J., A. Rattner, and J. Nathans. 2005. The rod photoreceptor-specific nuclear receptor Nr2e3 represses transcription of multiple cone-specific genes. *J Neurosci* 25:118–129.

Chen, J., and A. P. Sampath. 2013. Structure and function of rod and cone photoreceptors. In *Retina*, edited by S. Ryan, 342–359. Elsevier.

Chen, J., and A. Streit. 2013. Induction of the inner ear: stepwise specification of otic fate from multipotent progenitors. *Hear Res* 297:3–12.

Chen, J., M. Tambalo, M. Barembaum, R. Ranganathan, M. Simoes-Costa, M. E. Bronner, and A. Streit. 2017. A systems-level approach reveals new gene regulatory modules in the developing ear. *Development* 144:1531–1543.

Chen, P., J. E. Johnson, H. Y. Zoghbi, and N. Segil. 2002. The role of Math1 in inner ear development: uncoupling the establishment of the sensory primordium from hair cell fate determination. *Development* 129:2495–2505.

Chen, S., Q. L. Wang, Z. Nie, H. Sun, G. Lennon, N. G. Copeland, D. J. Gilbert, N. A. Jenkins, and D. J. Zack. 1997. Crx, a novel Otx-like paired-homeodomain protein, binds to and transactivates photoreceptor cell-specific genes. *Neuron* 19:1017–1030.

Chen, X., H. Fang, and J. E. Schwob. 2004. Multipotency of purified, transplanted globose basal cells in olfactory epithelium. *J Comp Neurol* 469:457–474.

Chen, Y. C., M. Auer-Grumbach, S. Matsukawa, M. Zitzelsberger, A. C. Themistocleous, T. M. Strom, C. Samara, A. W. Moore, L. T. Cho, G. T. Young, C. Weiss, M. Schabhuttl, R. Stucka, A. B. Schmid, Y. Parman, L. Graul-Neumann, W. Heinritz, E. Passarge, R. M. Watson, J. M. Hertz, U. Moog, M. Baumgartner, E. M. Valente, D. Pereira, C. M. Restrepo, I. Katona, M. Dusl, C. Stendel, T. Wieland, F. Stafford, F. Reimann, K. von Au, C. Finke, P. J. Willems, M. S. Nahorski, S. S. Shaikh, O. P. Carvalho, A. K. Nicholas, G. Karbani, M. A. McAleer, M. R. Cilio, J. C. McHugh, S. M. Murphy, A. D. Irvine, U. B. Jensen, R. Windhager, J. Weis, C. Bergmann, B. Rautenstrauss, J. Baets, P. De Jonghe, M. M. Reilly, R. Kropatsch, I. Kurth, R. Chrast, T. Michiue, D. L. Bennett, C. G. Woods, and J. Senderek. 2015. Transcriptional regulator PRDM12 is essential for human pain perception. *Nat Genet* 47:803–808.

Chen, Y., L. Shi, L. Zhang, R. Li, J. Liang, W. Yu, L. Sun, X. Yang, Y. Wang, Y. Zhang, and Y. Shang. 2008. The molecular mechanism governing the oncogenic potential of SOX2 in breast cancer. *J Biol Chem* 283:17969–17978.

Chen, Y., H. Yu, Y. Zhang, W. Li, N. Lu, W. Ni, Y. He, J. Li, S. Sun, Z. Wang, and H. Li. 2013. Cotransfection of Pax2 and Math1 promote in situ cochlear hair cell regeneration after neomycin insult. *Sci Rep* 3:2996.

Chen, Z. F., S. Rebelo, F. White, A. B. Malmberg, H. Baba, D. Lima, C. J. Woolf, A. I. Basbaum, and D. J. Anderson. 2001. The paired homeodomain protein DRG11 is required for the projection of cutaneous sensory afferent fibers to the dorsal spinal cord. *Neuron* 31:59–73.

Cheng, H., T. S. Aleman, A. V. Cideciyan, R. Khanna, S. G. Jacobson, and A. Swaroop. 2006. In vivo function of the orphan nuclear receptor NR2E3 in establishing photoreceptor identity during mammalian retinal development. *Hum Mol Genet* 15:2588–2602.

Cheng, H., H. Khanna, E. C. Oh, D. Hicks, K. P. Mitton, and A. Swaroop. 2004. Photoreceptor-specific nuclear receptor NR2E3 functions as a transcriptional activator in rod photoreceptors. *Hum Mol Genet* 13:1563–1575.

Cheng, L., A. Arata, R. Mizuguchi, Y. Qian, A. Karunaratne, P. A. Gray, S. Arata, S. Shirasawa, M. Bouchard, P. Luo, C. L. Chen, M. Busslinger, M. Goulding, H. Onimaru, and Q. Ma. 2004. Tlx3 and Tlx1 are post-mitotic selector genes determining glutamatergic over GABAergic cell fates. *Nat Neurosci* 7:510–517.

Chess, A., I. Simon, H. Cedar, and R. Axel. 1994. Allelic inactivation regulates olfactory receptor gene expression. *Cell* 78:823–834.

Cheung, L., P. Le Tissier, S. G. Goldsmith, M. Treier, R. Lovell-Badge, and K. Rizzoti. 2018. NOTCH activity differentially affects alternative cell fate acquisition and maintenance. *Elife* 7:e33318.

Cheung, L. Y. M., A. S. George, S. R. McGee, A. Z. Daly, M. L. Brinkmeier, B. S. Ellsworth, and S. A. Camper. 2018. Single-cell RNA sequencing reveals novel markers of male pituitary stem cells and hormone-producing cell types. *Endocrinology* 159:3910–3924.

Cheung, M., A. Tai, P. J. Lu, and K. S. Cheah. 2019. Acquisition of multipotent and migratory neural crest cells in vertebrate evolution. *Curr Opin Genet Dev* 57:84–90.

Chiappe, M. E., A. S. Kozlov, and A. J. Hudspeth. 2007. The structural and functional differentiation of hair cells in a lizard's basilar papilla suggests an operational principle of amniote cochleas. *J Neurosci* 27:11978–11985.

Chitnis, A. B., D. D. Nogare, and M. Matsuda. 2012. Building the posterior lateral line system in zebrafish. *Dev Neurobiol* 72:234–255.

Chiu, I. M., L. B. Barrett, E. K. Williams, D. E. Strochlic, S. Lee, A. D. Weyer, S. Lou, G. S. Bryman, D. P. Roberson, N. Ghasemlou, C. Piccoli, E. Ahat, V. Wang, E. J. Cobos, C. L. Stucky, Q. Ma, S. D. Liberles, and C. J. Woolf. 2014. Transcriptional profiling at whole population and single cell levels reveals somatosensory neuron molecular diversity. *Elife* 3:e04660.

Chiu, W. T., R. Charney Le, I. L. Blitz, M. B. Fish, Y. Li, J. Biesinger, X. Xie, and K. W. Cho. 2014. Genome-wide view of TGFbeta/Foxh1 regulation of the early mesendoderm program. *Development* 141:4537–4547.

Chng, Z., A. Teo, R. A. Pedersen, and L. Vallier. 2010. SIP1 mediates cell-fate decisions between neuroectoderm and mesendoderm in human pluripotent stem cells. *Cell Stem Cell* 6:59–70.

Cho, H. H., F. Cargnin, Y. Kim, B. Lee, R. J. Kwon, H. Nam, R. Shen, A. P. Barnes, J. W. Lee, S. Lee, and S. K. Lee. 2014. Isl1 directly controls a cholinergic neuronal identity in the developing forebrain and spinal cord by forming cell type-specific complexes. *PLoS Genet* 10:e1004280.

Cho, J. H., and M. J. Tsai. 2004. The role of BETA2/NeuroD1 in the development of the nervous system. *Mol Neurobiol* 30:35–47.

Chonko, K. T., I. Jahan, J. Stone, M. C. Wright, T. Fujiyama, M. Hoshino, B. Fritzsch, and S. M. Maricich. 2013. Atoh1 directs hair cell differentiation and survival in the late embryonic mouse inner ear. *Dev Biol* 381:401–410.

Chow, R. L., C. R. Altmann, R. A. Lang, and A. Hemmati-Brivanlou. 1999. Pax6 induces ectopic eyes in a vertebrate. *Development* 126:4213–4222.

Chow, R. L., and R. A. Lang. 2001. Early eye development in vertebrates. *Annu Rev Cell Dev Biol* 17:255–296.

Christ, S., U. W. Biebel, S. Hoidis, S. Friedrichsen, K. Bauer, and J. W. Smolders. 2004. Hearing loss in athyroid pax8 knockout mice and effects of thyroxine substitution. *Audiol Neurootol* 9:88–106.

Christensen, K. L., A. N. Patrick, E. L. McCoy, and H. L. Ford. 2008. The six family of homeobox genes in development and cancer. *Adv Cancer Res* 101:93–126.
Christophorou, N. A., A. P. Bailey, S. Hanson, and A. Streit. 2009. Activation of Six1 target genes is required for sensory placode formation. *Dev Biol* 336:327–336.
Christophorou, N. A., M. Mende, L. Lleras-Forero, T. Grocott, and A. Streit. 2010. Pax2 coordinates epithelial morphogenesis and cell fate in the inner ear. *Dev Biol* 345:180–190.
Chuah, M. I., and C. Au. 1991. Olfactory Schwann cells are derived from precursor cells in the olfactory epithelium. *J Neurosci Res* 29:172–180.
Churcher, A. M., and J. S. Taylor. 2011. The antiquity of chordate odorant receptors is revealed by the discovery of orthologs in the cnidarian *Nematostella vectensis*. *Genome Biol Evol* 3:36–43.
Cirillo, L. A., F. R. Lin, I. Cuesta, D. Friedman, M. Jarnik, and K. S. Zaret. 2002. Opening of compacted chromatin by early developmental transcription factors HNF3 (FoxA) and GATA-4. *Mol Cell* 9:279–289.
Clack, J. A., and E. Allin. 2004. The evolution of single- and multiple-ossicle ears in fishes and tetrapods. In *Evolution of the Vertebrate Auditory System*, edited by G. A. Manley, J. C. Fay and A. N. Popper, 128–163. New York: Springer.
Clevers, H., and R. Nusse. 2012. Wnt/beta-catenin signaling and disease. *Cell* 149:1192–1205.
Clowney, E. J., M. A. LeGros, C. P. Mosley, F. G. Clowney, E. C. Markenskoff-Papadimitriou, M. Myllys, G. Barnea, C. A. Larabell, and S. Lomvardas. 2012. Nuclear aggregation of olfactory receptor genes governs their monogenic expression. *Cell* 151:724–737.
Coate, T. M., and M. W. Kelley. 2013. Making connections in the inner ear: recent insights into the development of spiral ganglion neurons and their connectivity with sensory hair cells. *Semin Cell Dev Biol* 24:460–469.
Coffin, A., M. Kelley, G. A. Manley, and A. N. Popper. 2005. Evolution of sensory hair cells. In *Evolution of the vertebrate auditory system*, edited by G. A. Manley, A. N. Popper and R. R. Fay. New York: Springer.
Coletta, R. D., K. L. Christensen, D. S. Micalizzi, P. Jedlicka, M. Varella-Garcia, and H. L. Ford. 2008. Six1 overexpression in mammary cells induces genomic instability and is sufficient for malignant transformation. *Cancer Res* 68:2204–2213.
Coletta, R. D., K. Christensen, K. J. Reichenberger, J. Lamb, D. Micomonaco, L. Huang, D. M. Wolf, C. Müller-Tidow, T. R. Golub, K. Kawakami, and H. L. Ford. 2004. The Six1 homeoprotein stimulates tumorigenesis by reactivation of cyclin A1. *Proc Natl Acad Sci USA* 101:6478–6483.
Collazo, A., S. E. Fraser, and P. M. Mabee. 1994. A dual embryonic origin for vertebrate mechanoreceptors. *Science* 264:426–430.
Collin, J., R. Queen, D. Zerti, B. Dorgau, R. Hussain, J. Coxhead, S. Cockell, and M. Lako. 2019. Deconstructing retinal organoids: single cell RNA-Seq reveals the cellular components of human pluripotent stem cell-derived retina. *Stem Cells* 37:593–598.
Collin, S. P. 2009. Early evolution of vertebrate photoreception: lessons from lampreys and lungfishes. *Integr Zool* 4:87–98.
Collin, S. P., M. A. Knight, W. L. Davies, I. C. Potter, D. M. Hunt, and A. E. Trezise. 2003. Ancient colour vision: multiple opsin genes in the ancestral vertebrates. *Curr Biol* 13:R864–R865.
Collinson, J. M., R. E. Hill, and J. D. West. 2000. Different roles for Pax6 in the optic vesicle and facial epithelium mediate early morphogenesis of the murine eye. *Development* 127:945–956.
Collinson, J. M., J. C. Quinn, R. E. Hill, and J. D. West. 2003. The roles of Pax6 in the cornea, retina, and olfactory epithelium of the developing mouse embryo. *Dev Biol* 255:303–312.
Contin, M. A., D. M. Verra, and M. E. Guido. 2006. An invertebrate-like phototransduction cascade mediates light detection in the chicken retinal ganglion cells. *FASEB J* 20:2648–2650.

Cook, P. J., B. G. Ju, F. Telese, X. Wang, C. K. Glass, and M. G. Rosenfeld. 2009. Tyrosine dephosphorylation of H2AX modulates apoptosis and survival decisions. *Nature* 458:591–596.

Coolen, M., K. Sii-Felice, O. Bronchain, A. Mazabraud, F. Bourrat, S. Retaux, M. P. Felder-Schmittbuhl, S. Mazan, and J. L. Plouhinec. 2005. Phylogenomic analysis and expression patterns of large Maf genes in *Xenopus tropicalis* provide new insights into the functional evolution of the gene family in osteichthyans. *Dev Genes Evol* 215:327–339.

Coppola, E., M. Rallu, J. Richard, S. Dufour, D. Riethmacher, F. Guillemot, C. Goridis, and J. F. Brunet. 2010. Epibranchial ganglia orchestrate the development of the cranial neurogenic crest. *Proc Natl Acad Sci USA* 107:2066–2071.

Corbo, J. C., and C. L. Cepko. 2005. A hybrid photoreceptor expressing both rod and cone genes in a mouse model of enhanced S-cone syndrome. *PLoS Genet* 1:e11.

Cornesse, Y., T. Pieler, and T. Hollemann. 2005. Olfactory and lens placode formation is controlled by the hedgehog-interacting protein (Xhip) in Xenopus.*Dev Biol* 277:296–315.

Cosgrove, D., and M. Zallocchi. 2014. Usher protein functions in hair cells and photoreceptors. *Int J Biochem Cell Biol* 46:80–89.

Costa, A., and D. Henrique. 2015. Transcriptome profiling of induced hair cells (iHCs) generated by combined expression of Gfi1, Pou4f3 and Atoh1 during embryonic stem cell differentiation. *Genom Data* 6:77–80.

Costa, A., L. M. Powell, S. Lowell, and A. P. Jarman. 2017. Atoh1 in sensory hair cell development: constraints and cofactors. *Semin Cell Dev Biol* 65:60–68.

Costa, A., L. M. Powell, A. Soufi, S. Lowell, and A. P. Jarman. 2019. Atoh1 is repurposed from neuronal to hair cell determinant by Gfi1 acting as a coactivator without redistributing Atoh1's genomic binding sites. *Biorxiv* https://doi.org/10.1101/767574.

Couly, G. F., and N. M. Le Douarin. 1985. Mapping the early neural primordium in quail-chick chimeras. I. Developmental relationships between placodes, facial ectoderm, and prosencephalon. *Dev Biol* 110:422–439.

Couly, G. F., and N. M. Le Douarin. 1987. Mapping of the early neural primordium in quail chick chimeras. II. The prosencephalic neural plate and neural folds: implications for the genesis of cephalic human congenital abnormalities. *Dev Biol* 120:198–214.

Couly, G., and N. M. Le Douarin. 1990. Head morphogenesis in embryonic avian chimeras: evidence for a segmental pattern in the ectoderm corresponding to the neuromeres. *Development* 108:543–558.

Cui, W., S. I. Tomarev, J. Piatigorsky, A. B. Chepelinsky, and M. K. Duncan. 2004. Mafs, Prox1, and Pax6 can regulate chicken betaB1-crystallin gene expression. *J Biol Chem* 279:11088–11095.

Cuschieri, A., and L. H. Bannister. 1975. The development of the olfactory mucosa in the mouse: electron microscopy. *J Anat* 119:471–498.

Cvekl, A., and R. Ashery-Padan. 2014. The cellular and molecular mechanisms of vertebrate lens development. *Development* 141:4432–4447.

Cvekl, A., and M. K. Duncan. 2007. Genetic and epigenetic mechanisms of gene regulation during lens development. *Prog Retin Eye Res* 26:555–597.

Cvekl, A., and J. Piatigorsky. 1996. Lens development and crystallin gene expression: many roles for Pax-6. *Bioessays* 18:621–630.

Cvekl, A., and X. Zhang. 2017. Signaling and gene regulatorynetworks in mammalian lens development. *Trends Genet* 33:677–702.

D'Amico-Martel, A., and D. M. Noden. 1980. An autoradiographic analysis of the development of the chick trigeminal ganglion. *J Embryol Exp Morph* 55:167–182.

D'Amico-Martel, A., and D. M. Noden. 1983. Contributions of placodal and neural crest cells to avian cranial peripheral ganglia. *Am J Anat* 166:445–468.

D'Autreaux, F., E. Coppola, M. R. Hirsch, C. Birchmeier, and J. F. Brunet. 2011. Homeoprotein Phox2b commands a somatic-to-visceral switch in cranial sensory pathways. *Proc Natl Acad Sci USA* 108:20018–20023.

Dabdoub, A., C. Puligilla, J. M. Jones, B. Fritzsch, K. S. Cheah, L. H. Pevny, and M. W. Kelley. 2008. Sox2 signaling in prosensory domain specification and subsequent hair cell differentiation in the developing cochlea. *Proc Natl Acad Sci USA* 105:18396–18401.

Dalle Nogare, D., and A. B. Chitnis. 2017. A framework for understanding morphogenesis and migration of the zebrafish posterior lateral line primordium. *Mech Dev* 148:69–78.

Dallos, P. 2008. Cochlear amplification, outer hair cells and prestin. *Curr Opin Neurobiol* 18:370–376.

Dallos, P., and B. Fakler. 2002. Prestin, a new type of motor protein. *Nat Rev Mol Cell Biol* 3:104–111.

Dalton, R. P., and S. Lomvardas. 2015. Chemosensory receptor specificity and regulation. *Annu Rev Neurosci* 38:3313–3349.

Dalton, R. P., D. B. Lyons, and S. Lomvardas. 2013. Co-opting the unfolded protein response to elicit olfactory receptor feedback. *Cell* 155:321–332.

Dang, P., S. A. Fisher, D. J. Stefanik, J. Kim, and J. A. Raper. 2018. Coordination of olfactory receptor choice with guidance receptor expression and function in olfactory sensory neurons. *PLoS Genet* 14:e1007164.

Darian-Smith, I. 1973. The trigeminal system. In *Somatosensory system. Handbook of sensory physiology*, Vol. 2, edited by Iggo A., 271–314. Berlin, Heidelberg: Springer.

Dasen, J. S., J. P. Barbera, T. S. Herman, S. O. Connell, L. Olson, B. Ju, J. Tollkuhn, S. H. Baek, D. W. Rose, and M. G. Rosenfeld. 2001. Temporal regulation of a paired-like homeodomain repressor/TLE corepressor complex and a related activator is required for pituitary organogenesis. *Genes Dev* 15:3193–3207.

Dasen, J. S., S. M. O'Connell, S. E. Flynn, M. Treier, A. S. Gleiberman, D. P. Szeto, F. Hooshmand, A. K. Aggarwal, and M. G. Rosenfeld. 1999. Reciprocal interactions of Pit1 and GATA2 mediate signaling gradient-induced determination of pituitary cell types. *Cell* 97:587–598.

Dasen, J. S., and M. G. Rosenfeld. 2001. Signaling and transcriptional mechanisms in pituitary development. *Annu Rev Neurosci* 24:327–355.

Dash, S., and P. A. Trainor. 2020. The development, patterning and evolution of neural crest cell differentiation into cartilage and bone. *Bone* 137:115409.

Date-Ito, A., H. Ohara, M. Ichikawa, Y. Mori, and K. Hagino-Yamagishi. 2008. *Xenopus* V1R vomeronasal receptor family is expressed in the main olfactory system. *Chem Senses* 33:339–346.

Dattani, M. T., J. P. Martinez-Barbera, P. Q. Thomas, J. M. Brickman, R. Gupta, I. L. Martensson, H. Toresson, M. Fox, J. K. Wales, P. C. Hindmarsh, S. Krauss, R. S. Beddington, and I. C. Robinson. 1998. Mutations in the homeobox gene HESX1/Hesx1 associated with septo-optic dysplasia in human and mouse. *Nat Genet* 19:125–133.

Daudet, N., L. Ariza-McNaughton, and J. Lewis. 2007. Notch signalling is needed to maintain, but not to initiate, the formation of prosensory patches in the chick inner ear. *Development* 134:2369–2378.

Daudet, N., and J. Lewis. 2005. Two contrasting roles for Notch activity in chick inner ear development: specification of prosensory patches and lateral inhibition of hair-cell differentiation. *Development* 132:541–551.

Dauger, S., A. Pattyn, F. Lofaso, C. Gaultier, C. Goridis, J. Gallego, and J. F. Brunet. 2003. Phox2b controls the development of peripheral chemoreceptors and afferent visceral pathways. *Development* 130:6635–6642.

David, R., K. Ahrens, D. Wedlich, and G. Schlosser. 2001. *Xenopus* Eya1 demarcates all neurogenic placodes as well as migrating hypaxial muscle precursors. *Mech Dev* 103:189–192.

Davidson, E. H. 2010. Emerging properties of animal gene regulatory networks. *Nature* 468:911–920.

Davies, W. I., S. P. Collin, and D. M. Hunt. 2012. Molecular ecology and adaptation of visual photopigments in craniates. *Mol Ecol* 21:3121–3158.

Davies, W. I., L. Zheng, S. Hughes, T. K. Tamai, M. Turton, S. Halford, R. G. Foster, D. Whitmore, and M. W. Hankins. 2011. Functional diversity of melanopsins and their global expression in the teleost retina. *Cell Mol Life Sci* 68:4115–4132.

Davies, W. L., M. W. Hankins, and R. G. Foster. 2010. Vertebrate ancient opsin and melanopsin: divergent irradiance detectors. *Photochem Photobiol Sci* 9:1444–1457.

Davis, J. A., and R. R. Reed. 1996. Role of Olf-1 and Pax-6 transcription factors in neurodevelopment. *J Neurosci* 16:5082–5094.

Davis, R. L., and D. L. Turner. 2001. Vertebrate hairy and Enhancer of split related proteins: transcriptional repressors regulating cellular differentiation and embryonic patterning. *Oncogene* 20:8342–8357.

Davis, S. W., B. S. Ellsworth, M. I. Perez Millan, P. Gergics, V. Schade, N. Foyouzi, M. L. Brinkmeier, A. H. Mortensen, and S. A. Camper. 2013. Pituitary gland development and disease: from stem cell to hormone production. *Curr Top Dev Biol* 106:1–47.

Day, R. N., S. Koike, M. Sakai, M. Muramatsu, and R. A. Maurer. 1990. Both Pit-1 and the estrogen receptor are required for estrogen responsiveness of the rat prolactin gene. *Mol Endocrinol* 4:1964–1971.

De Camilli, P., and R. Jahn. 1990. Pathways to regulated exocytosis in neurons. *Annu Rev Physiol* 52:625–645.

de Croze, N., F. Maczkowiak, and A. H. Monsoro-Burq. 2011. Reiterative AP2a activity controls sequential steps in the neural crest gene regulatory network. *Proc Natl Acad Sci USA* 108:155–160.

de Jong, W. W., J. A. Leunissen, and C. E. Voorter. 1993. Evolution of the alpha-crystallin/small heat-shock protein family. *Mol Biol Evol* 10:103–126.

de la Rosa-Prieto, C., D. Saiz-Sánchez, I. Ubeda-Banon, L. Argandona-Palacios, S. Garcia-Munozguren, and A. Martinez-Marcos. 2010. Neurogenesis in subclasses of vomeronasal sensory neurons in adult mice. *Dev Neurobiol* 70:961–970.

De Miguel, E., M. C. Rodicio, and R. Anadón. 1989. Ganglion cells and retinopetal fibers of the larval lamprey retina: an HRP ultrastructural study. *Neurosci Lett* 106:1–6.

De Robertis, E. M., and H. Kuroda. 2004. Dorsal-ventral patterning and neural induction in *Xenopus* embryos. *Annu Rev Cell Dev Biol* 20:285–308.

DeCarvalho, A. C., S. L. Cappendijk, and J. M. Fadool. 2004. Developmental expression of the POU domain transcription factor Brn-3b (Pou4f2) in the lateral line and visual system of zebrafish. *Dev Dyn* 229:869–876.

Decatur, W. A., J. A. Hall, J. J. Smith, W. Li, and S. A. Sower. 2013. Insight from the lamprey genome: glimpsing early vertebrate development via neuroendocrine-associated genes and shared synteny of gonadotropin-releasing hormone (GnRH). *Gen Comp Endocrinol* 192:237–245.

Dee, C. T., C. S. Hirst, Y. H. Shih, V. B. Tripathi, R. K. Patient, and P. J. Scotting. 2008. Sox3 regulates both neural fate and differentiation in the zebrafish ectoderm. *Dev Biol* 320:301.

Dehal, P., Y. Satou, R. K. Campbell, J. Chapman, B. Degnan, A. De Tomaso, B. Davidson, A. Di Gregorio, M. Gelpke, D. M. Goodstein, N. Harafuji, K. E. Hastings, I. Ho, K. Hotta, W. Huang, T. Kawashima, P. Lemaire, D. Martinez, I. A. Meinertzhagen, S. Necula, M. Nonaka, N. Putnam, S. Rash, H. Saiga, M. Satake, A. Terry, L. Yamada, H. G. Wang, S. Awazu, K. Azumi, J. Boore, M. Branno, S. Chin-Bow, R. DeSantis, S. Doyle, P. Francino, D. N. Keys, S. Haga, H. Hayashi, K. Hino, K. S. Imai, K. Inaba, S. Kano, K. Kobayashi, M. Kobayashi, B. I. Lee, K. W. Makabe, C. Manohar, G. Matassi, M.

Medina, Y. Mochizuki, S. Mount, T. Morishita, S. Miura, A. Nakayama, S. Nishizaka, H. Nomoto, F. Ohta, K. Oishi, I. Rigoutsos, M. Sano, A. Sasaki, Y. Sasakura, E. Shoguchi, T. Shin-i, A. Spagnuolo, D. Stainier, M. M. Suzuki, O. Tassy, N. Takatori, M. Tokuoka, K. Yagi, F. Yoshizaki, S. Wada, C. Zhang, P. D. Hyatt, F. Larimer, C. Detter, N. Doggett, T. Glavina, T. Hawkins, P. Richardson, S. Lucas, Y. Kohara, M. Levine, N. Satoh, and D. S. Rokhsar. 2002. The draft genome of *Ciona intestinalis*: insights into chordate and vertebrate origins. *Science* 298:2157–2167.

DeHamer, M. K., J. L. Guevara, K. Hannon, B. B. Olwin, and A. L. Calof. 1994. Genesis of olfactory receptor neurons in vitro: regulation of progenitor cell divisions by fibroblast growth factors. *Neuron* 13:1083–1097.

Del Punta, K., T. Leinders-Zufall, I. Rodriguez, D. Jukam, C. J. Wysocki, S. Ogawa, F. Zufall, and P. Mombaerts. 2002. Deficient pheromone responses in mice lacking a cluster of vomeronasal receptor genes. *Nature* 419:70–74.

Del Rio-Tsonis, K., and P. A. Tsonis. 2003. Eye regeneration at the molecular age. *Dev Dyn* 226:211–224.

Delaune, E., P. Lemaire, and L. Kodjabachian. 2005. Neural induction in *Xenopus* requires early FGF signalling in addition to BMP inhibition. *Development* 132:299–310.

Demski, L. S. 1993. Terminal nerve complex. *Acta Anat* 148:81–95.

Denef, C. 2008. Paracrinicity: the story of 30 years of cellular pituitary crosstalk. *J Neuroendocrinol* 20:1–70.

Deng, M., H. Yang, X. Xie, G. Liang, and L. Gan. 2014. Comparative expression analysis of POU4F1, POU4F2 and ISL1 in developing mouse cochleovestibular ganglion neurons. *Gene Expr Patterns* 15:31–37.

Dennis, D. J., S. Han, and C. Schuurmans. 2019. bHLH transcription factors in neural development, disease, and reprogramming. *Brain Res* 1705:48–65.

Deretic, D., R. H. Aebersold, H. D. Morrison, and D. S. Papermaster. 1994. Alpha A- and alpha B-crystallin in the retina. Association with the post-Golgi compartment of frog retinal photoreceptors. *J Biol Chem* 269:16853–16861.

Devnath, S., and K. Inoue. 2008. An insight to pituitary folliculo-stellate cells. *J Neuroendocrinol* 20:687–691.

Dhanesh, S. B., C. Subashini, and J. James. 2016. Hes1: the maestro in neurogenesis. *Cell Mol Life Sci* 73:4019–4042.

Dickinson, A., and H. Sive. 2007. Positioning the extreme anterior in *Xenopus*: cement gland, primary mouth and anterior pituitary. *Semin Cell Dev Biol* 18:525–533.

Dietschi, Q., J. Tuberosa, L. Rosingh, G. Loichot, M. Ruedi, A. Carleton, and I. Rodriguez. 2017. Evolution of immune chemoreceptors into sensors of the outside world. *Proc Natl Acad Sci USA* 114:7397–7402.

Ding, Y., G. Colozza, E. A. Sosa, Y. Moriyama, S. Rundle, L. Salwinski, and E. M. De Robertis. 2018. Bighead is a Wnt antagonist secreted by the *Xenopus* Spemann organizer that promotes Lrp6 endocytosis. *Proc Natl Acad Sci USA* 115:E9135–E9144.

Ding, Y. Q., J. Yin, H. M. Xu, M. F. Jacquin, and Z. F. Chen. 2003. Formation of whisker-related principal sensory nucleus-based lemniscal pathway requires a paired homeodomain transcription factor, Drg11. *J Neurosci* 23:7246–7254.

Donner, A. L., V. Episkopou, and R. L. Maas. 2007. Sox2 and Pou2f1 interact to control lens and olfactory placode development. *Dev Biol* 303:784–799.

Dores, R. M., and A. J. Baron. 2011. Evolution of POMC: origin, phylogeny, posttranslational processing, and the melanocortins. *Ann NY Acad Sci* 1220:34–48.

Drysdale, T. A., and R. P. Elinson. 1992. Cell migration and induction in the development of the surface ectodermal pattern of the *Xenopus laevis* tadpole. *Dev Growth Differ.* 34:51–59.

Dubreuil, V., M. R. Hirsch, C. Jouve, J. F. Brunet, and C. Goridis. 2002. The role of Phox2b in synchronizing pan-neuronal and type-specific aspects of neurogenesis. *Development* 129:5241–5253.

Dubreuil, V., M. R. Hirsch, A. Pattyn, J. F. Brunet, and C. Goridis. 2000. The Phox2b transcription factor coordinately regulates neuronal cell cycle exit and identity. *Development* 127:5191–5201.

Dude, C. M., C. Y. Kuan, J. R. Bradshaw, N. D. Greene, F. Relaix, M. R. Stark, and C. V. Baker. 2009. Activation of Pax3 target genes is necessary but not sufficient for neurogenesis in the ophthalmic trigeminal placode. *Dev Biol* 326:314–326.

Dufourcq, P., S. Rastegar, U. Strähle, and P. Blader. 2004. Parapineal specific expression of gfi1 in the zebrafish epithalamus. *Gene Expr Patterns* 4:53–57.

Dufourcq, P., M. Roussigne, P. Blader, F. Rosa, N. Peyrieras, and S. Vriz. 2006. Mechanosensory organ regeneration in adults: the zebrafish lateral line as a model. *Mol Cell Neurosci* 33:180–187.

Duggan, C. D., S. Demaria, A. Baudhuin, D. Stafford, and J. Ngai. 2008. Foxg1 is required for development of the vertebrate olfactory system. *J Neurosci* 28:5229–5239.

Dulac, C., and R. Axel. 1995. A novel family of genes encoding putative pheromone receptors in mammals. *Cell* 83:195–206.

Dulac, C., and A. T. Torello. 2003. Molecular detection of pheromone signals in mammals: from genes to behaviour. *Nat Rev Neurosci* 4:551–562.

Duncan, J. S., and B. Fritzsch. 2012. Evolution of sound and balance perception: innovations that aggregate single hair cells into the ear and transform a gravistatic sensor into the organ of corti. *Anat Rec* 295:1760–1774.

Duncan, J. S., and B. Fritzsch. 2013. Continued expression of GATA3 is necessary for cochlear neurosensory development. *PLoS One* 8:e62046.

Duncan, M. K., L. Kos, N. A. Jenkins, D. J. Gilbert, N. G. Copeland, and S. I. Tomarev. 1997. Eyes absent: a gene family found in several metazoan phyla. *Mamm Genome* 8:479–485.

Dupin, E., G. W. Calloni, J. M. Coelho-Aguiar, and N. M. Le Douarin. 2018. The issue of the multipotency of the neural crest cells. *Dev Biol* 444 Suppl 1:S47–S59.

Duran Alonso, M. B., I. López Hernandez, M. A. de la Fuente, J. Garcia-Sancho, F. Giráldez, and T. Schimmang. 2018. Transcription factor induced conversion of human fibroblasts towards the hair cell lineage. *PLoS One* 13:e0200210.

Durruthy-Durruthy, R., A. Gottlieb, B. H. Hartman, J. Waldhaus, R. D. Laske, R. Altman, and S. Heller. 2014. Reconstruction of the mouse otocyst and early neuroblast lineage at single-cell resolution. *Cell* 157:964–978.

Dutta, S., J. E. Dietrich, G. Aspock, R. D. Burdine, A. Schier, M. Westerfield, and Z. M. Varga. 2005. pitx3 defines an equivalence domain for lens and anterior pituitary placode. *Development* 132:1579–1590.

Dutton, K., L. Abbas, J. Spencer, C. Brannon, C. Mowbray, M. Nikaido, R. N. Kelsh, and T. T. Whitfield. 2009. A zebrafish model for Waardenburg syndrome type IV reveals diverse roles for Sox10 in the otic vesicle. *Dis Model Mech* 2:68–83.

Dvorakova, M., I. Jahan, I. Macova, T. Chumak, R. Bohuslavova, J. Syka, B. Fritzsch, and G. Pavlinkova. 2016. Incomplete and delayed Sox2 deletion defines residual ear neurosensory development and maintenance. *Sci Rep* 6:38253.

Dvorakova, M., I. Macova, R. Bohuslavova, M. Anderova, B. Fritzsch, and G. Pavlinkova. 2020. Early ear neuronal development, but not olfactory or lens development, can proceed without SOX2. *Dev Biol* 457:43–56.

Dvoryanchikov, G., D. Hernandez, J. K. Roebber, D. L. Hill, S. D. Roper, and N. Chaudhari. 2017. Transcriptomes and neurotransmitter profiles of classes of gustatory and somatosensory neurons in the geniculate ganglion. *Nat Commun* 8:760.

Dykes, I. M., J. Lanier, S. R. Eng, and E. E. Turner. 2010. Brn3a regulates neuronal subtype specification in the trigeminal ganglion by promoting Runx expression during sensory differentiation. *Neural Dev* 5:3.

Dykes, I. M., L. Tempest, S. I. Lee, and E. E. Turner. 2011. Brn3a and Islet1 act epistatically to regulate the gene expression program of sensory differentiation. *J Neurosci* 31:9789–9799.

Eagleson, G., B. Ferreiro, and W. A. Harris. 1995. Fate of the anterior neural ridge and the morphogenesis of the *Xenopus* forebrain. *J Neurobiol* 28:146–158.

Eagleson, G. W., and R. D. Dempewolf. 2002. The role of the anterior neural ridge and Fgf-8 in early forebrain patterning and regionalization in *Xenopus laevis. Comp Biochem Physiol B Biochem Mol Biol* 132:179–189.

Eagleson, G. W., and W. A. Harris. 1990. Mapping of the presumptive brain regions in the neural plate of *Xenopus laevis. J Neurobiol* 21:427–440.

Eagleson, G. W., B. G. Jenks, and A. P. Van Overbeeke. 1986. The pituitary adrenocorticotropes originate from neural ridge tissue in *Xenopus laevis. J Embryol Exp Morphol* 95:1–14.

Eakin, R.M. 1963. Lines of evolution of photoreceptors. In *General physiology of cell specialization*, edited by D. Mazia and A. Tyler, 393–425. New York: McGraw-Hill.

Eakin, R.M. 1979. Evolutionary significance of photoreceptors: in retrospect. *Am Zool* 19:647–653.

Eames, B. F., D. M. Medeiros, and I. Adameyko. 2020. *Evolving neural crest cells.* Boca Raton: CRC Press.

Eatock, R. A., and A. Lysakowski. 2006. Mammalian vestibular hair cells. In *Vertebrate hair cells*, edited by R. A. Eatock, R. R. Fay and A. N. Popper, 348–442. New York: Springer.

Eatock, R. A., A. Rusch, A. Lysakowski, and M. Saeki. 1998. Hair cells in mammalian utricles. *Otolaryngol Head Neck Surg* 119:172–181.

Eckler, M. J., W. L. McKenna, S. Taghvaei, S. K. McConnell, and B. Chen. 2011. Fezf1 and Fezf2 are required for olfactory development and sensory neuron identity. *J Comp Neurol* 519:1829–1846.

Eddison, M., I. Leroux, and J. Lewis. 2000. Notch signaling in the development of the inner ear: lessons from *Drosophila. Proc Nat Acad Sci USA* 97:11692–11699.

Edlund, R. K., O. Birol, and A. K. Groves. 2015. The role of foxi family transcription factors in the development of the ear and jaw. *Curr Top Dev Biol* 111:461–495.

Eijkelkamp, N., K. Quick, and J. N. Wood. 2013. Transient receptor potential channels and mechanosensation. *Annu Rev Neurosci* 36:519–546.

Eilertsen, M., O. Drivenes, R. B. Edvardsen, C. A. Bradley, L. O. Ebbesson, and J. V. Helvik. 2014. Exorhodopsin and melanopsin systems in the pineal complex and brain at early developmental stages of Atlantic halibut (*Hippoglossus hippoglossus*). *J Comp Neurol* 522:4003–4022.

Eisthen, H. L. 1992. Phylogeny of the vomeronasal system and of receptor cell types in the olfactory and vomeronasal epithelia of vertebrates. *Microsc Res Tech* 23:1–21.

Eisthen, H. L. 1997. Evolution of vertebrate olfactory systems. *Brain Behav Evol* 50:222–233.

Eisthen, H. L., R. J. Delay, C. R. Wirsig-Wiechmann, and V. E. Dionne. 2000. Neuromodulatory effects of gonadotropin releasing hormone on olfactory receptor neurons. *J Neurosci* 20:3947–3955.

Ekström, P., and H. Meissl. 2003. Evolution of photosensory pineal organs in new light: the fate of neuroendocrine photoreceptors. *Philos Trans R Soc Lond B Biol Sci* 358:1679–700.

El Amraoui, A., and P. M. Dubois. 1993. Experimental evidence for an early commitment of gonadotropin-releasing hormone neurons, with special regard to their origin from the ectoderm of nasal cavity presumptive territory. *Neuroendocrinology* 57:991–1002.

El Hashash, A. H., D. Al Alam, G. Turcatel, O. Rogers, X. Li, S. Bellusci, and D. Warburton. 2011. Six1 transcription factor is critical for coordination of epithelial, mesenchymal and vascular morphogenesis in the mammalian lung. *Dev Biol* 353:242–258.

Elinson, R. P. 1997. Amphibians. In *Embryology. Constructing the organism*, edited by S. F. Gilbert and A. M. Raunio, 409–436. Sunderland: Sinauer.

Elkon, R., B. Milon, L. Morrison, M. Shah, S. Vijayakumar, M. Racherla, C. C. Leitch, L. Silipino, S. Hadi, M. Weiss-Gayet, E. Barras, C. D. Schmid, A. Ait-Lounis, A. Barnes, Y. Song, D. J. Eisenman, E. Eliyahu, G. I. Frolenkov, S. E. Strome, B. Durand, N. A. Zaghloul, S. M. Jones, W. Reith, and R. Hertzano. 2015. RFX transcription factors are essential for hearing in mice. *Nat Commun* 6:8549.

Elkouby, Y. M., S. Elias, E. S. Casey, S. A. Blythe, N. Tsabar, P. S. Klein, H. Root, K. J. Liu, and D. Frank. 2010. Mesodermal Wnt signaling organizes the neural plate via Meis3. *Development* 137:1531–1541.

Elliott, K. L., B. Fritzsch, and J. S. Duncan. 2018. Evolutionary and developmental biology provide insights into the regeneration of organ of Corti hair cells. *Front Cell Neurosci* 12:252.

Ellsworth, B. S., D. L. Butts, and S. A. Camper. 2008. Mechanisms underlying pituitary hypoplasia and failed cell specification in Lhx3-deficient mice. *Dev Biol* 313:118–129.

Ellsworth, B. S., N. Egashira, J. L. Haller, D. L. Butts, J. Cocquet, C. M. Clay, R. Y. Osamura, and S. A. Camper. 2006. FOXL2 in the pituitary: molecular, genetic and developmental analysis. *Mol Endocrinol* 20:2796–2805.

Elsaesser, R., G. Montani, R. Tirindelli, and J. Paysan. 2005. Phosphatidyl-inositide signalling proteins in a novel class of sensory cells in the mammalian olfactory epithelium. *Eur J Neurosci* 21:2692–2700.

Elsaesser, R., and J. Paysan. 2007. The sense of smell, its signalling pathways, and the dichotomy of cilia and microvilli in olfactory sensory cells. *BMC Neurosci* 8 Suppl 3:S1.

Embry, A. C., J. L. Glick, M. E. Linder, and P. J. Casey. 2004. Reciprocal signaling between the transcriptional co-factor Eya2 and specific members of the Galphai family. *Mol Pharmacol* 66:1325–1331.

Emerson, M. M., N. Surzenko, J. J. Goetz, J. Trimarchi, and C. L. Cepko. 2013. Otx2 and Onecut1 promote the fates of cone photoreceptors and horizontal cells and repress rod photoreceptors. *Dev Cell* 26:59–72.

Eng, S. R., I. M. Dykes, J. Lanier, N. Fedtsova, and E. E. Turner. 2007. POU-domain factor Brn3a regulates both distinct and common programs of gene expression in the spinal and trigeminal sensory ganglia. *Neural Develop* 2:3.

Eng, S. R., K. Gratwick, J. M. Rhee, N. Fedtsova, L. Gan, and E. E. Turner. 2001. Defects in sensory axon growth precede neuronal death in Brn3a-deficient mice. *J Neurosci* 21:541–549.

Engelen, E., U. Akinci, J. C. Bryne, J. Hou, C. Gontan, M. Moen, D. Szumska, C. Kockx, W. van Ijcken, D. H. Dekkers, J. Demmers, E. J. Rijkers, S. Bhattacharya, S. Philipsen, L. H. Pevny, F. G. Grosveld, R. J. Rottier, B. Lenhard, and R. A. Poot. 2011. Sox2 cooperates with Chd7 to regulate genes that are mutated in human syndromes. *NatGenet* 43:607–611.

Enright, J. M., K. A. Lawrence, T. Hadzic, and J. C. Corbo. 2015. Transcriptome profiling of developing photoreceptor subtypes reveals candidate genes involved in avian photoreceptor diversification. *J Comp Neurol* 523:649–668.

Ericson, J., S. Norlin, T. M. Jessell, and T. Edlund. 1998. Integrated FGF and BMP signaling controls the progression of progenitor cell differentiation and the emergence of pattern in the embryonic anterior pituitary. *Development* 125:1005–1015.

Erkman, L., R. J. McEvilly, L. Luo, A. K. Ryan, F. Hooshmand, S. M. O'Connell, E. M. Keithley, D. H. Rapaport, A. F. Ryan, and M. G. Rosenfeld. 1996. Role of transcription factors Brn-3.1 and Brn-3.2 in auditory and visual system development. *Nature* 381:603–606.

Ernsberger, U. 2015. Can the 'neuron theory' be complemented by a universal mechanism for generic neuronal differentiation. *Cell Tissue Res* 359:343–384.

Esteve, P., and P. Bovolenta. 1999. cSix4, a member of the six gene family of transcription factors, is expressed during placode and somite development. *Mech Dev* 85:161–165.

Evsen, L., S. Sugahara, M. Uchikawa, H. Kondoh, and D. K. Wu. 2013. Progression of neurogenesis in the inner ear requires inhibition of Sox2 transcription by neurogenin1 and neurod1. *J Neurosci* 33:3879–3890.

Ezin, A. M., S. E. Fraser, and M. Bronner-Fraser. 2009. Fate map and morphogenesis of presumptive neural crest and dorsal neural tube. *Dev Biol* 330:221–236.

Faber, S. C., P. Dimanlig, H. P. Makarenkova, S. Shirke, K. Ko, and R. A. Lang. 2001. FGF receptor signalin plays a role in lens induction. *Development* 128:4425–4438.

Fain, G. L., R. Hardie, and S. B. Laughlin. 2010. Phototransduction and the evolution of photoreceptors. *Curr Biol* 20:R114–R124.

Falcon, J., L. Besseau, M. Fuentes, S. Sauzet, E. Magnanou, and G. Boeuf. 2009. Structural and functional evolution of the pineal melatonin system in vertebrates. *Ann N Y Acad Sci* 1163:101–111.

Falk, N., M. Losl, N. Schroder, and A. Giessl. 2015. Specialized cilia in mammalian sensory systems. *Cells* 4:500–519.

Fan, X., L. F. Brass, M. Poncz, F. Spitz, P. Maire, and D. R. Manning. 2000. The alpha subunits of Gz and Gi interact with the eyes absent transcription cofactor Eya2, preventing its interaction with the six class of homeodomain-containing proteins. *J Biol Chem* 275:32129–32134.

Farabaugh, S. M., D. S. Micalizzi, P. Jedlicka, R. Zhao, and H. L. Ford. 2011. Eya2 is required to mediate the pro-metastatic functions of Six1 via the induction of TGF-beta signaling, epithelial-mesenchymal transition, and cancer stem cell properties. *Oncogene* 31:552–562.

Farah, M. H., J. M. Olson, H. B. Sucic, R. I. Hume, S. J. Tapscott, and D. L. Turner. 2000. Generation of neurons by transient expression of neural bHLH proteins in mammalian cells. *Development* 127:693–702.

Farrell, J. A., Y. Wang, S. J. Riesenfeld, K. Shekhar, A. Regev, and A. F. Schier. 2018. Single-cell reconstruction of developmental trajectories during zebrafish embryogenesis. *Science* 360:eaar3131.

Faucherre, A., J. Pujol-Marti, K. Kawakami, and H. López-Schier. 2009. Afferent neurons of the zebrafish lateral line are strict selectors of hair-cell orientation. *PLoS One* 4:e4477.

Fauquier, T., N. C. Guerineau, R. A. McKinney, K. Bauer, and P. Mollard. 2001. Folliculostellate cell network: a route for long-distance communication in the anterior pituitary. *Proc Natl Acad Sci USA* 98:8891–8896.

Fauquier, T., K. Rizzoti, M. Dattani, R. Lovell-Badge, and I. C. Robinson. 2008. SOX2-expressing progenitor cells generate all of the major cell types in the adult mouse pituitary gland. *Proc Natl Acad Sci USA* 105:2907–2912.

Faure, S., Barbara P. de Santa, D. J. Roberts, and M. Whitman. 2002. Endogenous patterns of BMP signaling during early chick development. *Dev Biol* 244:44–65.

Fautrez, J. 1942. La signification de la partie céphalique du bourrelet de la plaque médullaire chez les urodèles. Localisation des ébauches présomptives des microplacodes des nerfs crâniens et de la tête au stade neurale. *Bull Cl Sci Acad R Belg* 28:391–403.

Favaro, R., M. Valotta, A. L. Ferri, E. Latorre, J. Mariani, C. Giachino, C. Lancini, V. Tosetti, S. Ottolenghi, V. Taylor, and S. K. Nicolis. 2009. Hippocampal development and neural stem cell maintenance require Sox2-dependent regulation of Shh. *Nat Neurosci* 12:1248–1256.

Favor, J., R. Sandulache, A. Neuhäuser-Klaus, W. Pretsch, B. Chatterjee, E. Senft, W. Wurst, V. Blanquet, P. Grimes, R. Spörle, and K. Schughart. 1996. The mouse Pax2 1Neu mutation is identical to a human Pax2 mutation in a family with renal-coloboma syndrome and results in developmental defects of the brain, ear, eye and kidney. *Proc Natl Acad Sci USA* 93:13870–13875.

Fay, R. R., and A. N. Popper. 2000. Evolution of hearing in vertebrates: the inner ears and processing. *Hear Res* 149:1–10.

Fedtsova, N. G., and E. E. Turner. 1995. Brn-3.0 expression identifies early post-mitotic CNS neurons and sensory neural precursors. *Mech Dev* 53:291–304.

Feijoo, C. G., M. P. Saldias, J. F. De la Paz, J. L. Gómez-Skarmeta, and M. L. Allende. 2009. Formation of posterior cranial placode derivatives requires the Iroquois transcription factor irx4a. *Mol Cell Neurosci* 40:328–337.

Feinstein, P., T. Bozza, I. Rodriguez, A. Vassalli, and P. Mombaerts. 2004. Axon guidance of mouse olfactory sensory neurons by odorant receptors and the beta2 adrenergic receptor. *Cell* 117:833–846.

Fekete, D. M., and D. K. Wu. 2002. Revisiting cell fate specification in the inner ear. *Curr Opin Neurobiol* 12:35–42.

Feledy, J. A., M. J. Beanan, J. J. Sandoval, J. S. Goodrich, J. H. Lim, M. Matsuo-Takasaki, S. M. Sato, and T. D. Sargent. 1999. Inhibitory patterning of the anterior neural plate in *Xenopus* by homeodomain factors Dlx3 and Msx1. *Dev Biol* 212:455–464.

Feng, Y., and Q. Xu. 2010. Pivotal role of hmx2 and hmx3 in zebrafish inner ear and lateral line development. *Dev Biol* 339:507–518.

Fernandes, A. M., K. Fero, A. B. Arrenberg, S. A. Bergeron, W. Driever, and H. A. Burgess. 2012. Deep brain photoreceptors control light-seeking behavior in zebrafish larvae. *Curr Biol* 22:2042–2047.

Fernàndez, C., J. M. Goldberg, and R. A. Baird. 1990. The vestibular nerve of the chinchilla. III. Peripheral innervation patterns in the utricular macula. *J Neurophysiol* 63:767–780.

Fernholm, B., and K. Holmberg. 1975. The eyes in three genera of hagfish (*Eptatretus*, *Paramyxine* and *Myxine*)-a case of degenerative evolution. *Vision Res* 15:253–259.

Ferrando, S., and L. Gallus. 2013. Is the olfactory system of cartilaginous fishes a vomeronasal system? *Front Neuroanat* 7:37.

Ferrando, S., C. Gambardella, S. Ravera, S. Bottero, T. Ferrando, L. Gallus, V. Manno, A. P. Salati, P. Ramoino, and G. Tagliafierro. 2009. Immunolocalization of G protein alpha subunits in the olfactory system of the cartilaginous fish *Scyliorhinus canicula*. *Anat Rec* 292:1771–1779.

Ferri, A. L., M. Cavallaro, D. Braida, A. Di Cristofano, A. Canta, A. Vezzani, S. Ottolenghi, P. P. Pandolfi, M. Sala, S. DeBiasi, and S. K. Nicolis. 2004. Sox2 deficiency causes neurodegeneration and impaired neurogenesis in the adult mouse brain. *Development* 131:3805–3819.

Fettiplace, R., and C. M. Hackney. 2006. The sensory and motor roles of auditory hair cells. *Nat Rev Neurosci* 7:19–29.

Fettiplace, R., and K. X. Kim. 2014. The physiology of mechanoelectrical transduction channels in hearing. *Physiol Rev* 94:951–986.

Feuda, R., S. C. Hamilton, J. O. McInerney, and D. Pisani. 2012. Metazoan opsin evolution reveals a simple route to animal vision. *Proc Natl Acad Sci USA* 109:18868–18872.

Finger, T. E. 1997. Evolution of taste and solitary chemoreceptor cell systems. *Brain Behav Evol* 50:234–243.

Fischer, F. P. 1992. Quantitative analysis of the innervation of the chicken basilar papilla. *Hear Res* 61:167–178.

Fisher, M., and R. M. Grainger. 2004. Lens induction and determination. In *Development of the ocular lens*, edited by F. J. Lovicu and M. L. Robinson, 27–47. New York: Cambridge University Press.

Flasse, L. C., D. G. Stern, J. L. Pirson, I. Manfroid, B. Peers, and M. L. Voz. 2013. The bHLH transcription factor Ascl1a is essential for the specification of the intestinal secretory cells and mediates Notch signaling in the zebrafish intestine. *Dev Biol* 376:187–197.

Fletcher, R. B., D. Das, L. Gadye, K. N. Street, A. Baudhuin, A. Wagner, M. B. Cole, Q. Flores, Y. G. Choi, N. Yosef, E. Purdom, S. Dudoit, D. Risso, and J. Ngai. 2017. Deconstructing olfactory stem cell trajectories at single-cell resolution. *Cell Stem Cell* 20:817–830.e8.

Fletcher, R. B., M. S. Prasol, J. Estrada, A. Baudhuin, K. Vranizan, Y. G. Choi, and J. Ngai. 2011. p63 regulates olfactory stem cell self-renewal and differentiation. *Neuron* 72:748–759.

Flici, H., C. E. Schnitzler, R. C. Millane, G. Govinden, A. Houlihan, S. D. Boomkamp, S. Shen, A. D. Baxevanis, and U. Frank. 2017. An evolutionarily conserved SoxB-Hdac2 crosstalk regulates neurogenesis in a cnidarian. *Cell Rep* 18:1395–1409.

Flock, A. 1967. Ultrastructure and function in the lateral line organs. In *Lateral line detectors*, edited by P. Cahn, 163–197. Bloomington: Indiana University Press.

Flock, A., and J. Wersäll. 1962. A study of the orientation of the sensory hairs of the receptor cells in the lateral line organ of fish, with special reference to the function of the receptors. *J Cell Biol* 15:19–27.

Fode, C., G. Gradwohl, X. Morin, A. Dierich, M. Lemeur, C. Goridis, and F. Guillemot. 1998. The bHLH protein neurogenin 2 is a determination factor for epibranchial placode-derived sensory neurons. *Neuron* 20:483–494.

Ford, H. L., E. N. Kabingu, E. A. Bump, G. L. Mutter, and A. B. Pardee. 1998. Abrogation of the G2 cell cycle checkpoint associated with overexpression of HSIX1: a possible mechanism of breast carcinogenesis. *Proc Natl Acad Sci USA* 95:12608–12613.

Forget, A., L. Bihannic, S. M. Cigna, C. Lefevre, M. Remke, M. Barnat, S. Dodier, H. Shirvani, A. Mercier, A. Mensah, M. Garcia, S. Humbert, M. D. Taylor, A. Lasorella, and O. Ayrault. 2014. Shh signaling protects Atoh1 from degradation mediated by the E3 ubiquitin ligase Huwe1 in neural precursors. *Dev Cell* 29:649–661.

Forni, P. E., C. Taylor-Burds, V. S. Melvin, T. Williams, and S. Wray. 2011. Neural crest and ectodermal cells intermix in the nasal placode to give rise to GnRH-1 neurons, sensory neurons, and olfactory ensheathing cells. *J Neurosci* 31:6915–6927.

Forni, P. E., and S. Wray. 2015. GnRH, anosmia and hypogonadotropic hypogonadism–where are we? *Front Neuroendocrinol* 36:165–177.

Fortin, J., U. Boehm, C. X. Deng, M. Treier, and D. J. Bernard. 2014. Follicle-stimulating hormone synthesis and fertility depend on SMAD4 and FOXL2. *FASEB J* 28:3396–3410.

Fortin, J., P. Lamba, Y. Wang, and D. J. Bernard. 2009. Conservation of mechanisms mediating gonadotrophin-releasing hormone 1 stimulation of human luteinizing hormone beta subunit transcription. *Mol Hum Reprod* 15:77–87.

Frederikse, P. H., C. Kasinathan, and N. J. Kleiman. 2012. Parallels between neuron and lens fiber cell structure and molecular regulatory networks. *Dev Biol* 368:255–260.

Frederikse, P., and C. Kasinathan. 2017. Lens biology is a dimension of neurobiology. *Neurochem Res* 42:933–942.

Freitas, R., G. Zhang, J. S. Albert, D. H. Evans, and M. J. Cohn. 2006. Developmental origin of shark electrosensory organs. *Evol Dev* 8:74–80.

Freter, S., S. J. Fleenor, R. Freter, K. J. Liu, and J. Begbie. 2013. Cranial neural crest cells form corridors prefiguring sensory neuroblast migration. *Development* 140:3595–3600.

Freter, S., Y. Muta, S. S. Mak, S. Rinkwitz, and R. K. Ladher. 2008. Progressive restriction of otic fate: the role of FGF and Wnt in resolving inner ear potential. *Development* 135:3415–3424.

Freter, S., Y. Muta, P. O'Neill, V. S. Vassilev, S. Kuraku, and R. K. Ladher. 2012. Pax2 modulates proliferation during specification of the otic and epibranchial placodes. *Dev Dyn* 241:1716–1728.

Freyer, L., V. Aggarwal, and B. E. Morrow. 2011. Dual embryonic origin of the mammalian otic vesicle forming the inner ear. *Development* 138:5403–5414.

Friedle, H., and W. Knöchel. 2002. Cooperative interaction of Xvent-2 and GATA-2 in the activation of the ventral homeobox gene Xvent-1B. *J Biol Chem* 277:23872–23881.

Friedman, R. A., L. Makmura, E. Biesiada, X. Wang, and E. M. Keithley. 2005. Eya1 acts upstream of Tbx1, Neurogenin 1, NeuroD and the neurotrophins BDNF and NT-3 during inner ear development. *Mech Dev* 122:625–634.

Frisdal, A., and P. A. Trainor. 2014. Development and evolution of the pharyngeal apparatus. *Wiley Interdiscip Rev Dev Biol* 3:403–418.

Fritzsch, B. 1989. Diversity and regression in the amphibian lateral line and electrosensory system. In *The mechanosensory lateral line*, edited by S. Coombs, P. Görner and H. Münz, 99–114. New York: Springer.

Fritzsch, B., K. F. Barald, and M. I. Lomax. 1998. Early embryology of the vertebrate ear. In *Development of the auditory system*, edited by E. W. Rubel, A. N. Popper and R. R. Fay, 80–145. New York: Springer.

Fritzsch, B., and K. W. Beisel. 2004. Keeping sensory cells and evolving neurons to connect them to the brain: molecular conservation and novelties in vertebrate ear development. *Brain Behav Evol* 64:182–197.

Fritzsch, B., K. W. Beisel, and N. A. Bermingham. 2000. Developmental evolutionary biology of the vertebrate ear: conserving mechanoelectric transduction and developmental pathways in diverging morphologies. *Neuroreport* 11:R35–R44.

Fritzsch, B., K. W. Beisel, K. Jones, I. Farinas, A. Maklad, J. Lee, and L. F. Reichardt. 2002. Development and evolution of inner ear sensory epithelia and their innervation. *J Neurobiol* 53:143–156.

Fritzsch, B., K. W. Beisel, S. Pauley, and G. Soukup. 2007. Molecular evolution of the vertebrate mechanosensory cell and ear. *Int J Dev Biol* 51:663–678.

Fritzsch, B., and K. L. Elliott. 2017. Gene, cell, and organ multiplication drives inner ear evolution. *Dev Biol* 431:3–15.

Fritzsch, B., N. Pan, I. Jahan, J. S. Duncan, B. J. Kopecky, K. L. Elliott, J. Kersigo, and T. Yang. 2013. Evolution and development of the tetrapod auditory system: an organ of Corti-centric perspective. *Evol Dev* 15:63–79.

Fritzsch, B., S. Pauley, and K. W. Beisel. 2006. Cells, molecules and morphogenesis: the making of the vertebrate ear. *Brain Res* 1091:151–171.

Fritzsch, B., and H. Straka. 2014. Evolution of vertebrate mechanosensory hair cells and inner ears: toward identifying stimuli that select mutation driven altered morphologies. *J Comp Physiol A Neuroethol Sens Neural Behav Physiol* 200:5–18.

Frolenkov, G. I., I. A. Belyantseva, T. B. Friedman, and A. J. Griffith. 2004. Genetic insights into the morphogenesis of inner ear hair cells. *Nat Rev Genet* 5:489–498.

Fu, Y., H. Zhong, M. H. Wang, D. G. Luo, H. W. Liao, H. Maeda, S. Hattar, L. J. Frishman, and K. W. Yau. 2005. Intrinsically photosensitive retinal ganglion cells detect light with a vitamin A-based photopigment, melanopsin. *Proc Natl Acad Sci USA* 102:10339–10344.

Fu, Z., T. Ogura, W. Luo, and W. Lin. 2018. ATP and odor mixture activate TRPM5-expressing microvillous cells and potentially induce acetylcholine release to enhance supporting cell endocytosis in mouse main olfactory epithelium. *Front Cell Neurosci* 12:71.

Fudge, D. S., J. V. McCuaig, S. Van Stralen, J. F. Hess, H. Wang, R. T. Mathias, and P. G. FitzGerald. 2011. Intermediate filaments regulate tissue size and stiffness in the murine lens. *Invest Ophthalmol Vis Sci* 52:3860–3867.

Fujimoto, M., H. Izu, K. Seki, K. Fukuda, T. Nishida, S. Yamada, K. Kato, S. Yonemura, S. Inouye, and A. Nakai. 2004. HSF4 is required for normal cell growth and differentiation during mouse lens development. *EMBO J* 23:4297–4306.

Furukawa, T., E. M. Morrow, and C. L. Cepko. 1997. Crx, a novel otx-like homeobox gene, shows photoreceptor-specific expression and regulates photoreceptor differentiation. *Cell* 91:531–541.

Furukawa, T., E. M. Morrow, T. Li, F. C. Davis, and C. L. Cepko. 1999. Retinopathy and attenuated circadian entrainment in Crx-deficient mice. *Nat Genet* 23:466–470.

Furuta, Y., and B. L. Hogan. 1998. BMP4 is essential for lens induction in the mouse embryo. *Genes Dev* 12:3764–3775.

Gale, J., and D. Jagger. 2010. Cochlear supporting cells. In *Oxford handbook of auditory science: the ear*, edited by P. Fuchs, 307–328. Oxford: Oxford Univ. Press.

Gallagher, B. C., J. J. Henry, and R. M. Grainger. 1996. Inductive processes leading to inner ear formation during *Xenopus* development. *Dev Biol* 175:95–107.

Galvez, H., G. Abelló, and F. Giráldez. 2017. Signaling and transcription factors during inner ear development: the generation of hair cells and otic neurons. *Front Cell Dev Biol* 5:21.

Gammill, L. S., and H. Sive. 2001. Otx2 expression in the ectoderm activates anterior neural determination and is required for *Xenopus* cement gland formation. *Dev Biol* 240:223–236.

Gamse, J. T., Y. C. Shen, C. Thisse, B. Thisse, P. A. Raymond, M. E. Halpern, and J. O. Liang. 2002. Otx5 regulates genes that show circadian expression in the zebrafish pineal complex. *Nat Genet* 30:117–121.

Gan, L., S. W. Wang, Z. Huang, and W. H. Klein. 1999. POU domain factor Brn-3b is essential for retinal ganglion cell differentiation and survival but not for initial cell fate specification. *Dev Biol* 210:469–480.

Gans, C., and R. G. Northcutt. 1983. Neural crest and the origin of vertebrates: a new head. *Science* 220:268–274.

Gao, W. Q. 2003. Hair cell development in higher vertebrates. *Curr Top Dev Biol* 57:293–319.

Garcia-Castro, M. I., C. Marcelle, and M. Bronner-Fraser. 2002. Ectodermal Wnt function as a neural crest inducer. *Science* 297:848–851.

Garcia, C. M., J. Huang, B. P. Madakashira, Y. Liu, R. Rajagopal, L. Dattilo, M. L. Robinson, and D. C. Beebe. 2011. The function of FGF signaling in the lens placode. *Dev Biol* 351:176–185.

Garnett, A. T., T. A. Square, and D. M. Medeiros. 2012. BMP, Wnt and FGF signals are integrated through evolutionarily conserved enhancers to achieve robust expression of Pax3 and Zic genes at the zebrafish neural plate border. *Development* 139:4220–4231.

Gaston-Massuet, C., C. L. Andoniadou, M. Signore, E. Sajedi, S. Bird, J. M. Turner, and J. P. Martinez-Barbera. 2008. Genetic interaction between the homeobox transcription factors HESX1 and SIX3 is required for normal pituitary development. *Dev Biol* 324:322–333.

Gatto, G., K. M. Smith, S. E. Ross, and M. Goulding. 2019. Neuronal diversity in the somatosensory system: bridging the gap between cell type and function. *Curr Opin Neurobiol* 56:167–174.

Gau, P., A. Curtright, L. Condon, D. W. Raible, and A. Dhaka. 2017. An ancient neurotrophin receptor code; a single Runx/Cbfbeta complex determines somatosensory neuron fate specification in zebrafish. *PLoS Genet* 13:e1006884.

Geary, L., and C. LaBonne. 2018. FGF mediated MAPK and PI3K/Akt signals make distinct contributions to pluripotency and the establishment of neural crest. *Elife* 7:e33845.

Ghanbari, H., H. C. Seo, A. Fjose, and A. W. Br„ndli. 2001. Molecular cloning and embryonic expression of *Xenopus* six homeobox genes. *Mech Dev* 101:271–277.

Ghysen, A., and C. Dambly-Chaudière. 2004. Development of the zebrafish lateral line. *Curr Opin Neurobiol* 14:67–73.

Ghysen, A., and C. Dambly-Chaudière. 2007. The lateral line microcosmos. *Genes Dev* 21:2118–2130.

Gibbs, M. A. 2004. Lateral line receptors: where do they come from developmentally and where is our research going? *Brain Behav Evol* 64:163–181.

Gillis, J. A., M. S. Modrell, R. G. Northcutt, K. C. Catania, C. A. Luer, and C. V. Baker. 2012. Electrosensory ampullary organs are derived from lateral line placodes in cartilaginous fishes. *Development* 139:3142–3146.

Gilmour, D., H. Knaut, H. M. Maischein, and C. Nüsslein-Volhard. 2004. Towing of sensory axons by their migrating target cells in vivo. *Nat Neurosci* 7:491–492.
Gilmour, D. T., H. M. Maischein, and C. Nüsslein-Volhard. 2002. Migration and function of a glial subtype in the vertebrate peripheral nerveous system. *Neuron* 34:577–588.
Glavic, A., J. L. Gómez-Skarmeta, and R. Mayor. 2002. The homeoprotein Xirol is required for midbrain-hindbrain boundary formation. *Development* 129:1609–1621.
Glavic, A., Honoré S. Maris, Feijoo C. Gloria, F. Bastidas, M. L. Allende, and R. Mayor. 2004. Role of BMP signaling and the homeoprotein Iroquois in the specification of the cranial placodal field. *Dev Biol* 272:89–103.
Glover, J. C., K. L. Elliott, A. Erives, V. V. Chizhikov, and B. Fritzsch. 2018. Wilhelm His' lasting insights into hindbrain and cranial ganglia development and evolution. *Dev Biol* 444 Suppl 1:S14–S24.
Goldberg, L. B., P. K. Aujla, and L. T. Raetzman. 2011. Persistent expression of activated Notch inhibits corticotrope and melanotrope differentiation and results in dysfunction of the HPA axis. *Dev Biol* 358:23–32.
Goldsmith, S., R. Lovell-Badge, and K. Rizzoti. 2016. SOX2 is sequentially required for progenitor proliferation and lineage specification in the developing pituitary. *Development* 143:2376–2388.
Goldstein, B. J., H. Fang, S. L. Youngentob, and J. E. Schwob. 1998. Transplantation of multipotent progenitors from the adult olfactory epithelium. *Neuroreport* 9:1611–1617.
Gómez-Skarmeta, J., E. Calle-Mustienes, and J. Modolell. 2001. The Wnt-activated Xirol gene encodes a repressor that is essential for neural development and downregulates Bmp4. *Development* 128:551–560.
Gompel, N., C. Dambly-Chaudière, and A. Ghysen. 2001. Neuronal differences prefigure somatotopy in the zebrafish lateral line. *Development* 128:387–393.
Gonzalez, A., R. Morona, J. M. López, N. Moreno, and R. G. Northcutt. 2010. Lungfishes, like tetrapods, possess a vomeronasal system. *Front Neuroanat* 4:130.
Goodrich, E. S. 1930. *Studies on the structure and development of vertebrates.* London: Macmillan.
Goodrich, L. V. 2016. Early development of the spiral ganglion. In *The primary auditory neurons of the mammalian cochlea*, edited by A. Dabdoub, B. Fritzsch, A. N. Popper and R. R. Fay, 11–48. New York: Springer.
Goodyear, R. J., C. J. Kros, and G. P. Richardson. 2006. The development of hair cells in the inner ear. In *Vertebrate hair cells*, edited by R. A. Eatock, R. R. Fay and A. N. Popper, 20–94. New York: Springer.
Goodyear, R. J., and G. P. Richardson. 2018. Structure, function, and development of the tectorial membrane: an extracellular matrix essential for hearing. *Curr Topics Dev Biol* 130:217–244.
Goodyer, C. G., J. J. Tremblay, F. W. Paradis, A. Marcil, C. Lanctot, Y. Gauthier, and J. Drouin. 2003. Pitx1 in vivo promoter activity and mechanisms of positive autoregulation. *Neuroendocrinology* 78:129–137.
Gorbman, A. 1983. Early development of the hagfish pituitary gland: evidence for the endodermal origin of the adenohypophysis. *Amer Zool* 23:639–654.
Gordon, M. K., J. S. Mumm, R. A. Davis, J. D. Holcomb, and A. L. Calof. 1995. Dynamics of MASH1 expression in vitro and in vivo suggest a non-stem cell site of MASH1 action in the olfactory receptor neuron lineage. *Mol Cell Neurosci* 6:363–379.
Görner, P. 1961. Beitrag zum Bau und zur Arbeitsweise des Seitenorgans von *Xenopus laevis. Verh Zool Ges* 22E:193–198.
Gou, Y., J. Guo, K. Maulding, and B. B. Riley. 2018. sox2 and sox3 cooperate to regulate otic/epibranchial placode induction in zebrafish. *Dev Biol* 435:84–95.
Gou, Y., S. Vemaraju, E. M. Sweet, H. J. Kwon, and B. B. Riley. 2018. sox2 and sox3 play unique roles in development of hair cells and neurons in the zebrafish inner ear. *Dev Biol* 435:73–83.
Graf, T., and T. Enver. 2009. Forcing cells to change lineages. *Nature* 462:587–594.

Graham, A. 2008. Deconstructing the pharyngeal metamere. *J Exp Zool B Mol Dev Evol* 310:336–344.

Graham, A., and J. Begbie. 2000. Neurogenic placodes: a common front. *Trends Neurosci* 23:313–316.

Graham, A., T. Butts, A. Lumsden, and C. Kiecker. 2014. What can vertebrates tell us about segmentation? *Evodevo* 5:24.

Graham, A., I. Heyman, and A. Lumsden. 1993. Even-numbered rhombomeres control the apoptotic elimination of neural crest cells from odd-numbered rhombomeres in the chick hindbrain. *Development* 119:233–245.

Graham, D. M., K. Y. Wong, P. Shapiro, C. Frederick, K. Pattabiraman, and D. M. Berson. 2008. Melanopsin ganglion cells use a membrane-associated rhabdomeric phototransduction cascade. *J Neurophysiol* 99:2522–2532.

Graham, V., J. Khudyakov, P. Ellis, and L. Pevny. 2003. SOX2 functions to maintain neural progenitor identity. *Neuron* 39:749–765.

Grainger, R. M., J. E. Mannion, T. L. Cook, Jr., and C. A. Zygar. 1997. Defining intermediate stages in cell determination: acquisition of a lens-forming bias in head ectoderm during lens determination. *Dev Genet* 20:246–257.

Grant, K. A., D. W. Raible, and T. Piotrowski. 2005. Regulation of latent sensory hair cell precursors by glia in the zebrafish lateral line. *Neuron* 45:69–80.

Grati, M., and B. Kachar. 2011. Myosin VIIa and sans localization at stereocilia upper tip-link density implicates these Usher syndrome proteins in mechanotransduction. *Proc Natl Acad Sci USA* 108:11476–11481.

Graw, J. 2010. Eye development. *Curr Top Dev Biol* 90:343–386.

Graziadei, P. P., and G. A. Graziadei. 1979. Neurogenesis and neuron regeneration in the olfactory system of mammals. I. Morphological aspects of differentiation and structural organization of the olfactory sensory neurons. *J Neurocytol* 8:1–18.

Green, S. A., M. Simoes-Costa, and M. E. Bronner. 2015. Evolution of vertebrates as viewed from the crest. *Nature* 520:474–482.

Greer, P. L., D. M. Bear, J. M. Lassance, M. L. Bloom, T. Tsukahara, S. L. Pashkovski, F. K. Masuda, A. C. Nowlan, R. Kirchner, H. E. Hoekstra, and S. R. Datta. 2016. A family of non-GPCR chemosensors defines an alternative logic for mammalian olfaction. *Cell* 165:1734–1748.

Greiling, T. M., and J. I. Clark. 2012. New insights into the mechanism of lens development using zebra fish. *Int Rev Cell Mol Biol* 296:1–61.

Griep, A. E., and P. Zhang. 2004. Lens cell proliferation: the cell cycle. In *Development of the ocular lens*, edited by F. J. Lovicu and M. L. Robinson, 191–213. Cambridge: Cambridge University Press.

Grifone, R., J. Demignon, J. Giordani, C. Niro, E. Souil, F. Bertin, C. Laclef, P. X. Xu, and P. Maire. 2007. Eya1 and Eya2 proteins are required for hypaxial somitic myogenesis in the mouse embryo. *Dev Biol* 302:602–616.

Grifone, R., J. Demignon, C. Houbron, E. Souil, C. Niro, M. J. Seller, G. Hamard, and P. Maire. 2005. Six1 and Six4 homeoproteins are required for Pax3 and Mrf expression during myogenesis in the mouse embryo. *Development* 132:2235–2249.

Grimm, C., B. Chatterjee, J. Favor, T. Immervoll, J. Loster, N. Klopp, R. Sandulache, and J. Graw. 1998. Aphakia (ak), a mouse mutation affecting early eye development: fine mapping, consideration of candidate genes and altered Pax6 and Six3 gene expression pattern. *Dev Genet* 23:299–316.

Grocott, T., M. Tambalo, and A. Streit. 2012. The peripheral sensory nervous system in the vertebrate head: a gene regulatory perspective. *Dev Biol* 370:3–23.

Groves, A. K., and M. Bronner-Fraser. 2000. Competence, specification and commitment in otic placode induction. *Development* 127:3489–3499.

Groves, A. K., and D. M. Fekete. 2012. Shaping sound in space: the regulation of inner ear patterning. *Development* 139:245–257.

Groves, A. K., and C. LaBonne. 2014. Setting appropriate boundaries: fate, patterning and competence at the neural plate border. *Dev Biol* 389:2–12.

Groves, A. K., K. D. Zhang, and D. M. Fekete. 2013. The genetics of hair cell development and regeneration. *Annu Rev Neurosci* 36:361–381.

Grus, W. E., and J. Zhang. 2009. Origin of the genetic components of the vomeronasal system in the common ancestor of all extant vertebrates. *Mol Biol Evol* 26:407–419.

Gubbels, S. P., D. W. Woessner, J. C. Mitchell, A. J. Ricci, and J. V. Brigande. 2008. Functional auditory hair cells produced in the mammalian cochlea by in utero gene transfer. *Nature* 455:537–541.

Guillemot, F. 2007. Spatial and temporal specification of neural fates by transcription factor codes. *Development* 134:3771–3780.

Guillemot, F., and B. A. Hassan. 2017. Beyond proneural: emerging functions and regulations of proneural proteins. *Curr Opin Neurobiol* 42:93–101.

Guillemot, F., and A. L. Joyner. 1993. Dynamic expression of the murine Achaete-Scute homologue Mash-1 in the developing nervous system. *Mech Dev* 42:171–185.

Guner, B., A. T. Ozacar, J. E. Thomas, and R. O. Karlstrom. 2008. Graded Hh and Fgf signaling independently regulate pituitary cell fates and help establish the PD and PI of the zebrafish adenohypophysis. *Endocrinology* 149:4435–4451.

Guo, Z., A. Packard, R. C. Krolewski, M. T. Harris, G. L. Manglapus, and J. E. Schwob. 2010. Expression of pax6 and sox2 in adult olfactory epithelium. *J Comp Neurol* 518:4395–4418.

Guo, Z., L. Zhang, Z. Wu, Y. Chen, F. Wang, and G. Chen. 2014. In vivo direct reprogramming of reactive glial cells into functional neurons after brain injury and in an Alzheimer's disease model. *Cell Stem Cell* 14:188–202.

Gupta, D., S. A. Harvey, N. Kaminski, and S. K. Swamynathan. 2011. Mouse conjunctival forniceal gene expression during postnatal development and its regulation by Kruppel-like factor 4. *Invest Ophthalmol Vis Sci* 52:4951–4962.

Guruharsha, K. G., M. W. Kankel, and S. Artavanis-Tsakonas. 2012. The Notch signalling system: recent insights into the complexity of a conserved pathway. *Nat Rev Genet* 13:654–666.

Guthrie, S. 2007. Patterning and axon guidance of cranial motor neurons. *Nat Rev Neurosci* 8:859–871.

Haddon, C., Y. J. Jiang, L. Smithers, and J. Lewis. 1998. Delta-Notch signalling and the patterning of sensory cell differentiation in the zebrafish ear: evidence from the mind bomb mutant. *Development* 125:4637–4644.

Haddon, C., and J. Lewis. 1996. Early ear development in the embryo of the zebrafish, *Danio rerio*. *J Comp Neurol* 365:113–128.

Hadrys, T., T. Braun, S. Rinkwitz-Brandt, H. H. Arnold, and E. Bober. 1998. Nkx5-1 controls semicircular canal formation in the mouse inner ear. *Development* 125:33–39.

Haeberle, H., M. Fujiwara, J. Chuang, M. M. Medina, M. V. Panditrao, S. Bechstedt, J. Howard, and E. A. Lumpkin. 2004. Molecular profiling reveals synaptic release machinery in Merkel cells. *Proc Natl Acad Sci USA* 101:14503-14508.

Hafler, B. P., N. Surzenko, K. T. Beier, C. Punzo, J. M. Trimarchi, J. H. Kong, and C. L. Cepko. 2012. Transcription factor Olig2 defines subpopulations of retinal progenitor cells biased toward specific cell fates. *Proc Natl Acad Sci USA* 109:7882–7887.

Hagey, D. W., and J. Muhr. 2014. Sox2 acts in a dose-dependent fashion to regulate proliferation of cortical progenitors. *Cell Rep* 9:1908–1920.

Haggis, A. J. 1956. Analysis of the determination of the olfactory placode in *Amblystoma punctatum*. *J Embryol Exp Morphol* 4:120–138.

Hagino-Yamagishi, K., K. Moriya, H. Kubo, Y. Wakabayashi, N. Isobe, S. Saito, M. Ichikawa, and K. Yazaki. 2004. Expression of vomeronasal receptor genes in *Xenopus laevis*. *J Comp Neurol* 472:246–256.

Haider, N. B., J. K. Naggert, and P. M. Nishina. 2001. Excess cone cell proliferation due to lack of a functional NR2E3 causes retinal dysplasia and degeneration in rd7/rd7 mice. *Hum Mol Genet* 10:1619–1626.

Haldin, C. E., S. Nijjar, K. Masse, M. W. Barnett, and E. A. Jones. 2003. Isolation and growth factor inducibility of the *Xenopus laevis* Lmx1b gene. *Int J Dev Biol* 47:253–262.

Halfter, W., S. Dong, B. Schurer, C. Ring, G. J. Cole, and A. Eller. 2005. Embryonic synthesis of the inner limiting membrane and vitreous body. *Invest Ophthalmol Vis Sci* 46:2202–2209.

Hall, B. K. 2000. The neural crest as a fourth germ layer and vertebrates as quadroblastic not triploblastic. *Evol Dev* 2:3–5.

Halpern, M., L. S. Shapiro, and C. Jia. 1995. Differential localization of G proteins in the opossum vomeronasal system. *Brain Res* 677:157–161.

Hamburger, V. 1961. Experimental analysis of the dual origin of the trigeminal ganglion in the chick embryo. *J Exp Zool* 148:91–123.

Han, D. W., N. Tapia, A. Hermann, K. Hemmer, S. Hoing, M. J. Arauzo-Bravo, H. Zaehres, G. Wu, S. Frank, S. Moritz, B. Greber, J. H. Yang, H. T. Lee, J. C. Schwamborn, A. Storch, and H. R. Schöler. 2012. Direct reprogramming of fibroblasts into neural stem cells by defined factors. *Cell Stem Cell* 10:465–472.

Hanchate, N. K., K. Kondoh, Z. Lu, D. Kuang, X. Ye, X. Qiu, L. Pachter, C. Trapnell, and L. B. Buck. 2015. Single-cell transcriptomics reveals receptor transformations during olfactory neurogenesis. *Science* 350:1251–1255.

Hans, S., J. Christison, D. Liu, and M. Westerfield. 2007. Fgf-dependent otic induction requires competence provided by Foxi1 and Dlx3b. *BMC Dev Biol* 7:5.

Hans, S., A. Irmscher, and M. Brand. 2013. Zebrafish Foxi1 provides a neuronal ground state during inner ear induction preceding the Dlx3b/4b-regulated sensory lineage. *Development* 140:1936–1945.

Hans, S., D. Liu, and M. Westerfield. 2004. Pax8 and Pax2a function synergistically in otic specification, downstream of the Foxi1 and Dlx3b transcription factors. *Development* 131:5091–5102.

Hansen, A., K. T. Anderson, and T. E. Finger. 2004. Differential distribution of olfactory receptor neurons in goldfish: structural and molecular correlates. *J Comp Neurol* 477:347–359.

Hansen, A., and T. E. Finger. 2000. Phyletic distribution of crypt-type olfactory receptor neurons in fishes. *Brain Behav Evol* 55:100–110.

Hansen, A., and T. E. Finger. 2008. Is TrpM5 a reliable marker for chemosensory cells? Multiple types of microvillous cells in the main olfactory epithelium of mice. *BMC Neurosci* 9:115.

Hansen, A., J. O. Reiss, C. L. Gentry, and G. D. Burd. 1998. Ultrastructure of the olfactory organ in the clawed frog, *Xenopus laevis*, during larval development and metamorphosis. *J Comp Neurol* 398:273–288.

Hansen, A., S. H. Rolen, K. Anderson, Y. Morita, J. Caprio, and T. E. Finger. 2003. Correlation between olfactory receptor cell type and function in the channel catfish. *J Neurosci* 23:9328–9339.

Hansen, A., and E. Zeiske. 1993. Development of the olfactory organ in the zebrafish, *Brachydanio rerio*. *J Comp Neurol* 333:289–300.

Hansen, A., and E. Zeiske. 1998. The peripheral olfactory organ of the zebrafish, *Danio rerio*: an ultrastructural study. *Chem Senses* 23:39–48.

Hansen, A., and B. S. Zielinski. 2005. Diversity in the olfactory epithelium of bony fishes: development, lamellar arrangement, sensory neuron cell types and transduction components. *J Neurocytol* 34:183–208.

Hardwick, L. J. A., R. Azzarelli, and A. Philpott. 2018. Cell cycle-dependent phosphorylation and regulation of cellular differentiation. *Biochem Soc Trans* 46:1083–1091.

Harlow, D. E., and L. A. Barlow. 2007. Embryonic origin of gustatory cranial sensory neurons. *Dev Biol* 310:317–328.

Hartman, B. H., T. A. Reh, and O. Bermingham-McDonogh. 2010. Notch signaling specifies prosensory domains via lateral induction in the developing mammalian inner ear. *Proc Natl Acad Sci USA* 107:15792–15797.

Haruta, M., M. Kosaka, Y. Kanegae, I. Saito, T. Inoue, R. Kageyama, A. Nishida, Y. Honda, and M. Takahashi. 2001. Induction of photoreceptor-specific phenotypes in adult mammalian iris tissue. *Nat Neurosci* 4:1163–1164.

Hashiguchi, Y., and M. Nishida. 2007. Evolution of trace amine associated receptor (TAAR) gene family in vertebrates: lineage-specific expansions and degradations of a second class of vertebrate chemosensory receptors expressed in the olfactory epithelium. *Mol Biol Evol* 24:2099–2107.

Hashiguchi, Y., Y. Furuta, and M. Nishida. 2008. Evolutionary patterns and selective pressures of odorant/pheromone receptor gene families in teleost fishes. *PLoS One* 3:e4083.

Hashimoto, H., M. Itoh, Y. Yamanaka, S. Yamashita, T. Shimizu, L. Solnica-Krezel, M. Hibi, and T. Hirano. 2000. Zebrafish Dkk1 functions in forebrain specification and axial mesendoderm formation. *Dev Biol* 217:138–152.

Hassan, B. A., and H. J. Bellen. 2000. Doing the MATH: is the mouse a good model for fly development? *Genes Dev* 14:1852–1865.

Hatakeyama, J., Y. Bessho, K. Katoh, S. Ookawara, M. Fujioka, F. Guillemot, and R. Kageyama. 2004. Hes genes regulate size, shape and histogenesis of the nervous system by control of the timing of neural stem cell differentiation. *Development* 131:5539–5550.

Hatakeyama, J., and R. Kageyama. 2004. Retinal cell fate determination and bHLH factors. *Semin Cell Dev Biol* 15:83–89.

Hattar, S., H. W. Liao, M. Takao, D. M. Berson, and K. W. Yau. 2002. Melanopsin-containing retinal ganglion cells: architecture, projections, and intrinsic photosensitivity. *Science* 295:1065–1070.

Hausken, K. N., B. Tizon, M. Shpilman, S. Barton, W. Decatur, D. Plachetzki, S. Kavanaugh, S. Ul-Hasan, B. Levavi-Sivan, and S. A. Sower. 2018. Cloning and characterization of a second lamprey pituitary glycoprotein hormone, thyrostimulin (GpA2/GpB5). *Gen Comp Endocrinol* 264:16–27.

Hawley, S. H., K. Wunnenberg-Stapleton, C. Hashimoto, M. N. Laurent, T. Watabe, B. W. Blumberg, and K. W. Cho. 1995. Disruption of BMP signals in embryonic *Xenopus* ectoderm leads to direct neural induction. *Genes Dev* 9:2923–2935.

He, D. Z., S. Lovas, Y. Ai, Y. Li, and K. W. Beisel. 2014. Prestin at year 14: progress and prospect. *Hear Res* 311:25–35.

Heasman, J. 2006. Patterning the early *Xenopus* embryo. *Development* 133:1205–1217.

Heeg-Truesdell, E., and C. LaBonne. 2006. Neural induction in *Xenopus* requires inhibition of Wnt-beta-catenin signaling. *Dev Biol* 298:71–86.

Heller, N., and A. W. Brändli. 1999. *Xenopus* Pax-2/5/8 orthologues: novel insights into Pax gene evolution and identification of Pax-8 as the earliest marker for otic and pronephric cell lineages. *Dev Genet* 24:208–219.

Helms, A. W., A. L. Abney, N. Ben-Arie, H. Y. Zoghbi, and J. E. Johnson. 2000. Autoregulation and multiple enhancers control Math1 expression in the developing nervous system. *Development* 127:1185–1196.

Helms, J. A., and R. A. Schneider. 2003. Cranial skeletal biology. *Nature* 423:326–331.

Hemmati-Brivanlou, A., and G. H. Thomsen. 1995. Ventral mesodermal patterning in *Xenopus* embryos: expression patterns and activities of BMP-2 and BMP-4. *Dev Genet* 17:78–89.

Hennig, A. K., G. H. Peng, and S. Chen. 2008. Regulation of photoreceptor gene expression by Crx-associated transcription factor network. *Brain Res* 1192:114–133.

Henry, J. J., and R. M. Grainger. 1987. Inductive interactions in the spatial and temporal restriction of lens-forming potential in embryonic ectoderm of *Xenopus laevis*. *Dev Biol* 124:200–214.

Henry, J. J., and R. M. Grainger. 1990. Early tissue interactions leading to embryonic lens formation in *Xenopus laevis*. *Dev Biol* 141:149–163.

Hermesz, E., S. Mackem, and K. A. Mahon. 1996. Rpx: a novel anterior-restricted homeobox gene progressively activated in the prechordal plate, anterior neural plate and Rathke's pouch of the mouse embryo. *Development* 122:41–52.

Hernandez, P. P., F. A. Olivari, A. F. Sarrazin, P. C. Sandoval, and M. L. Allende. 2007. Regeneration in zebrafish lateral line neuromasts: expression of the neural progenitor cell marker sox2 and proliferation-dependent and-independent mechanisms of hair cell renewal. *Dev Neurobiol* 67:637–654.

Heron, P. M., A. J. Stromberg, P. Breheny, and T. S. McClintock. 2013. Molecular events in the cell types of the olfactory epithelium during adult neurogenesis. *Mol Brain* 6:49.

Herrada, G., and C. Dulac. 1997. A novel family of putative pheromone receptors in mammals with a topographically organized and sexually dimorphic distribution. *Cell* 90:763–773.

Herrick, C. J. 1899. The cranial and first spinal nerves of *Menidia*; a contribution upon the nerve components of the bony fishes. *J Comp Neurol* 9:153–455.

Hertzano, R., A. A. Dror, M. Montcouquiol, Z. M. Ahmed, B. Ellsworth, S. Camper, T. B. Friedman, M. W. Kelley, and K. B. Avraham. 2007. Lhx3, a LIM domain transcription factor, is regulated by Pou4f3 in the auditory but not in the vestibular system. *Eur J Neurosci* 25:999–1005.

Hertzano, R., M. Montcouquiol, S. Rashi-Elkeles, R. Elkon, R. Yucel, W. N. Frankel, G. Rechavi, T. Moroy, T. B. Friedman, M. W. Kelley, and K. B. Avraham. 2004. Transcription profiling of inner ears from Pou4f3(ddl/ddl) identifies Gfi1 as a target of the Pou4f3 deafness gene. *Hum Mol Genet* 13:2143–2153.

Herzog, W., C. Sonntag, B. Walderich, J. Odenthal, H. M. Maischein, and M. Hammerschmidt. 2004. Genetic analysis of adenohypophysis formation in zebrafish. *Mol Endocrinol* 18:1185–1195.

Herzog, W., X. Zeng, Z. Lele, C. Sonntag, J. W. Ting, C. Y. Chang, and M. Hammerschmidt. 2003. Adenohypophysis formation in the zebrafish and its dependence on sonic hedgehog. *Dev Biol* 254:36–49.

Hikasa, H., and S. Y. Sokol. 2013. Wnt signaling in vertebrate axis specification. *Cold Spring Harb Perspect Biol* 5:a007955.

Hilding, D. A., and R. D. Ginzberg. 1977. Pigmentation of the stria vascularis. The contribution of neural crest melanocytes. *Acta Otolaryngol* 84:24–37.

Hindley, C., F. Ali, G. McDowell, K. Cheng, A. Jones, F. Guillemot, and A. Philpott. 2012. Post-translational modification of Ngn2 differentially affects transcription of distinct targets to regulate the balance between progenitor maintenance and differentiation. *Development* 139:1718–1723.

Hintze, M., R. S. Prajapati, M. Tambalo, N. A. D. Christophorou, M. Anwar, T. Grocott, and A. Streit. 2017. Cell interactions, signals and transcriptional hierarchy governing placode progenitor induction. *Development* 144:2810–2823.

Hirata, H., S. Yoshiura, T. Ohtsuka, Y. Bessho, T. Harada, K. Yoshikawa, and R. Kageyama. 2002. Oscillatory expression of the bHLH factor Hes1 regulated by a negative feedback loop. *Science* 298:840–843.

Hirata, T., M. Nakazawa, O. Muraoka, R. Nakayama, Y. Suda, and M. Hibi. 2006. Zinc-finger genes Fez and Fez-like function in the establishment of diencephalon subdivisions. *Development* 133:3993–4004.

Hirokawa, N. 1978. The ultrastructure of the basilar papilla of the chick. *J Comp Neurol* 181:361–374.

Hirota, J., and P. Mombaerts. 2004. The LIM-homeodomain protein Lhx2 is required for complete development of mouse olfactory sensory neurons. *Proc Natl Acad Sci USA* 101:8751–8755.

Hirota, J., M. Omura, and P. Mombaerts. 2007. Differential impact of Lhx2 deficiency on expression of class I and class II odorant receptor genes in mouse. *Mol Cell Neurosci* 34:679–688.

His, W. 1868. *Untersuchungen über die erste Anlage des Wirbelthierleibes: die erste Entwickelung des Hühnchens im Ei*. Leipzig: F.C.W. Vogel.

Ho, H. Y., K. H. Chang, J. Nichols, and M. Li. 2009. Homeodomain protein Pitx3 maintains the mitotic activity of lens epithelial cells. *Mech Dev* 126:18–29.

Ho, Y., P. Hu, M. T. Peel, S. Chen, P. G. Camara, D. J. Epstein, H. Wu, and S. A. Liebhaber. 2020. Single-cell transcriptomic analysis of adult mouse pituitary reveals sexual dimorphism and physiologic demand-induced cellular plasticity. *Protein Cell* 11:565–583.

Hobert, O. 2008. Regulatory logic of neuronal diversity: terminal selector genes and selector motifs. *Proc Natl Acad Sci USA* 105:20067–20071.

Hobert, O. 2011. Regulation of terminal differentiation programs in the nervous system. *Annu Rev Cell Dev Biol* 27:681–696.

Hobert, O., I. Carrera, and N. Stefanakis. 2010. The molecular and gene regulatory signature of a neuron. *Trends Neurosci* 33:435–445.

Hobert, O., and H. Westphal. 2000. Functions of LIM-homeobox genes. *Trends Genet* 16:75–83.

Hockman, D., A. J. Burns, G. Schlosser, K. P. Gates, B. Jevans, A. Mongera, S. Fisher, G. Unlu, E. W. Knapik, C. K. Kaufman, C. Mosimann, L. I. Zon, J. J. Lancman, P. D. S. Dong, H. Lickert, A. S. Tucker, and C. V. Baker. 2017. Evolution of the hypoxia-sensitive cells involved in amniote respiratory reflexes. *Elife* 6:e21231.

Hodson, D. J., N. Romano, M. Schaeffer, P. Fontanaud, C. Lafont, T. Fiordelisio, and P. Mollard. 2012. Coordination of calcium signals by pituitary endocrine cells in situ. *Cell Calcium* 51:222–230.

Holbrook, E. H., K. E. Szumowski, and J. E. Schwob. 1995. An immunochemical, ultrastructural, and developmental characterization of the horizontal basal cells of rat olfactory epithelium. *J Comp Neurol* 363:129–146.

Holland, L. Z., R. Albalat, K. Azumi, E. Benito-Gutierrez, M. J. Blow, M. Bronner-Fraser, F. Brunet, T. Butts, S. Candiani, L. J. Dishaw, D. E. Ferrier, J. Garcia-Fernandez, J. J. Gibson-Brown, C. Gissi, A. Godzik, F. Hallbook, D. Hirose, K. Hosomichi, T. Ikuta, H. Inoko, M. Kasahara, J. Kasamatsu, T. Kawashima, A. Kimura, M. Kobayashi, Z. Kozmik, K. Kubokawa, V. Laudet, G. W. Litman, A. C. McHardy, D. Meulemans, M. Nonaka, R. P. Olinski, Z. Pancer, L. A. Pennacchio, M. Pestarino, J. P. Rast, I. Rigoutsos, M. Robinson-Rechavi, G. Roch, H. Saiga, Y. Sasakura, M. Satake, Y. Satou, M. Schubert, N. Sherwood, T. Shiina, N. Takatori, J. Tello, P. Vopalensky, S. Wada, A. Xu, Y. Ye, K. Yoshida, F. Yoshizaki, J. K. Yu, Q. Zhang, C. M. Zmasek, P. J. de Jong, K. Osoegawa, N. H. Putnam, D. S. Rokhsar, N. Satoh, and P. W. Holland. 2008. The amphioxus genome illuminates vertebrate origins and cephalochordate biology. *Genome Res* 18:1100–1111.

Holland, L. Z., J. E. Carvalho, H. Escriva, V. Laudet, M. Schubert, S. M. Shimeld, and J. K. Yu. 2013. Evolution of bilaterian central nervous systems: a single origin? *Evodevo* 4:27.

Hollemann, T., and T. Pieler. 1999. Xpitx-1: a homeobox gene expressed during pituitary and cement gland formation of *Xenopus* embryos. *Mech Dev* 88:249–252.

Holmberg, J., E. Hansson, M. Malewicz, M. Sandberg, T. Perlmann, U. Lendahl, and J. Muhr. 2008. SoxB1 transcription factors and Notch signaling use distinct mechanisms to regulate proneural gene function and neural progenitor differentiation. *Development* 135:1843–1851.

Holmberg, K. 1971. The hagfish retina: electron microscopic study comparing receptor and epithelial cells in the Pacific hagfish, *Polistotrema stouti*, with those in the Atlantic hagfish, *Myxine glutinosa*. *Z Zellforsch Mikrosk Anat* 121:249–269.

Holmberg, K., P. Ohman, and T. Dreyfert. 1977. ERG-recordings from the retina of the river lamprey (*Lampetra fluviatilis*). *Vision Res* 17:715–717.

Holthues, H., L. Engel, R. Spessert, and L. Vollrath. 2005. Circadian gene expression patterns of melanopsin and pinopsin in the chick pineal gland. *Biochem Biophys Res Commun* 326:160–165.

Holzschuh, J., N. Wada, C. Wada, A. Schaffer, Y. Javidan, A. Tallafuss, L. Bally-Cuif, and T. F. Schilling. 2005. Requirements for endoderm and BMP signaling in sensory neurogenesis in zebrafish. *Development* 132:3731–3742.

Hong, C. S., and J. P. Saint-Jeannet. 2007. The activity of Pax3 and Zic1 regulates three distinct cell fates at the neural plate border. *Mol Biol Cell* 18:2192–2202.

Honoré, S. M., M. J. Aybar, and R. Mayor. 2003. Sox10 is required for the early development of the prospective neural crest in *Xenopus* embryos. *Dev Biol* 260:79–96.

Horton, S., A. Meredith, J. A. Richardson, and J. E. Johnson. 1999. Correct coordination of neuronal differentiation events in ventral forebrain requires the bHLH factor MASH1. *Mol Cell Neurosci* 14:355–369.

Horwitz, J. 1992. Alpha-crystallin can function as a molecular chaperone. *Proc Natl Acad Sci USA* 89:10449–10453.

Houart, C., M. Westerfield, and S. W. Wilson. 1998. A small population of anterior cells patterns the forebrain during zebrafish gastrulation. *Nature* 391:788–792.

Hrabovszky, E., and Z. Liposits. 2008. Novel aspects of glutamatergic signalling in the neuroendocrine system. *J Neuroendocrinol* 20:743–751.

Hu, M., D. Krause, M. Greaves, S. Sharkis, M. Dexter, C. Heyworth, and T. Enver. 1997. Multilineage gene expression precedes commitment in the hemopoietic system. *Genes Dev* 11:774–785.

Hu, Q., L. Zhang, J. Wen, S. Wang, M. Li, R. Feng, X. Yang, and L. Li. 2010. The EGF receptor-sox2-EGF receptor feedback loop positively regulates the self-renewal of neural precursor cells. *Stem Cells* 28:279–286.

Huang, C., J. A. Chan, and C. Schuurmans. 2014. Proneural bHLH genes in development and disease. *Curr Top Dev Biol* 110:75–127.

Huang, E. J., K. Zang, A. Schmidt, A. Saulys, M. Xiang, and L. F. Reichardt. 1999. POU domain factor Brn-3a controls the differentiatio and survival of trigeminal neurons by regulating Trk receptor expresssion. *Development* 126:2869–2882.

Huang, J., R. Rajagopal, Y. Liu, L. K. Dattilo, O. Shaham, R. Ashery-Padan, and D. C. Beebe. 2011. The mechanism of lens placode formation: a case of matrix-mediated morphogenesis. *Dev Biol* 355:32–42.

Huang, X., C. S. Hong, M. O'Donnell, and J. P. Saint-Jeannet. 2005. The doublesex-related gene, XDmrt4, is required for neurogenesis in the olfactory system. *Proc Natl Acad Sci USA* 102:11349–11354.

Huard, J. M., S. L. Youngentob, B. J. Goldstein, M. B. Luskin, and J. E. Schwob. 1998. Adult olfactory epithelium contains multipotent progenitors that give rise to neurons and non-neural cells. *J Comp Neurol* 400:469–486.

Huber, K., S. Combs, U. Ernsberger, C. Kalcheim, and K. Unsicker. 2002. Generation of neuroendocrine chromaffin cells from sympathoadrenal progenitors: beyond the glucocorticoid hypothesis. *Ann N Y Acad Sci* 971:554–559.

Hudspeth, A. J. 2014. Integrating the active process of hair cells with cochlear function. *Nat Rev Neurosci* 15:600–614.

Hudspeth, A. J., Y. Choe, A. D. Mehta, and P. Martin. 2000. Putting ion channels to work: mechanoelectrical transduction, adaptation, and amplification by hair cells. *Proc Natl Acad Sci USA* 97:11765–11772.

Hunt, P., M. Gulisano, M. Cook, M. H. Sham, A. Faiella, D. Wilkinson, E. Boncinelli, and R. Krumlauf. 1991. A distinct Hox code for the branchial region of the vertebrate head. *Nature* 353:861–864.

Hussain, A., L. R. Saraiva, and S. I. Korsching. 2009. Positive Darwinian selection and the birth of an olfactory receptor clade in teleosts. *Proc Natl Acad Sci USA* 106:4313–4318.

Hutcheson, D. A., and M. L. Vetter. 2001. The bHLH factors Xath5 and XNeuroD can upregulate the expression of XBrn3d, a POU-homeodomain transcription factor. *Dev Biol* 232:327–338.

Iida, H., Y. Ishii, and H. Kondoh. 2017. Intrinsic lens potential of neural retina inhibited by Notch signaling as the cause of lens transdifferentiation. *Dev Biol* 421:118–125.

Ikeda, K., S. Ookawara, S. Sato, Z. Ando, R. Kageyama, and K. Kawakami. 2007. Six1 is essential for early neurogenesis in the development of olfactory epithelium. *Dev Biol* 311:53–68.

Ikeda, K., Y. Watanabe, H. Ohto, and K. Kawakami. 2002. Molecular interaction and synergistic activation of a promoter by Six, Eya, and Dach proteins mediated through CREB binding protein. *Mol Cell Biol* 22:6759–6766.

Ikeda, R., K. Pak, E. Chavez, and A. F. Ryan. 2015. Transcription factors with conserved binding sites near ATOH1 on the POU4F3 gene enhance the induction of cochlear hair cells. *Mol Neurobiol* 51:672–684.

Ikeda, Y. 1937. Beiträge zur entwicklungsmechanischen Stütze der Kupfferschen Theorie der Sinnesplakoden. *Roux Arch Entw Mech* 136:672–675.

Ikeda, Y. 1938. Über die wechselseitigen Beziehungen der Sinnesorgane untereinander in ihrer normalen und experimentell bedingten Entwicklung. *Arch Anat Inst Sendai* 21:1–44.

Imayoshi, I., A. Isomura, Y. Harima, K. Kawaguchi, H. Kori, H. Miyachi, T. Fujiwara, F. Ishidate, and R. Kageyama. 2013. Oscillatory control of factors determining multipotency and fate in mouse neural progenitors. *Science* 342:1203–1208.

Imayoshi, I., and R. Kageyama. 2014. Oscillatory control of bHLH factors in neural progenitors. *Trends Neurosci* 37:531–538.

Inoue, K., E. F. Couch, K. Takano, and S. Ogawa. 1999. The structure and function of folliculo-stellate cells in the anterior pituitary gland. *Arch Histol Cytol* 62:205–218.

Inoue, K., K. Ito, M. Osato, B. Lee, S. C. Bae, and Y. Ito. 2007. The transcription factor Runx3 represses the neurotrophin receptor TrkB during lineage commitment of dorsal root ganglion neurons. *J Biol Chem* 282:24175–24184.

Inoue, K., S. Ozaki, T. Shiga, K. Ito, T. Masuda, N. Okado, T. Iseda, S. Kawaguchi, M. Ogawa, S. C. Bae, N. Yamashita, S. Itohara, N. Kudo, and Y. Ito. 2002. Runx3 controls the axonal projection of proprioceptive dorsal root ganglion neurons. *Nat Neurosci* 5:946–954.

Ishibashi, M., S. L. Ang, K. Shiota, S. Nakanishi, R. Kageyama, and F. Guillemot. 1995. Targeted disruption of mammalian hairy and Enhancer of split homolog-1 (HES-1) leads to up-regulation of neural helix-loop- helix factors, prematur e neurogenesis, and severe neural tube defects. *Gene Develop* 9:3136–3148.

Ishibashi, S., and K. Yasuda. 2001. Distinct roles of maf genes during *Xenopus* lens development *Mech Dev* 101:155–166.

Ishii, T., and P. Mombaerts. 2011. Coordinated coexpression of two vomeronasal receptor V2R genes per neuron in the mouse. *Mol Cell Neurosci* 46:397–408.

Ishii, Y., M. Abu-Elmagd, and P. J. Scotting. 2001. Sox3 expression defines a common primordium for the epibranchial placodes in chick. *Dev Biol* 236:344–353.

Isoldi, M. C., M. D. Rollag, A. M. Castrucci, and I. Provencio. 2005. Rhabdomeric phototransduction initiated by the vertebrate photopigment melanopsin. *Proc Natl Acad Sci USA* 102:1217–12121.

Itoh, M., and A. B. Chitnis. 2001. Expression of proneural and neurogenic genes in the zebrafish lateral line primordium correlates with selection of hair cell fate in neuromasts. *Mech Dev* 102:263–266.

Itoh, M., T. Kudoh, M. Dedekian, C. H. Kim, and A. B. Chitnis. 2002. A role for iro1 and iro7 in the establishment of an anteroposterior compartment of the ectoderm adjacent to the midbrain-hindbrain boundary. *Development* 129:2317–2327.

Iwafuchi-Doi, M. 2019. The mechanistic basis for chromatin regulation by pioneer transcription factors. *Wiley Interdiscip Rev Syst Biol Med* 11:e1427.

Iwafuchi-Doi, M., Y. Yoshida, D. Onichtchouk, M. Leichsenring, W. Driever, T. Takemoto, M. Uchikawa, Y. Kamachi, and H. Kondoh. 2011. The Pou5f1/Pou3f-dependent but SoxB-independent regulation of conserved enhancer N2 initiates Sox2 expression during epiblast to neural plate stages in vertebrates. *Dev Biol* 352:354–366.

Iwafuchi-Doi, M., and K. S. Zaret. 2016. Cell fate control by pioneer transcription factors. *Development* 143:1833–1837.

Iwai, N., Z. Zhou, D. R. Roop, and R. R. Behringer. 2008. Horizontal basal cells are multipotent progenitors in normal and injured adult olfactory epithelium. *Stem Cells* 26:1298–1306.

Jacobson, A. G. 1963. The determination and positioning of the nose, lens and ear. III. Effects of reversing the antero-posterior axis of epidermis, neural plate and neural fold. *J Exp Zool* 154:293–303.

Jacobson, C. O. 1959. The localization of the presumptive cerebral regions in the neural plate of the axolotl larva. *J Embryol Exp Morphol* 7:1–21.

Jacobson, G., and B. Meister. 1996. Molecular components of the exocytotic machinery in the rat pituitary gland. *Endocrinology* 137:5344–5356.

Jacquin, M. F., J. J. Arends, C. Xiang, L. A. Shapiro, C. E. Ribak, and Z. F. Chen. 2008. In DRG11 knock-out mice, trigeminal cell death is extensive and does not account for failed brainstem patterning. *J Neurosci* 28:3577–3585.

Jahan, I., N. Pan, and B. Fritzsch. 2015. Opportunities and limits of the one gene approach: the ability of Atoh1 to differentiate and maintain hair cells depends on the molecular context. *Front Cell Neurosci* 9:26.

Jahan, I., N. Pan, J. Kersigo, and B. Fritzsch. 2010. Neurod1 suppresses hair cell differentiation in ear ganglia and regulates hair cell subtype development in the cochlea. *PLoS One* 5:e11661.

Janesick, A., J. Shiotsugu, M. Taketani, and B. Blumberg. 2012. RIPPLY3 is a retinoic acid-inducible repressor required for setting the borders of the pre-placodal ectoderm. *Development* 139:1213–1224.

Jaurena, M. B., H. Juraver-Geslin, A. Devotta, and J. P. Saint-Jeannet. 2015. Zic1 controls placode progenitor formation non-cell autonomously by regulating retinoic acid production and transport. *Nat Commun* 6:7476.

Jayasena, C. S., T. Ohyama, N. Segil, and A. K. Groves. 2008. Notch signaling augments the canonical Wnt pathway to specify the size of the otic placode. *Development* 135:2251–2261.

Jékely, G. 2013. Global view of the evolution and diversity of metazoan neuropeptide signaling. *Proc Natl Acad Sci USA* 110:8702–8707.

Jemc, J., and I. Rebay. 2007. The eyes absent family of phosphotyrosine phosphatases: properties and roles in developmental regulation of transcription. *Annu Rev Biochem* 76:513–538.

Jeon, S. J., M. Fujioka, S. C. Kim, and A. S. Edge. 2011. Notch signaling alters sensory or neuronal cell fate specification of inner ear stem cells. *J Neurosci* 31:8351–8358.

Jeong, J. Y., Z. Einhorn, P. Mathur, L. Chen, S. Lee, K. Kawakami, and S. Guo. 2007. Patterning the zebrafish diencephalon by the conserved zinc-finger protein Fezl. *Development* 134:127–136.

Jeong, J. Y., Z. Einhorn, S. Mercurio, S. Lee, B. Lau, M. Mione, S. W. Wilson, and S. Guo. 2006. Neurogenin1 is a determinant of zebrafish basal forebrain dopaminergic neurons and is regulated by the conserved zinc finger protein Tof/Fezl. *Proc Natl Acad Sci USA* 103:5143–5148.

Johnson, J. L., and M. R. Leroux. 2010. cAMP and cGMP signaling: sensory systems with prokaryotic roots adopted by eukaryotic cilia. *Trends Cell Biol* 20:435–444.

Johnson, K. R., S. A. Cook, L. C. Erway, A. N. Matthews, L. P. Sanford, N. E. Paradies, and R. A. Friedman. 1999. Inner ear and kidney anomalies caused by IAP insertion in an intron of the Eya1 gene in a mouse model of BOR syndrome. *Hum Mol Genet* 8:645–643.

Johnson, M. A., L. Tsai, D. S. Roy, D. H. Valenzuela, C. Mosley, A. Magklara, S. Lomvardas, S. D. Liberles, and G. Barnea. 2012. Neurons expressing trace amine-associated receptors project to discrete glomeruli and constitute an olfactory subsystem. *Proc Natl Acad Sci USA* 109:13410–13415.

Johnston, J. 1902. An attempt to define the primitive functional divisions of the central nervous system. *J Comp Neurol* 12:87–106.

Jones, J. M., and M. E. Warchol. 2009. Expression of the Gata3 transcription factor in the acoustic ganglion of the developing avian inner ear. *J Comp Neurol* 516:507–518.

Jørgensen, J. M. 2005. Morphology of electroreceptive sensory organs. In *Electroreception*, edited by T. H. Bullock, C. D. Hopkins, A. N. Popper and R. R. Fay, 47–67. New York: Springer.

Julian, L. M., A. C. McDonald, and W. L. Stanford. 2017. Direct reprogramming with SOX factors: masters of cell fate. *Curr Opin Genet Dev* 46:24–36.

Kageyama, R., T. Ohtsuka, and T. Kobayashi. 2008. Roles of Hes genes in neural development. *Dev Growth Differ* 50 Suppl 1:S97–103.

Kageyama, R., T. Ohtsuka, H. Shimojo, and I. Imayoshi. 2009. Dynamic regulation of Notch signaling in neural progenitor cells. *Curr Opin Cell Biol* 21:733–740.

Kaji, T., and K. B. Artinger. 2004. dlx3b and dlx4b function in the development of Rohon-Beard sensory neurons and trigeminal placode in the zebrafish neurula. *Dev Biol* 276:523–540.

Kalatzis, V., I. Sahly, A. El-Amraoui, and C. Petit. 1998. Eya1 expression in the developing ear and kidney: towards the understanding of the pathogenesis of branchio-oto-renal (BOR) syndrome. *Dev Dyn* 213:486–499.

Kam, J. W., R. Raja, and J. F. Cloutier. 2014. Cellular and molecular mechanisms regulating embryonic neurogenesis in the rodent olfactory epithelium. *Int J Dev Neurosci* 37:76–86.

Kan, L., N. Israsena, Z. Zhang, M. Hu, L. R. Zhao, A. Jalali, V. Sahni, and J. A. Kessler. 2004. Sox1 acts through multiple independent pathways to promote neurogenesis. *Dev Biol* 269:580–594.

Kan, L., A. Jalali, L. R. Zhao, X. Zhou, T. McGuire, I. Kazanis, V. Episkopou, A. G. Bassuk, and J. A. Kessler. 2007. Dual function of Sox1 in telencephalic progenitor cells. *Dev Biol* 310:85–98.

Kanekar, S., M. Perron, R. Dorsky, W. A. Harris, L. Y. Jan, Y. N. Jan, and M. L. Vetter. 1997. Xath5 participates in a network of bhlh genes in the developing *Xenopus* retina. *Neuron* 19:981–994.

Kardong, K. V. 2009. *Vertebrates. Comparative anatomy, function, evolution*. 5th ed. New York: McGraw Hill.

Karis, A., I. Pata, J. H. van Doorninck, F. Grosveld, C. I. de Zeeuw, D. de Caprona, and B. Fritzsch. 2001. Transcription factor GATA-3 alters pathway selection of olivocochlear neurons and affects morphogenesis of the ear. *J Comp Neurol* 429:615–630.

Karlstrom, R. O., W. S. Talbot, and A. F. Schier. 1999. Comparative synteny cloning of zebrafish you-too: mutations in the Hedgehog target gli2 affect ventral forebrain patterning. *Genes Dev* 13:388–393.

Kasberg, A. D., E. W. Brunskill, and S. Steven Potter. 2013. SP8 regulates signaling centers during craniofacial development. *Dev Biol* 381:312–323.

Katoh, H., S. Shibata, K. Fukuda, M. Sato, E. Satoh, N. Nagoshi, T. Minematsu, Y. Matsuzaki, C. Akazawa, Y. Toyama, M. Nakamura, and H. Okano. 2011. The dual origin of the peripheral olfactory system: placode and neural crest. *Mol Brain* 4:34.

Kaupp, U. B. 2010. Olfactory signalling in vertebrates and insects: differences and commonalities. *Nat Rev Neurosci* 11:188–200.

Kawai, T., Y. Oka, and H. Eisthen. 2009. The role of the terminal nerve and GnRH in olfactory system neuromodulation. *Zoolog Sci* 26:669–680.

Kawakami, K., S. Sato, H. Ozaki, and K. Ikeda. 2000. Six family genes – structure and function as transcription factors and their roles in development. *Bioessays* 22:616–626.

Kawamura, K., and S. Kikuyama. 1992. Evidence that hypophysis and hypothalamus constitute a single entity from the primary stage of histogenesis. *Development* 115:1–9.

Kawasaki, M. 2005. Physiology of tuberous electrosensory systems In *Electroreception*, edited by T. H. Bullock, C. D. Hopkins, A. N. Popper and R. R. Fay, 154–194. New York: Springer.

Kawauchi, H., and S. A. Sower. 2006. The dawn and evolution of hormones in the adenohypophysis. *Gen Comp Endocrinol* 148:3–14.

Kawauchi, S., C. L. Beites, C. E. Crocker, H. H. Wu, A. Bonnin, R. Murray, and A. L. Calof. 2004. Molecular signals regulating proliferation of stem and progenitor cells in mouse olfactory epithelium. *Dev Neurosci* 26:166–180.

Kawauchi, S., J. Kim, R. Santos, H. H. Wu, A. D. Lander, and A. L. Calof. 2009. Foxg1 promotes olfactory neurogenesis by antagonizing Gdf11. *Development* 136:1453–1464.

Kawauchi, S., J. Shou, R. Santos, J. M. Hebert, S. K. McConnell, I. Mason, and A. L. Calof. 2005. Fgf8 expression defines a morphogenetic center required for olfactory neurogenesis and nasal cavity development in the mouse. *Development* 132:5211–5223.

Kawauchi, S., S. Takahashi, O. Nakajima, H. Ogino, M. Morita, M. Nishizawa, K. Yasuda, and M. Yamamoto. 1999. Regulation of lens fiber cell differentiation by transcription factor c-Maf. *J Biol Chem* 274:19254–19260.

Kay, J. N., K. C. Finger-Baier, T. Roeser, W. Staub, and H. Baier. 2001. Retinal ganglion cell genesis requires lakritz, a Zebrafish atonal homolog. *Neuron* 30:725–736.

Kazanskaya, O., A. Glinka, and C. Niehrs. 2000. The role of *Xenopus* dickkopf1 in prechordal plate specification and neural patterning. *Development* 127:4981–4992.

Kee, Y., and M. Bronner-Fraser. 2005. To proliferate or to die: role of Id3 in cell cycle progression and survival of neural crest progenitors. *Genes Dev* 19:744–755.

Kelberman, D., S. C. de Castro, S. Huang, J. A. Crolla, R. Palmer, J. W. Gregory, D. Taylor, L. Cavallo, M. F. Faienza, R. Fischetto, J. C. Achermann, J. P. Martinez-Barbera, K. Rizzoti, R. Lovell-Badge, I. C. Robinson, D. Gerrelli, and M. T. Dattani. 2008. SOX2 plays a critical role in the pituitary, forebrain, and eye during human embryonic development. *J Clin Endocrinol Metab* 93:1865–1873.

Kelberman, D., K. Rizzoti, A. Avilion, M. Bitner-Glindzicz, S. Cianfarani, J. Collins, W. K. Chong, J. M. Kirk, J. C. Achermann, R. Ross, D. Carmignac, R. Lovell-Badge, I. C. Robinson, and M. T. Dattani. 2006. Mutations withinSox2/SOX2 are associated with abnormalities in the hypothalamo-pituitary-gonadal axis in mice and humans. *J Clin Invest* 116 (9):2442–2455.

Kelberman, D., K. Rizzoti, R. Lovell-Badge, I. C. Robinson, and M. T. Dattani. 2009. Genetic regulation of pituitary gland development in human and mouse. *Endocr Rev* 30:790–829.

Kelly, M. C., Q. Chang, A. Pan, X. Lin, and P. Chen. 2012. Atoh1 directs the formation of sensory mosaics and induces cell proliferation in the postnatal mammalian cochlea in vivo. *J Neurosci* 32:6699–6710.

Kelly, M. C., and P. Chen. 2009. Development of form and function in the mammalian cochlea. *Curr Opin Neurobiol* 19:395–401.

Kenyon, K. L., S. A. Moody, and M. Jamrich. 1999. A novel fork head gene mediates early steps during *Xenopus* lens formation. *Development* 126:5107–5116.

Keynes, R. J., and C. D. Stern. 1984. Segmentation in the vertebrate nervous system. *Nature* 310:786–789.

Khan, M., E. Vaes, and P. Mombaerts. 2011. Regulation of the probability of mouse odorant receptor gene choice. *Cell* 147:907–921.

Khan, S. Y., S. F. Hackett, M. C. Lee, N. Pourmand, C. C. Talbot, Jr., and S. A. Riazuddin. 2015. Transcriptome profiling of developing murine lens through RNA sequencing. *Invest Ophthalmol Vis Sci* 56:4919–4926.

Khatri, S. B., R. K. Edlund, and A. K. Groves. 2014. Foxi3 is necessary for the induction of the chick otic placode in response to FGF signaling. *Dev Biol* 391 (2):158–169.

Khonsari, R. H., M. Seppala, A. Pradel, H. Dutel, G. Clement, O. Lebedev, S. Ghafoor, M. Rothova, A. Tucker, J. G. Maisey, C. M. Fan, M. Kawasaki, A. Ohazama, P. Tafforeau, B. Franco, J. Helms, C. J. Haycraft, A. David, P. Janvier, M. T. Cobourne, and P. T. Sharpe. 2013. The buccohypophyseal canal is an ancestral vertebrate trait maintained by modulation in sonic hedgehog signaling. *BMC Biol* 11:27.

Khudyakov, J., and M. Bronner-Fraser. 2009. Comprehensive spatiotemporal analysis of early chick neural crest network genes. *Dev Dyn* 238:716–723.

Kiecker, C., and A. Lumsden. 2005. Compartments and their boundaries in vertebrate brain development. *Nat Rev Neurosci* 6:553–564.

Kiecker, C., and C. Niehrs. 2001. A morphogen gradient of Wnt/beta-catenin signalling regulates anteroposterior neural patterning in *Xenopus*. *Development* 128:4189–4201.

Kiernan, A. E., A. L. Pelling, K. K. Leung, A. S. Tang, D. M. Bell, C. Tease, R. Lovell-Badge, K. P. Steel, and K. S. Cheah. 2005. Sox2 is required for sensory organ development in the mammalian inner ear. *Nature* 434:1031–1035.

Kil, S. H., and A. Collazo. 2001. Origins of inner ear sensory organs revealed by fate map and time-lapse analyses. *Dev Biol* 233:365–379.

Kim, H. S., H. Seo, C. Yang, J. F. Brunet, and K. S. Kim. 1998. Noradrenergic-specific transcription of the dopamine beta-hydroxylase gene requires synergy of multiple cis-acting elements including at least two Phox2a-binding sites. *J Neurosci* 18:8247–8260.

Kim, W. Y., B. Fritzsch, A. Serls, L. A. Bakel, E. J. Huang, L. F. Reichardt, D. S. Barth, and J. E. Lee. 2001. NeuroD-null mice are deaf due to a severe loss of the inner ear sensory neurons during development. *Development* 128:417–426.

Kimelman, D. 2006. Mesoderm induction: from caps to chips. *Nat Rev Genet* 7:360–372.

Kimura, A., D. Singh, E. F. Wawrousek, M. Kikuchi, M. Nakamura, and T. Shinohara. 2000. Both PCE-1/RX and OTX/CRX interactions are necessary for photoreceptor-specific gene expression. *J Biol Chem* 275:1152–1160.

Kirjavainen, A., M. Sulg, F. Heyd, K. Alitalo, S. Yla-Herttuala, T. Moroy, T. V. Petrova, and U. Pirvola. 2008. Prox1 interacts with Atoh1 and Gfi1, and regulates cellular differentiation in the inner ear sensory epithelia. *Dev Biol* 322:33–45.

Kishi, M., K. Mizuseki, N. Sasai, H. Yamazaki, K. Shiota, S. Nakanishi, and Y. Sasai. 2000. Requirement of Sox2-mediated signaling for differentiation of early *Xenopus* neuroectoderm. *Development* 127:791–800.

Klein, S. L., and P. P. C. Graziadei. 1983. The differentiation of the olfactory placode in *Xenopus laevis*: a light and electron microscope study. *J Comp Neurol* 217:7–30.

Klemenz, R., E. Frohli, R. H. Steiger, R. Schafer, and A. Aoyama. 1991. Alpha B-crystallin is a small heat shock protein. *Proc Natl Acad Sci USA* 88:3652–3656.

Klisch, T. J., Y. Xi, A. Flora, L. Wang, W. Li, and H. Y. Zoghbi. 2011. In vivo Atoh1 targetome reveals how a proneural transcription factor regulates cerebellar development. *Proc Natl Acad Sci USA* 108:3288–3293.

Klum, S., C. Zaouter, Z. Alekseenko, A. K. Bjorklund, D. W. Hagey, J. Ericson, J. Muhr, and M. Bergsland. 2018. Sequentially acting SOX proteins orchestrate astrocyte- and oligodendrocyte-specific gene expression. *EMBO Rep* 19:e46635.

Knabe, W., B. Obermayer, H. J. Kuhn, G. Brunnett, and S. Washausen. 2009. Apoptosis and proliferation in the trigeminal placode. *Brain Struct Funct* 214:49–65.
Knouff, R. A. 1927. The origin of the cranial ganglia of *Rana*. *J Comp Neurol* 44:259–361.
Knouff, R. A. 1935. The developmental pattern of ectodermal placodes in *Rana pipiens*. *J Comp Neurol* 62:17–71.
Kobayashi, D., M. Kobayashi, K. Matsumoto, T. Ogura, M. Nakafuku, and K. Shimamura. 2002. Early subdivisions in the neural plate define distinct competence for inductive signals. *Development* 129:83–93.
Kobayashi, M., K. Nishikawa, T. Suzuki, and M. Yamamoto. 2001. The homeobox protein Six3 interacts with the Groucho corepressor and acts as a transcriptional repressor in eye and forebrain formation. *Dev Biol* 232:315–326.
Kobayashi, M., H. Osanai, K. Kawakami, and M. Yamamoto. 2000. Expression of three zebrafish Six4 genes in the cranial sensory placodes and the developing somites. *Mech Dev* 98:151–155.
Kobayashi, T., Y. Iwamoto, K. Takashima, A. Isomura, Y. Kosodo, K. Kawakami, T. Nishioka, K. Kaibuchi, and R. Kageyama. 2015. Deubiquitinating enzymes regulate Hes1 stability and neuronal differentiation. *FEBS J* 282:2411–2423.
Kobayashi, T., and R. Kageyama. 2014. Expression dynamics and functions of Hes factors in development and diseases. *Curr Top Dev Biol* 110:263–283.
Kochhar, A., S. M. Fischer, W. J. Kimberling, and R. J. Smith. 2007. Branchio-oto-renal syndrome. *Am J Med Genet A* 143:1671–1678.
Kolterud, A., M. Alenius, L. Carlsson, and S. Bohm. 2004. The Lim homeobox gene Lhx2 is required for olfactory sensory neuron identity. *Development* 131:5319–5326.
Kondoh, H 2008. Shedding light on developmental gene regulation through the lens. *Dev Growth Differ* 50 Suppl 1:S57–S69.
Kondoh, H., and Y. Kamachi. 2010. SOX-partner code for cell specification: regulatory target selection and underlying molecular mechanisms. *Int J Biochem Cell Biol* 42:391–399.
Kondoh, H., M. Uchikawa, H. Yoda, H. Takeda, M. FurutaniSeiki, and R. O. Karlstrom. 2000. Zebrafish mutations in Gli-mediated hedgehog signaling lead to lens transdifferentiation from the adenohypophysis anlage. *Mech Dev* 96:165–174.
Konishi, Y., K. Ikeda, Y. Iwakura, and K. Kawakami. 2006. Six1 and Six4 promote survival of sensory neurons during early trigeminal gangliogenesis. *Brain Res* 1116:93–102.
Koo, S. K., J. K. Hill, C. H. Hwang, Z. S. Lin, K. J. Millen, and D. K. Wu. 2009. Lmx1a maintains proper neurogenic, sensory, and non-sensory domains in the mammalian inner ear. *Dev Biol* 333:14–25.
Köppl, C. 2011. Birds–same thing, but different? Convergent evolution in the avian and mammalian auditory systems provides informative comparative models. *Hear Res* 273:65–71.
Korzh, V., T. Edlund, and S. Thor. 1993. Zebrafish primary neurons initiate expression of the LIM homeodomain protein Isl-1 at the end of gastrulation. *Development* 118:417–425.
Köster, R. W., R. P. Kühnlein, and J. Wittbrodt. 2000. Ectopic Sox3 activity elicits sensory placode formation. *Mech Dev* 95:175–187.
Kouki, T., H. Imai, K. Aoto, K. Eto, S. Shioda, K. Kawamura, and S. Kikuyama. 2001. Developmental origin of the rat adenohypophysis prior to the formation of Rathke's pouch. *Development* 128:959–963.
Kozlowski, D. J., T. Murakami, R. K. Ho, and E. S. Weinberg. 1997. Regional cell movement and tissue patterning in the zebrafish embryo revealed by fate mapping with caged fluorescein. *Biochem Cell Biol* 75:551–562.
Kozlowski, D. J., T. T. Whitfield, N. A. Hukriede, W. K. Lam, and E. S. Weinberg. 2005. The zebrafish dog-eared mutation disrupts eya1, a gene required for cell survival and differentiation in the inner ear and lateral line. *Dev Biol* 277:27–41.

Kramer, I., M. Sigrist, J. C. de Nooij, I. Taniuchi, T. M. Jessell, and S. Arber. 2006. A role for Runx transcription factor signaling in dorsal root ganglion sensory neuron diversification. *Neuron* 49:379–393.

Kramer, P. R., and S. Wray. 2000. Midline nasal tissue influences nestin expression in nasal-placode-derived luteinizing hormone-releasing hormone neurons during development. *Dev Biol* 227:343–357.

Kriebel, M., F. Müller, and T. Hollemann. 2007. Xeya3 regulates survival and proliferation of neural progenitor cells within the anterior neural plate of *Xenopus* embryos. *Dev Dyn* 236:1526–1534.

Kriebitz, N. N., C. Kiecker, L. McCormick, A. Lumsden, A. Graham, and E. Bell. 2009. PRDC regulates placode neurogenesis in chick by modulating BMP signalling. *Dev Biol* 336:280–292.

Krimm, R. F 2007. Factors that regulate embryonic gustatory development. *BMC Neurosci* 8 Suppl 3:S4.

Krishnan, N., D. G. Jeong, S. K. Jung, S. E. Ryu, A. Xiao, C. D. Allis, S. J. Kim, and N. K. Tonks. 2009. Dephosphorylation of the C-terminal tyrosyl residue of the DNA damage-related histone H2A.X is mediated by the protein phosphatase Eyes Absent. *J Biol Chem* 284:16066–16070.

Kroll, K. L., A. N. Salic, L. M. Evans, and M. W. Kirschner. 1998. Geminin, a neuralizing molecule that demarcates the future neural plate at the onset of gastrulation. *Development* 125:3247–3258.

Krull, C. E., R. Lansford, N. W. Gale, A. Collazo, C. Marcelle, G. D. Yancopoulos, S. E. Fraser, and M. Bronner-Fraser. 1997. Interactions of Eph-related receptors and ligands confer rostrocaudal pattern to trunk neural crest migration. *Curr Biol* 7:571–580.

Kulesa, P., D. L. Ellies, and P. A. Trainor. 2004. Comparative analysis of neural crest cell death, migration, and function during vertebrate embryogenesis. *Dev Dyn* 229: 14–29.

Kumar, J. P. 2009. The sine oculis homeobox (SIX) family of transcription factors as regulators of development and disease. *Cell Mol Life Sci* 66:565–583.

Kupari, J., M. Haring, E. Agirre, G. Castelo-Branco, and P. Ernfors. 2019. An atlas of vagal sensory neurons and their molecular specialization. *Cell Rep* 27:2508–2523 e4.

Kuratani, S. 2003. Evolutionary developmental biology and vertebrate head segmentation: a perspective from developmental constraint. *Theory Biosci* 122:230–251.

Kuratani, S. 2005. Craniofacial development and the evolution of the vertebrates: the old problems on a new background. *Zoolog Sci* 22:1–19.

Kuratani, S. 2008. Is the vertebrate head segmented?-evolutionary and developmental considerations. *Integr Comp Biol* 48:647–657.

Kuratani, S., N. Adachi, N. Wada, Y. Oisi, and F. Sugahara. 2012. Developmental and evolutionary significance of the mandibular arch and prechordal/premandibular cranium in vertebrates: revising the heterotopy scenario of gnathostome jaw evolution. *J Anat* 222:41–55.

Kuratani, S. C., and N. A. Wall. 1992. Expression of Hox 2.1 protein in restricted populations of neural crest cells and pharyngeal ectoderm. *Dev Dyn* 195:15–28.

Kuratani, S., I. Matsuo, and S. Aizawa. 1997. Developmental patterning and evolution of the mammalian viscerocranium: genetic insights into comparative morphology. *Dev Dyn* 209:139–155.

Kuratani, S., T. Ueki, S. Aizawa, and S. Hirano. 1997. Peripheral development of cranial nerves in a cyclostome, *Lampetra japonica*: morphological distribution of nerve branches and the vertebrate body plan. *J Comp Neurol* 384:483–500.

Kuwabara, T., J. Hsieh, A. Muotri, G. Yeo, M. Warashina, D. C. Lie, L. Moore, K. Nakashima, M. Asashima, and F. H. Gage. 2009. Wnt-mediated activation of NeuroD1 and retro-elements during adult neurogenesis. *Nat Neurosci* 12:1097–1105.

Kwak, S. J., S. Vemaraju, S. J. Moorman, D. Zeddies, A. N. Popper, and B. B. Riley. 2006. Zebrafish pax5 regulates development of the utricular macula and vestibular function. *Dev Dyn* 235:3026–3038.

Kwan, K. M., H. Otsuna, H. Kidokoro, K. R. Carney, Y. Saijoh, and C. B. Chien. 2012. A complex choreography of cell movements shapes the vertebrate eye. *Development* 139:359–372.

Kwon, H. J., N. Bhat, E. M. Sweet, R. A. Cornell, and B. B. Riley. 2010. Identification of early requirements for preplacodal ectoderm and sensory organ development. *PLoS Genet* 6:e1001133.

La Bonne, C., and M. Bronner-Fraser. 1998. Neural crest induction in *Xenopus*: evidence for a two-signal model. *Development* 125:2403–2414.

La Mantia, A. S., N. Bhasin, K. Rhodes, and J. Heemskerk. 2000. Mesenchymal/epithelial induction mediates olfactory pathway formation. *Neuron* 28:411–425.

Laclef, C., G. Hamard, J. Demignon, E. Souil, C. Houbron, and P. Maire. 2003. Altered myogenesis in Six1-deficient mice. *Development* 130:2239–2252.

Laclef, C., E. Souil, J. Demignon, and P. Maire. 2003. Thymus, kidney and craniofacial abnormalities in Six1 deficient mice. *Mech Dev* 120:669–679.

Lacomme, M., L. Liaubet, F. Pituello, and S. Bel-Vialar. 2012. NEUROG2 drives cell cycle exit of neuronal precursors by specifically repressing a subset of cyclins acting at the G1 and S phases of the cell cycle. *Mol Cell Biol* 32:2596–2607.

Ladher, R. K. 2017. Changing shape and shaping change: inducing the inner ear. *Semin Cell Dev Biol* 65:39–46.

Ladher, R. K., K. U. Anakwe, A. L. Gurney, G. C. Schoenwolf, and P. H. Francis-West. 2000. Identification of synergistic signals initiating inner ear development. *Science* 290:1965–1967.

Ladher, R. K., P. O'Neill, and J. Begbie. 2010. From shared lineage to distinct functions: the development of the inner ear and epibranchial placodes. *Development* 137:1777–1785.

Ladich, F., and A. N. Popper. 2004. Parallel evolution of fish hearing organs. In *Evolution of the vertebrate auditory system*, edited by G. A. Manley, A. N. Popper and R. R. Fay, 95–127. New York: Springer.

Laframboise, A. J., X. Ren, S. Chang, R. Dubuc, and B. S. Zielinski. 2007. Olfactory sensory neurons in the sea lamprey display polymorphisms. *Neurosci Lett* 414:277–281.

Lagman, D., D. Ocampo Daza, J. Widmark, X. M. Abalo, G. Sundström, and D. Larhammar. 2013. The vertebrate ancestral repertoire of visual opsins, transducin alpha subunits and oxytocin/vasopressin receptors was established by duplication of their shared genomic region in the two rounds of early vertebrate genome duplications. *BMC Evol Biol* 13:238.

Lagutin, O. V., C. C. Zhu, D. Kobayashi, J. Topczewski, K. Shimamura, L. Puelles, H. R. Russell, P. J. McKinnon, L. Solnica-Krezel, and G. Oliver. 2003. Six3 repression of Wnt signaling in the anterior neuroectoderm is essential for vertebrate forebrain development. *Genes Dev* 17:368–379.

Lagutin, O., C. Q. C. Zhu, Y. Furuta, D. H. Rowitch, A. P. McMahon, and G. Oliver. 2001. Six3 promotes the formation of ectopic optic vesicle-like structures in mouse embryos. *Dev Dyn* 221:342–349.

Lamb, T. D. 2009. Evolution of vertebrate retinal photoreception. *Philos Trans R Soc Lond B Biol Sci* 364:2911–2924.

Lamb, T. D. 2013. Evolution of phototransduction, vertebrate photoreceptors and retina. *Prog Retin Eye Res* 36:52–119.

Lamb, T. D., S. P. Collin, and E. N. Pugh, Jr. 2007. Evolution of the vertebrate eye: opsins, photoreceptors, retina and eye cup. *Nat Rev Neurosci* 8:960–976.

Lamolet, B., A. M. Pulichino, T. Lamonerie, Y. Gauthier, T. Brue, A. Enjalbert, and J. Drouin. 2001. A pituitary cell-restricted T box factor, Tpit, activates POMC transcription in cooperation with Pitx homeoproteins. *Cell* 104:849–859.

Landacre, F. L. 1910. The origin of the cranial ganglia in *Ameiurus*. *J Comp Neurol* 20:309–411.

Lange, K. 2011. Fundamental role of microvilli in the main functions of differentiated cells: outline of an universal regulating and signaling system at the cell periphery. *J Cell Physiol* 226:896–927.

Langer, K. B., S. K. Ohlemacher, M. J. Phillips, C. M. Fligor, P. Jiang, D. M. Gamm, and J. S. Meyer. 2018. Retinal ganglion cell diversity and subtype specification from human pluripotent stem cells. *Stem Cell Rep* 10:1282–1293.

Lanier, J., I. M. Dykes, S. Nissen, S. R. Eng, and E. E. Turner. 2009. Brn3a regulates the transition from neurogenesis to terminal differentiation and represses non-neural gene expression in the trigeminal ganglion. *Dev Dyn* 238:3065–3079.

Lanigan, T. M., S. K. DeRaad, and A. F. Russo. 1998. Requirement of the MASH-1 transcription factor for neuroendocrine differentiation of thyroid C cells. *J Neurobiol* 34:126–134.

Larhammar, D., K. Nordström, and T. A. Larsson. 2009. Evolution of vertebrate rod and cone phototransduction genes. *Philos Trans R Soc Lond B Biol Sci* 364:2867–2880.

Lassiter, R. N., C. M. Dude, S. B. Reynolds, N. I. Winters, C. V. Baker, and M. R. Stark. 2007. Canonical Wnt signaling is required for ophthalmic trigeminal placode cell fate determination and maintenance. *Dev Biol* 308:392–406.

Lassiter, R. N., S. B. Reynolds, K. D. Marin, T. F. Mayo, and M. R. Stark. 2009. FGF signaling is essential for ophthalmic trigeminal placode cell delamination and differentiation. *Dev Dyn* 238:1073–1082.

Laub, F., R. Aldabe, V. Friedrich, Jr., S. Ohnishi, T. Yoshida, and F. Ramirez. 2001. Developmental expression of mouse Kruppel-like transcription factor KLF7 suggests a potential role in neurogenesis. *Dev Biol* 233:305–318.

Laub, F., L. Lei, H. Sumiyoshi, D. Kajimura, C. Dragomir, S. Smaldone, A. C. Puche, T. J. Petros, C. Mason, L. F. Parada, and F. Ramirez. 2005. Transcription factor KLF7 is important for neuronal morphogenesis in selected regions of the nervous system. *Mol Cell Biol* 25:5699–5711.

Lavoie, P. L., L. Budry, A. Balsalobre, and J. Drouin. 2008. Developmental dependence on NurRE and EboxNeuro for expression of pituitary proopiomelanocortin. *Mol Endocrinol* 22:1647–1657.

Lawoko-Kerali, G., M. N. Rivolta, P. Lawlor, D. I. Cacciabue-Rivolta, C. Langton-Hewer, J. H. van Doorninck, and M. C. Holley. 2004. GATA3 and NeuroD distinguish auditory and vestibular neurons during development of the mammalian inner ear. *Mech Dev* 121:287–299.

Le Douarin, N. M., and C. Kalcheim. 1999. *The neural crest*. Cambridge: Cambrdige University Press.

Le Tissier, P., T. Fiordelisio Coll, and P. Mollard. 2018. The processes of anterior pituitary hormone pulse generation. *Endocrinology* 159:3524–3535.

Lebel, M., P. Agarwal, C. W. Cheng, M. G. Kabir, T. Y. Chan, V. Thanabalasingham, X. Zhang, D. R. Cohen, M. Husain, S. H. Cheng, B. G. Bruneau, and C. C. Hui. 2003. The Iroquois homeobox gene Irx2 is not essential for normal development of the heart and midbrain-hindbrain boundary in mice. *Mol Cell Biol* 23:8216–8225.

Lecaudey, V., I. Anselme, R. Dildrop, U. Ruther, and S. Schneider-Maunoury. 2005. Expression of the zebrafish Iroquois genes during early nervous system formation and patterning. *J Comp Neurol* 492:289–302.

Ledent, V. 2002. Postembryonic development of the posterior lateral line in zebrafish. *Development* 129:597–604.

Lee, H. K., H. S. Lee, and S. A. Moody. 2014. Neural transcription factors: from embryos to neural stem cells. *Mol Cells* 37:705–712.

Lee, J. E., S. M. Hollenberg, L. Snider, D. L. Turner, N. Lipnick, and H. Weintraub. 1995. Conversion of *Xenopus* ectoderm into neurons by NeuroD, a basic helix-loop-helix protein. *Science* 268:836–844.

Lee, S., J. M. Cuvillier, B. Lee, R. Shen, J. W. Lee, and S. K. Lee. 2012. Fusion protein Isl1-Lhx3 specifies motor neuron fate by inducing motor neuron genes and concomitantly suppressing the interneuron programs. *Proc Natl Acad Sci USA* 109:3383–3388.

Lee, S. K., and S. L. Pfaff. 2003. Synchronization of neurogenesis and motor neuron specification by direct coupling of bHLH and homeodomain transcription factors. *Neuron* 38:731–745.

Lee, V. H., L. T. Lee, and B. K. Chow. 2008. Gonadotropin-releasing hormone: regulation of the GnRH gene. *FEBS J* 275:5458–5478.

Leger, S., and M. Brand. 2002. Fgf8 and Fgf3 are required for zebrafish ear placode induction, maintenance and inner ear patterning. *Mech Dev* 119:91–108.

Lei, L., F. Laub, M. Lush, M. Romero, J. Zhou, B. Luikart, L. Klesse, F. Ramirez, and L. F. Parada. 2005. The zinc finger transcription factor Klf7 is required for TrkA gene expression and development of nociceptive sensory neurons. *Genes Dev* 19:1354–1364.

Lei, L., L. Ma, S. Nef, T. Thai, and L. F. Parada. 2001. mKlf7, a potential transcriptional regulator of TrkA nerve growth factor receptor expression in sensory and sympathetic neurons. *Development* 128:1147–1158.

Lei, L., and L. F. Parada. 2007. Transcriptional regulation of Trk family neurotrophin receptors. *Cell Mol Life Sci* 64:522–532.

Lei, L., J. Zhou, L. Lin, and L. F. Parada. 2006. Brn3a and Klf7 cooperate to control TrkA expression in sensory neurons. *Dev Biol* 300:758–769.

Leung, C. T., P. A. Coulombe, and R. R. Reed. 2007. Contribution of olfactory neural stem cells to tissue maintenance and regeneration. *Nat Neurosci* 10:720–726.

Levanon, D., D. Bettoun, C. Harris-Cerruti, E. Woolf, V. Negreanu, R. Eilam, Y. Bernstein, D. Goldenberg, C. Xiao, M. Fliegauf, E. Kremer, F. Otto, O. Brenner, A. Lev-Tov, and Y. Groner. 2002. The Runx3 transcription factor regulates development and survival of TrkC dorsal root ganglia neurons. *EMBO J* 21:3454–3463.

Li, B., S. Kuriyama, M. Moreno, and R. Mayor. 2009. The posteriorizing gene Gbx2 is a direct target of Wnt signalling and the earliest factor in neural crest induction. *Development* 136:3267–3278.

Li, C. L., K. C. Li, D. Wu, Y. Chen, H. Luo, J. R. Zhao, S. S. Wang, M. M. Sun, Y. J. Lu, Y. Q. Zhong, X. Y. Hu, R. Hou, B. B. Zhou, L. Bao, H. S. Xiao, and X. Zhang. 2016. Somatosensory neuron types identified by high-coverage single-cell RNA-sequencing and functional heterogeneity. *Cell Res* 26:83–102.

Li, C. W., T. R. Van De Water, and R. J. Ruben. 1978. The fate mapping of the eleventh and twelfth day mouse otocyst: an in vitro study of the sites of origin of the embryonic inner ear sensory structures. *J Morphol* 157:249–267.

Li, H., H. Liu, C. Sage, M. Huang, Z. Y. Chen, and S. Heller. 2004. Islet-1 expression in the developing chicken inner ear. *J Comp Neurol* 477:1–10.

Li, J., T. Zhang, A. Ramakrishnan, B. Fritzsch, J. Xu, E. Y. M. Wong, Y. E. Loh, J. Ding, L. Shen, and P. X. Xu. 2020. Dynamic changes in cis-regulatory occupancy by Six1 and its cooperative interactions with distinct cofactors drive lineage-specific gene expression programs during progressive differentiation of the auditory sensory epithelium. *Nucleic Acids Res* 48:2880–2896.

Li, R., F. Wu, R. Ruonala, D. Sapkota, Z. Hu, and X. Mu. 2014. Isl1 and Pou4f2 form a complex to regulate target genes in developing retinal ganglion cells. *PLoS One* 9:e92105.

Li, S., E. B. Crenshaw, 3rd, E. J. Rawson, D. M. Simmons, L. W. Swanson, and M. G. Rosenfeld. 1990. Dwarf locus mutants lacking three pituitary cell types result from mutations in the POU-domain gene pit-1. *Nature* 347:528–533.

Li, S., S. M. Price, H. Cahill, D. K. Ryugo, M. M. Shen, and M. Xiang. 2002. Hearing loss caused by progressive degeneration of cochlear hair cells in mice deficient for the Barhl1 homeobox gene. *Development* 129:3523–3532.

Li, S., W. Qian, G. Jiang, and Y. Ma. 2016. Transcription factors in the development of inner ear hair cells. *Front Biosci (Landmark Ed)* 21:1118–1125.

Li, W., and R. A. Cornell. 2007. Redundant activities of Tfap2a and Tfap2c are required for neural crest induction and development of other non-neural ectoderm derivatives in zebrafish embryos. *Dev Biol* 304:338–354.

Li, X., K. A. Oghi, J. Zhang, A. Krones, K. T. Bush, C. K. Glass, S. K. Nigam, A. K. Aggarwal, R. Maas, D. W. Rose, and M. G. Rosenfeld. 2003. Eya protein phosphatase activity regulates Six1-Dach-Eya transcriptional effects in mammalian organogenesis. *Nature* 426:247–254.

Li, X., V. Perissi, F. Liu, D. W. Rose, and M. G. Rosenfeld. 2002. Tissue-specific regulation of retinal and pituitary precursor cell proliferation. *Science* 297:1180–1183.

Libants, S., K. Carr, H. Wu, J. H. Teeter, Y. W. Chung-Davidson, Z. Zhang, C. Wilkerson, and W. Li. 2009. The sea lamprey *Petromyzon marinus* genome reveals the early origin of several chemosensory receptor families in the vertebrate lineage. *BMC Evol Biol* 9:180.

Liberles, S. D., and L. B. Buck. 2006. A second class of chemosensory receptors in the olfactory epithelium. *Nature* 442:645–650.

Liberles, S. D., L. F. Horowitz, D. Kuang, J. J. Contos, K. L. Wilson, J. Siltberg-Liberles, D. A. Liberles, and L. B. Buck. 2009. Formyl peptide receptors are candidate chemosensory receptors in the vomeronasal organ. *Proc Natl Acad Sci USA* 106:9842–9847.

Lichtneckert, R., and H. Reichert. 2005. Insights into the urbilaterian brain: conserved genetic patterning mechanisms in insect and vertebrate brain development. *Heredity* 94:465–477.

Light, W., A. E. Vernon, A. Lasorella, A. Iavarone, and C. LaBonne. 2005. *Xenopus* Id3 is required downstream of Myc for the formation of multipotent neural crest progenitor cells. *Development* 132:1831–1841.

Lin, S. C., S. Li, D. W. Drolet, and M. G. Rosenfeld. 1994. Pituitary ontogeny of the Snell dwarf mouse reveals Pit-1-independent and Pit-1-dependent origins of the thyrotrope. *Development* 120:515–522.

Lin, W., E. A. Ezekwe, Jr., Z. Zhao, E. R. Liman, and D. Restrepo. 2008. TRPM5-expressing microvillous cells in the main olfactory epithelium. *BMC Neurosci* 9:114.

Linder, B., E. Mentele, K. Mansperger, T. Straub, E. Kremmer, and R. A. Rupp. 2007. CHD4/Mi-2beta activity is required for the positioning of the mesoderm/neuroectoderm boundary in *Xenopus*. *Genes Dev* 21:973–983.

Ling, F., B. Kang, and X. H. Sun. 2014. Id proteins: small molecules, mighty regulators. *Curr Top Dev Biol* 110:189–216.

Litsiou, A., S. Hanson, and A. Streit. 2005. A balance of FGF, BMP and WNT signalling positions the future placode territory in the head. *Development* 132:4051–4062.

Liu, D., H. Chu, L. Maves, Y. L. Yan, P. A. Morcos, J. H. Postlethwait, and M. Westerfield. 2003. Fgf3 and Fgf8 dependent and independent transcription factors are required for otic placode specification. *Development* 130:2213–2224.

Liu, H., J. L. Pecka, Q. Zhang, G. A. Soukup, K. W. Beisel, and D. Z. He. 2014. Characterization of transcriptomes of cochlear inner and outer hair cells. *J Neurosci* 34:11085–11095.

Liu, M., F. A. Pereira, S. D. Price, M. J. Chu, C. Shope, D. Himes, R. A. Eatock, W. E. Brownell, A. Lysakowski, and M. J. Tsai. 2000. Essential role of BETA2/NeuroD1 in development of the vestibular and auditory systems. *Gene Develop* 14:2839–2854.

Liu, W., O. V. Lagutin, M. Mende, A. Streit, and G. Oliver. 2006. Six3 activation of Pax6 expression is essential for mammalian lens induction and specification. *EMBO J* 25:5383–5395.

Liu, W., Z. Mo, and M. Xiang. 2001. The Ath5 proneural genes function upstream of Brn3 POU domain transcription factor genes to promote retinal ganglion cell development. *Proc Natl Acad Sci USA* 98:1649–1654.

Liu, Y., and Q. Ma. 2011. Generation of somatic sensory neuron diversity and implications on sensory coding. *Curr Opin Neurobiol* 21:52–60.

Liu, Z., T. Owen, J. Fang, R. S. Srinivasan, and J. Zuo. 2012. In vivo notch reactivation in differentiating cochlear hair cells induces sox2 and prox1 expression but does not disrupt hair cell maturation. *Dev Dyn* 241 (4):684–696.

Livesey, F. J., and C. L. Cepko. 2001. Vertebrate neural cell-fate determination: lessons from the retina. *Nat Rev Neurosci* 2:109–118.

Lleras-Forero, L., M. Tambalo, N. Christophorou, D. Chambers, C. Houart, and A. Streit. 2013. Neuropeptides: developmental signals in placode progenitor formation. *Dev Cell* 26:195–203.

Locket, N. A., and J. M. Jørgensen. 1998. The eyes of hagfishes. In *The biology of hagfishes*, edited by J. M. Jørgensen, J. P. Lomholt, R. E. Weber and H. Malte, 541–556. Dordrecht: Springer.

Lodato, M. A., C. W. Ng, J. A. Wamstad, A. W. Cheng, K. K. Thai, E. Fraenkel, R. Jaenisch, and L. A. Boyer. 2013. SOX2 co-occupies distal enhancer elements with distinct POU factors in ESCs and NPCs to specify cell state. *PLoS Genet* 9:e1003288.

Logan, C., R. J. Wingate, I. J. McKay, and A. Lumsden. 1998. Tlx-1 and Tlx-3 homeobox gene expression in cranial sensory ganglia and hindbrain of the chick embryo: markers of patterned connectivity. *J Neurosci* 18:5389–402.

Londin, E. R., L. Mentzer, and H. I. Sirotkin. 2007. Churchill regulates cell movement and mesoderm specification by repressing Nodal signaling. *BMC Dev Biol* 7:120.

Loosli, F., S. Winkler, and J. Wittbrodt. 1999. Six3 overexpression initiates the formation of ectopic retina. *Genes Dev* 13:649–654.

Lopes, C., Z. Liu, Y. Xu, and Q. Ma. 2012. Tlx3 and Runx1 act in combination to coordinate the development of a cohort of nociceptors, thermoceptors, and pruriceptors. *J Neurosci* 32:9706–9715.

Lopes, D. M., F. Denk, and S. B. McMahon. 2017. The molecular fingerprint of dorsal root and trigeminal ganglion neurons. *Front Mol Neurosci* 10:304.

López-Rios, J., K. Tessmar, F. Loosli, J. Wittbrodt, and P. Bovolenta. 2003. Six3 and Six6 activity is modulated by members of the groucho family. *Development* 130:185–195.

López-Schier, H., and A. J. Hudspeth. 2005. Supernumerary neuromasts in the posterior lateral line of zebrafish lacking peripheral glia. *Proc Natl Acad Sci USA* 102:1496–1501.

López-Schier, H., and A. J. Hudspeth. 2006. A two-step mechanism underlies the planar polarization of regenerating sensory hair cells. *Proc Natl Acad Sci USA* 103:18615–20.

López-Schier, H., C. J. Starr, J. A. Kappler, R. Kollmar, and A. J. Hudspeth. 2004. Directional cell migration establishes the axes of planar polarity in the posterior lateral-line organ of the zebrafish. *Dev Cell* 7:401–412.

López, F., R. Delgado, R. López, J. Bacigalupo, and D. Restrepo. 2014. Transduction for pheromones in the main olfactory epithelium is mediated by the Ca^2+-activated channel TRPM5. *J Neurosci* 34:3268–3278.

Louvi, A., and S. Artavanis-Tsakonas. 2006. Notch signalling in vertebrate neural development. *Nat Rev Neurosci* 7:93–102.

Lovicu, F. J., and J. W. McAvoy. 2005. Growth factor regulation of lens development. *Dev Biol* 280:1–14.

Lu, C. C., J. M. Appler, E. A. Houseman, and L. V. Goodrich. 2011. Developmental profiling of spiral ganglion neurons reveals insights into auditory circuit assembly. *J Neurosci* 31:10903–10918.

Lu, X., and C. W. Sipe. 2016. Developmental regulation of planar cell polarity and hair-bundle morphogenesis in auditory hair cells: lessons from human and mouse genetics. *Wiley Interdiscip Rev Dev Biol* 5:85–101.

Lujan, E., S. Chanda, H. Ahlenius, T. C. Südhof, and M. Wernig. 2012. Direct conversion of mouse fibroblasts to self-renewing, tripotent neural precursor cells. *Proc Natl Acad Sci USA* 109:2527–2532.

Lunde, K., H. G. Belting, and W. Driever. 2004. Zebrafish pou5f1/pou2, homolog of mammalian Oct4, functions in the endoderm specification cascade. *Curr Biol* 14:48–55.

Luo, T., Y. H. Lee, J. P. Saint-Jeannet, and T. D. Sargent. 2003. Induction of neural crest in *Xenopus* by transcription factor AP2alpha. *Proc Natl Acad Sci USA* 100:532–537.

Luo, T., M. Matsuo-Takasaki, M. L. Thomas, D. L. Weeks, and T. D. Sargent. 2002. Transcription factor AP-2 is an essential and direct regulator of epidermal development in *Xenopus*. *Dev Biol* 245:136–144.

Lush, M. E., D. C. Diaz, N. Koenecke, S. Baek, H. Boldt, M. K. St Peter, T. Gaitan-Escudero, A. Romero-Carvajal, E. M. Busch-Nentwich, A. G. Perera, K. E. Hall, A. Peak, J. S. Haug, and T. Piotrowski. 2019. scRNA-Seq reveals distinct stem cell populations that drive hair cell regeneration after loss of Fgf and Notch signaling. *Elife* 8:e44431.

Luu, P., F. Acher, H. O. Bertrand, J. Fan, and J. Ngai. 2004. Molecular determinants of ligand selectivity in a vertebrate odorant receptor. *J Neurosci* 24:10128–10137.

Lyons, D. B., W. E. Allen, T. Goh, L. Tsai, G. Barnea, and S. Lomvardas. 2013. An epigenetic trap stabilizes singular olfactory receptor expression. *Cell* 154:325–336.

Ma, E. Y., and D. W. Raible. 2009. Signaling pathways regulating zebrafish lateral line development. *Curr Biol* 19:R381–R386.

Ma, Q., D. J. Anderson, and B. Fritzsch. 2000. Neurogenin1 null mutant ears develop fewer, morphologically normal hair cells in smaller sensory epithelia devoid of innervation. *J Assoc Res Otolaryngol* 1:129–143.

Ma, Q. F., Z. F. Chen, I. D. Barrantes, J. L. de la Pompa, and D. J. Anderson. 1998. Neurogenin1 is essential for the determination of neuronal precursors for proximal cranial sensory ganglia. *Neuron* 20:469–482.

Ma, Q. F., C. Fode, F. Guillemot, and D. J. Anderson. 1999. Neurogenin1 and Neurogenin2 control two distinct waves of neurogenesis in developing dorsal root ganglia. *Gene Dev* 13:1717–1728.

Ma, Q. F., C. Kintner, and D. J. Anderson. 1996. Identification of neurogenin, a vertebrate neuronal determination gene. *Cell* 87:43–52.

Ma, Y. C., M. R. Song, J. P. Park, H. Y. Henry Ho, L. Hu, M. V. Kurtev, J. Zieg, Q. Ma, S. L. Pfaff, and M. E. Greenberg. 2008. Regulation of motor neuron specification by phosphorylation of neurogenin 2. *Neuron* 58:65–77.

Mackereth, M. D., S. J. Kwak, A. Fritz, and B. B. Riley. 2005. Zebrafish pax8 is required for otic placode induction and plays a redundant role with Pax2 genes in the maintenance of the otic placode. *Development* 132:371–382.

Macosko, E. Z., A. Basu, R. Satija, J. Nemesh, K. Shekhar, M. Goldman, I. Tirosh, A. R. Bialas, N. Kamitaki, E. M. Martersteck, J. J. Trombetta, D. A. Weitz, J. R. Sanes, A. K. Shalek, A. Regev, and S. A. McCarroll. 2015. Highly parallel genome-wide expression profiling of individual cells using nanoliter droplets. *Cell* 161:1202–1214.

Maczkowiak, F., S. Mateos, E. Wang, D. Roche, R. Harland, and A. H. Monsoro-Burq. 2010. The Pax3 and Pax7 paralogs cooperate in neural and neural crest patterning using distinct molecular mechanisms, in *Xenopus laevis* embryos. *Dev Biol* 340: 381–396.

Magarinos, M., J. Contreras, M. R. Aburto, and I. Varela-Nieto. 2012. Early development of the vertebrate inner ear. *Anat Rec* 295:1775–1790.

Magklara, A., A. Yen, B. M. Colquitt, E. J. Clowney, W. Allen, E. Markenscoff-Papadimitriou, Z. A. Evans, P. Kheradpour, G. Mountoufaris, C. Carey, G. Barnea, M. Kellis, and S. Lomvardas. 2011. An epigenetic signature for monoallelic olfactory receptor expression. *Cell* 145:555–570.

Maharana, S. K., and G. Schlosser. 2018. A gene regulatory network underlying the formation of pre-placodal ectoderm in *Xenopus laevis*. *BMC Biol* 16:79.

Maier, E. C., A. Saxena, B. Alsina, M. E. Bronner, and T. T. Whitfield. 2014. Sensational placodes: neurogenesis in the otic and olfactory systems. *Dev Biol* 389:50–67.

Maier, E., J. von Hofsten, H. Nord, M. Fernandes, H. Paek, J. M. Hebert, and L. Gunhaga. 2010. Opposing Fgf and Bmp activities regulate the specification of olfactory sensory and respiratory epithelial cell fates. *Development* 137:1601–1611.

Mancilla, H., and R. Mayor. 1996. Neural crest formation in *Xenopus laevis*: mechanisms of Xslug induction. *Dev Biol* 177:580–589.

Manley, G. A. 2011. Lizard auditory papillae: an evolutionary kaleidoscope. *Hear Res* 273:59–64.

Manley, G. A. 2017. Comparative auditory neuroscience: understanding the evolution and function of ears. *J Assoc Res Otolaryngol* 18:1–24.

Manley, G. A., and C. Köppl. 1998. Phylogenetic development of the cochlea and its innervation. *Curr Opin Neurobiol* 8:468–474.

Mansour, S. L. 1994. Targeted disruption of int-2 (fgf-3) causes developmental defects in the tail and inner ear. *Mol Reprod Dev* 39:62–67.

Mansour, S. L., and G. C. Schoenwolf. 2005. Morphogenesis of the inner ear. In *Development of the inner ear*, edited by M. W. Kelley, D. K. Wu, A. N. Popper and R. R. Fay, 43–84. New York: Springer.

Marchal, L., G. Luxardi, V. Thome, and L. Kodjabachian. 2009. BMP inhibition initiates neural induction via FGF signaling and Zic genes. *Proc Natl Acad Sci USA* 106:17437–17442.

Markenscoff-Papadimitriou, E., W. E. Allen, B. M. Colquitt, T. Goh, K. K. Murphy, K. Monahan, C. P. Mosley, N. Ahituv, and S. Lomvardas. 2014. Enhancer interaction networks as a means for singular olfactory receptor expression. *Cell* 159:543–557.

Marmigere, F., and P. Ernfors. 2007. Specification and connectivity of neuronal subtypes in the sensory lineage. *Nat Rev Neurosci* 8:114–127.

Maroon, H., J. Walshe, R. Mahmood, P. Kiefer, C. Dickson, and I. Mason. 2002. Fgf3 and Fgf8 are required together for formation of the otic placode and vesicle. *Development* 129:2099–2108.

Marquardt, T., R. Ashery-Padan, N. Andrejewski, R. Scardigli, F. Guillemot, and P. Gruss. 2001. Pax6 is required for the multipotent state of retinal progenitor cells. *Cell* 105:43–55.

Marquis, T. J., M. Nozaki, W. Fagerberg, and S. A. Sower. 2017. Comprehensive histological and immunological studies reveal a novel glycoprotein hormone and thyrostimulin expressing proto-glycotrope in the sea lamprey pituitary. *Cell Tissue Res* 367:311–338.

Marra, N. J., M. J. Stanhope, N. K. Jue, M. Wang, Q. Sun, P. Pavinski Bitar, V. P. Richards, A. Komissarov, M. Rayko, S. Kliver, B. J. Stanhope, C. Winkler, S. J. O'Brien, A. Antunes, S. Jørgensen, and M. S. Shivji. 2019. White shark genome reveals ancient elasmobranch adaptations associated with wound healing and the maintenance of genome stability. *Proc Natl Acad Sci USA* 116:4446–4455.

Martin, K., and A. K. Groves. 2006. Competence of cranial ectoderm to respond to Fgf signaling suggests a two-step model of otic placode induction. *Development* 133:877–887.

Martinez-Barbera, J. P., and R. S. Beddington. 2001. Getting your head around Hex and Hesx1: forebrain formation in mouse. *Int J Dev Biol* 45:327–336.

Martinez-Barbera, J. P., M. Signore, P. P. Boyl, E. Puelles, D. Acampora, R. Gogoi, F. Schubert, A. Lumsden, and A. Simeone. 2001. Regionalisation of anterior neuroectoderm and its competence in responding to forebrain and midbrain inducing activities depend on mutual antagonism between OTX2 and GBX2. *Development* 128:4789–4800.

Martinez-Morales, J. R., M. Signore, D. Acampora, A. Simeone, and P. Bovolenta. 2001. Otx genes are required for tissue specification in the developing eye. *Development* 128:2019–2030.

Martinez Arias, A., and B. Steventon. 2018. On the nature and function of organizers. *Development* 145:dev159525.

Martini, S., L. Silvotti, A. Shirazi, N. J. Ryba, and R. Tirindelli. 2001. Co-expression of putative pheromone receptors in the sensory neurons of the vomeronasal organ. *J Neurosci* 21:843–848.

Masland, R. H. 2012. The neuronal organization of the retina. *Neuron* 76:266–280.

Masserdotti, G., S. Gillotin, B. Sutor, D. Drechsel, M. Irmler, H. F. Jørgensen, S. Sass, F. J. Theis, J. Beckers, B. Berninger, F. Guillemot, and M. Götz. 2015. Transcriptional mechanisms of proneural factors and REST in regulating neuronal reprogramming of astrocytes. *Cell Stem Cell* 17:74–88.

Masuda, M., D. Dulon, K. Pak, L. M. Mullen, Y. Li, L. Erkman, and A. F. Ryan. 2011. Regulation of POU4F3 gene expression in hair cells by 5' DNA in mice. *Neuroscience* 197:48–64.

Masuda, M., K. Pak, E. Chavez, and A. F. Ryan. 2012. TFE2 and GATA3 enhance induction of POU4F3 and myosin VIIa positive cells in nonsensory cochlear epithelium by ATOH1. *Dev Biol* 372:68–80.

Matei, V., S. Pauley, S. Kaing, D. Rowitch, K. W. Beisel, K. Morris, F. Feng, K. Jones, J. Lee, and B. Fritzsch. 2005. Smaller inner ear sensory epithelia in Neurog 1 null mice are related to earlier hair cell cycle exit. *Dev Dyn* 234:633–650.

Mathers, P. H., A. Miller, T. Doniach, M. L. Dirksen, and M. Jamrich. 1995. Initiation of anterior head specific gene expresion in uncommitted ectoderm of *Xenopus laevis* by ammonium chloride. *Dev Biol* 171:641–654.

Mathias, R. T., T. W. White, and X. Gong. 2010. Lens gap junctions in growth, differentiation, and homeostasis. *Physiol Rev* 90:179–206.

Matos-Cruz, V., J. Blasic, B. Nickle, P. R. Robinson, S. Hattar, and M. E. Halpern. 2011. Unexpected diversity and photoperiod dependence of the zebrafish melanopsin system. *PLoS One* 6:e25111.

Matsuda, K., and H. Kondoh. 2014. Dkk1-dependent inhibition of Wnt signaling activates Hesx1 expression through its 5' enhancer and directs forebrain precursor development. *Genes Cells* 19:374–385.

Matsuda, M., and A. B. Chitnis. 2010. Atoh1a expression must be restricted by Notch signaling for effective morphogenesis of the posterior lateral line primordium in zebrafish. *Development* 137:3477–3487.

Matsunami, H., and L. B. Buck. 1997. A multigene family encoding a diverse array of putative pheromone receptors in mammals. *Cell* 90:775–784.

Matsuno, A., J. Itoh, S. Takekoshi, Y. Itoh, Y. Ohsugi, H. Katayama, T. Nagashima, and R. Y. Osamura. 2003. Dynamics of subcellular organelles, growth hormone, Rab3B, SNAP-25, and syntaxin in rat pituitary cells caused by growth hormone releasing hormone and somatostatin. *Microsc Res Tech* 62:232–239.

Matsuno, A., A. Mizutani, H. Okinaga, K. Takano, S. Yamada, S. M. Yamada, H. Nakaguchi, K. Hoya, M. Murakami, M. Takeuchi, M. Sugaya, J. Itoh, S. Takekoshi, and R. Y. Osamura. 2011. Functional molecular morphology of anterior pituitary cells, from hormone production to intracellular transport and secretion. *Med Mol Morphol* 44:63–70.

Matsuo-Takasaki, M., M. Matsumura, and Y. Sasai. 2005. An essential role of *Xenopus* Foxi1a for ventral specification of the cephalic ectoderm during gastrulation. *Development* 132:3885–3894.

Matsuo, I., S. Kuratani, C. Kimura, N. Takeda, and S. Aizawa. 1995. Mouse Otx2 functions in the formation and patterning of rostral head. *Gene Dev* 9:2646–2658.

Matsushima, D., W. Heavner, and L. H. Pevny. 2011. Combinatorial regulation of optic cup progenitor cell fate by SOX2 and PAX6. *Development* 138:443–454.

Matthews, G., and P. Fuchs. 2010. The diverse roles of ribbon synapses in sensory neurotransmission. *Nat Rev Neurosci* 11:812–822.

Maucksch, C., K. S. Jones, and B. Connor. 2013. Concise review: the involvement of SOX2 in direct reprogramming of induced neural stem/precursor cells. *Stem Cells Transl Med* 2:579–583.

Maulding, K., M. S. Padanad, J. Dong, and B. B. Riley. 2014. Mesodermal Fgf10b cooperates with other fibroblast growth factors during induction of otic and epibranchial placodes in zebrafish. *Dev Dyn* 243:1275–1285.

Mazzoni, E. O., S. Mahony, M. Closser, C. A. Morrison, S. Nedelec, D. J. Williams, D. An, D. K. Gifford, and H. Wichterle. 2013. Synergistic binding of transcription factors to cell-specific enhancers programs motor neuron identity. *Nat Neurosci* 16:1219–1227.

McCabe, K. L., and M. Bronner-Fraser. 2008. Essential role for PDGF signaling in ophthalmic trigeminal placode induction. *Development* 135:1863–1874.

McCabe, K. L., J. W. Sechrist, and M. Bronner-Fraser. 2009. Birth of ophthalmic trigeminal neurons initiates early in the placodal ectoderm. *J Comp Neurol* 514:161–173.

McCarroll, M. N., Z. R. Lewis, M. D. Culbertson, B. L. Martin, D. Kimelman, and A. V. Nechiporuk. 2012. Graded levels of Pax2a and Pax8 regulate cell differentiation during sensory placode formation. *Development* 139:2740–2750.

McCarroll, M. N., and A. V. Nechiporuk. 2013. Fgf3 and Fgf10a work in concert to promote maturation of the epibranchial placodes in zebrafish. *PLoS One* 8:e85087.

McCauley, H. A., and G. Guasch. 2015. Three cheers for the goblet cell: maintaining homeostasis in mucosal epithelia. *Trends Mol Med* 21:492–503.

McCoy, E. L., R. Iwanaga, P. Jedlicka, N. S. Abbey, L. A. Chodosh, K. A. Heichman, A. L. Welm, and H. L. Ford. 2009. Six1 expands the mouse mammary epithelial stem/progenitor cell pool and induces mammary tumors that undergo epithelial-mesenchymal transition. *J Clin Invest* 119:2663–2677.

McDevitt, D. S. 1972. Presence of lateral eye lens crystallins in the median eye of the american chameleon. *Science* 175:763–764.

McEvilly, R. J., L. Erkman, L. Luo, P. E. Sawchenko, A. F. Ryan, and M. G. Rosenfeld. 1996. Requirement for brn-3.0 in differentiation and survival of sensory and motor neurons. *Nature* 384:574–577.

McEwen, D. P., P. M. Jenkins, and J. R. Martens. 2008. Olfactory cilia: our direct neuronal connection to the external world. *Curr Top Dev Biol* 85:333–370.

McGrath, J., P. Roy, and B. J. Perrin. 2017. Stereocilia morphogenesis and maintenance through regulation of actin stability. *Semin Cell Dev Biol* 65:88–95.

McLarren, K. W., A. Litsiou, and A. Streit. 2003. DLX5 positions the neural crest and preplacode region at the border of the neural plate. *Dev Biol* 259:34–47.

McPherson, D. R. 2018. Sensory hair cells: an introduction to structure and physiology. *Integr Comp Biol* 58:282–300.

Mears, A. J., M. Kondo, P. K. Swain, Y. Takada, R. A. Bush, T. L. Saunders, P. A. Sieving, and A. Swaroop. 2001. Nrl is required for rod photoreceptor development. *Nat Genet* 29:447–452.

Medeiros, D. M. 2013. The evolution of the neural crest: new perspectives from lamprey and invertebrate neural crest-like cells. *Wiley Interdiscip Rev Dev Biol* 2:1–15.

Medina-Martinez, O., and M. Jamrich. 2007. Foxe view of lens development and disease. *Development* 134:1455–1463.

Medina-Martinez, O., R. Shah, and M. Jamrich. 2009. Pitx3 controls multiple aspects of lens development. *Dev Dyn* 238:2193–201.

Menco, B. P. 1997. Ultrastructural aspects of olfactory signaling. *Chem Senses* 22:295–311.

Menco, B. P., R. C. Bruch, B. Dau, and W. Danho. 1992. Ultrastructural localization of olfactory transduction components: the G protein subunit Golf alpha and type III adenylyl cyclase. *Neuron* 8:441–453.

Menco, B. P., V. M. Carr, P. I. Ezeh, E. R. Liman, and M. P. Yankova. 2001. Ultrastructural localization of G-proteins and the channel protein TRP2 to microvilli of rat vomeronasal receptor cells. *J Comp Neurol* 438:468–489.

Menco, B. P., and A. I. Farbman. 1985. Genesis of cilia and microvilli of rat nasal epithelia during pre-natal development. I. Olfactory epithelium, qualitative studies. *J Cell Sci* 78:283–310.

Menco, B. P., and E. E. Morrison. 2003. Morphology of the mammalian olfactory epithelium: form, fine structure, function, and pathology. In *Handbook of olfaction and gustation*, edited by R. L. Doty, 17–49. New York: Marcel Dekker.

Metcalfe, W. K. 1985. Sensory neuron growth cones comigrate with posterior lateral line primordial cells in zebrafish. *J Comp Neurol* 238:218–224.

Metzis, V., S. Steinhauser, E. Pakanavicius, M. Gouti, D. Stamataki, K. Ivanovitch, T. Watson, T. Rayon, S. N. Mousavy Gharavy, R. Lovell-Badge, N. M. Luscombe, and J. Briscoe. 2018. Nervous system regionalization entails axial allocation before neural differentiation. *Cell* 175:1105–1118 e17.

Meulemans, D., and M. Bronner-Fraser. 2004. Gene-regulatory interactions in neural crest evolution and development. *Dev Cell* 7:291–299.

Meunier, A., and J. Azimzadeh. 2016. Multiciliated cells in animals. *Cold Spring Harb Perspect Biol* 8:a028233.

Micalizzi, D. S., K. L. Christensen, P. Jedlicka, R. D. Coletta, A. E. Baron, J. C. Harrell, K. B. Horwitz, D. Billheimer, K. A. Heichman, A. L. Welm, W. P. Schiemann, and H. L. Ford. 2009. The Six1 homeoprotein induces human mammary carcinoma cells to undergo epithelial-mesenchymal transition and metastasis in mice through increasing TGF-beta signaling. *J Clin Invest* 119:2678–2690.

Milet, C., F. Maczkowiak, D. D. Roche, and A. H. Monsoro-Burq. 2013. Pax3 and Zic1 drive induction and differentiation of multipotent, migratory, and functional neural crest in *Xenopus* embryos. *Proc Natl Acad Sci USA* 110:5528–5533.

Miller, S. J., Z. D. Lan, A. Hardiman, J. Wu, J. J. Kordich, D. M. Patmore, R. S. Hegde, T. P. Cripe, J. A. Cancelas, M. H. Collins, and N. Ratner. 2010. Inhibition of eyes absent homolog 4 expression induces malignant peripheral nerve sheath tumor necrosis. *Oncogene* 29:368–379.

Millet, S., K. Campbell, D. J. Epstein, K. Losos, E. Harris, and A. L. Joyner. 1999. A role for Gbx2 in repression of Otx-2 and positioning the mid/hindbrain organizer. *Nature* 401:161–164.

Millimaki, B. B., E. M. Sweet, M. S. Dhason, and B. B. Riley. 2007. Zebrafish atoh1 genes: classic proneural activity in the inner ear and regulation by Fgf and Notch. *Development* 134:295–305.

Millimaki, B. B., E. M. Sweet, and B. B. Riley. 2010. Sox2 is required for maintenance and regeneration, but not initial development, of hair cells in the zebrafish inner ear. *Dev Biol* 338:262–269.

Minoux, M., and F. M. Rijli. 2010. Molecular mechanisms of cranial neural crest cell migration and patterning in craniofacial development. *Development* 137:2605–2621.

Mirabeau, O., and J. S. Joly. 2013. Molecular evolution of peptidergic signaling systems in bilaterians. *Proc Natl Acad Sci USA* 110:E2028–E2037.

Mishima, N., and S. Tomarev. 1998. Chicken Eyes absent 2 gene: isolation and expression pattern during development. *Int J Dev Biol* 42:1109–1115.

Mistretta, C. M. 1989. Anatomy and neurophysiology of the taste system in aged animals. *Ann N Y Acad Sci* 561:277–290.

Mistri, T. K., A. G. Devasia, L. T. Chu, W. P. Ng, F. Halbritter, D. Colby, B. Martynoga, S. R. Tomlinson, I. Chambers, P. Robson, and T. Wohland. 2015. Selective influence of Sox2 on POU transcription factor binding in embryonic and neural stem cells. *EMBO Rep* 16:1177–1191.

Mitton, K. P., P. K. Swain, S. Chen, S. Xu, D. J. Zack, and A. Swaroop. 2000. The leucine zipper of NRL interacts with the CRX homeodomain. A possible mechanism of transcriptional synergy in rhodopsin regulation. *J Biol Chem* 275:29794–29799.

Miyagi, S., H. Kato, and A. Okuda. 2009. Role of SoxB1 transcription factors in development. *Cell Mol Life Sci* 66:3675–3684.

Miyagi, S., S. Masui, H. Niwa, T. Saito, T. Shimazaki, H. Okano, M. Nishimoto, M. Muramatsu, A. Iwama, and A. Okuda. 2008. Consequence of the loss of Sox2 in the developing brain of the mouse. *FEBS Lett* 582:2811–2815.

Miyake, T., I. H. von Herbing, and B. K. Hall. 1997. Neural ectoderm, neural crest, and placodes: contribution of the otic placode to the ectodermal lining of the embryonic opercular cavity in atlantic cod (Teleostei). *J Morphol* 231:231–252.

Miyamichi, K., S. Serizawa, H. M. Kimura, and H. Sakano. 2005. Continuous and overlapping expression domains of odorant receptor genes in the olfactory epithelium determine the dorsal/ventral positioning of glomeruli in the olfactory bulb. *J Neurosci* 25:3586–3592.

Miyazono, K., S. Maeda, and T. Imamura. 2005. BMP receptor signaling: transcriptional targets, regulation of signals, and signaling cross-talk. *Cytokine Growth Factor Rev* 16:251–263.

Mizoguchi, T., S. Togawa, K. Kawakami, and M. Itoh. 2011. Neuron and sensory epithelial cell fate is sequentially determined by Notch signaling in zebrafish lateral line development. *J Neurosci* 31:15522–15530.

Mizuguchi, R., M. Sugimori, H. Takebayashi, H. Kosako, M. Nagao, S. Yoshida, Y. Nabeshima, K. Shimamura, and M. Nakafuku. 2001. Combinatorial roles of Olig2 and Neurogenin2 in the coordinated induction of pan-neuronal and subtype-specific properties of motoneurons. *Neuron* 31:757–771.

Mizuseki, K., M. Kishi, M. Matsui, S. Nakanishi, and Y. Sasai. 1998. *Xenopus* zic-related-1 and sox-2, two factors induced by chordin, have distinct activities in the initiation of neural induction. *Development* 125:579–587.

Modrell, M. S., W. E. Bemis, R. G. Northcutt, M. C. Davis, and C. V. Baker. 2011. Electrosensory ampullary organs are derived from lateral line placodes in bony fishes. *Nat Commun* 2:496.

Modrell, M. S., D. Hockman, B. Uy, D. Buckley, T. Sauka-Spengler, M. E. Bronner, and C. V. Baker. 2014. A fate-map for cranial sensory ganglia in the sea lamprey. *Dev Biol* 385:405–416.

Modrell, M. S., M. Lyne, A. R. Carr, H. H. Zakon, D. Buckley, A. S. Campbell, M. C. Davis, G. Micklem, and C. V. Baker. 2017. Insights into electrosensory organ development, physiology and evolution from a lateral line-enriched transcriptome. *Elife* 6:e24197.

Møller, A. R. 2012. *Hearing: anatomy, physiology, and disorders of the auditory system*. 3rd ed. San Diego: Plural Publishing.

Mombaerts, P. 2004. Genes and ligands for odorant, vomeronasal and taste receptors. *Nat Rev Neurosci* 5:263–278.

Mombaerts, P., F. Wang, C. Dulac, S. K. Chao, A. Nemes, M. Mendelsohn, J. Edmondson, and R. Axel. 1996. Visualizing an olfactory sensory map. *Cell* 87:675–686.

Monahan, K., and S. Lomvardas. 2015. Monoallelic expression of olfactory receptors. *Annu Rev Cell Dev Biol* 31:721–740.

Monahan, K., I. Schieren, J. Cheung, A. Mumbey-Wafula, E. S. Monuki, and S. Lomvardas. 2017. Cooperative interactions enable singular olfactory receptor expression in mouse olfactory neurons. *Elife* 6:e28620.

Monsoro-Burq, A. H., R. B. Fletcher, and R. M. Harland. 2003. Neural crest induction by paraxial mesoderm in *Xenopus* embryos requires FGF signals *Development* 130:3111–3124.

Monsoro-Burq, A. H., E. Wang, and R. Harland. 2005. Msx1 and Pax3 cooperate to mediate FGF8 and WNT signals during *Xenopus* neural crest induction. *Dev Cell* 8:167–178.

Monsoro-Burq, A. H. 2020. *Xenopus* embryo: neural induction. In *Encyclopedia of life sciences*, https://doi.org/10.1002/9780470015902.a0000731.pub3. Wiley and Sons, Chichester.

Montani, G., S. Tonelli, R. Elsaesser, J. Paysan, and R. Tirindelli. 2006. Neuropeptide Y in the olfactory microvillar cells. *Eur J Neurosci* 24:20–24.

Monzack, E. L., and L. L. Cunningham. 2013. Lead roles for supporting actors: critical functions of inner ear supporting cells. *Hear Res* 303:20–29.

Moody, S. A., and A. S. LaMantia. 2015. Transcriptional regulation of cranial sensory placode development. *Curr Top Dev Biol* 111:301–350.

Morshedian, A., and G. L. Fain. 2015. Single-photon sensitivity of lamprey rods with cone-like outer segments. *Curr Biol* 25:484–487.

Morshedian, A., and G. L. Fain. 2017. Light adaptation and the evolution of vertebrate photoreceptors. *J Physiol* 595:4947–4960.

Mu, X., P. D. Beremand, S. Zhao, R. Pershad, H. Sun, A. Scarpa, S. Liang, T. L. Thomas, and W. H. Klein. 2004. Discrete gene sets depend on POU domain transcription factor Brn3b/Brn-3.2/POU4f2 for their expression in the mouse embryonic retina. *Development* 131:1197–1210.

Mu, X., X. Fu, P. D. Beremand, T. L. Thomas, and W. H. Klein. 2008. Gene regulation logic in retinal ganglion cell development: Isl1 defines a critical branch distinct from but overlapping with Pou4f2. *Proc Natl Acad Sci USA* 105:6942–6947.

Müller, U., and A. Littlewood-Evans. 2001. Mechanisms that regulate mechanosensory hair cell differentiation. *Trends Cell Biol* 11:334–342.

Mulvaney, J., and A. Dabdoub. 2012. Atoh1, an essential transcription factor in neurogenesis and intestinal and inner ear development: function, regulation, and context dependency. *J Assoc Res Otolaryngol* 13:281–293.

Munger, S. D., T. Leinders-Zufall, and F. Zufall. 2009. Subsystem organization of the mammalian sense of smell. *Annu Rev Physiol* 71:115–140.

Muñoz-Sanjuán, I., and A. H. Brivanlou. 2002. Neural induction, the default model and embryonic stem cells. *Nat Rev Neurosci* 3:271–280.

Murakami, S., and Y. Arai. 1994. Direct evidence for the migration of LHRH neurons from the nasal region to the forebrain in the chick embryo: a carbocyanine dye analysis. *Neurosci Res* 19:331–338.

Muranishi, Y., K. Terada, and T. Furukawa. 2012. An essential role for Rax in retina and neuroendocrine system development. *Dev Growth Differ* 54:341–348.

Muranishi, Y., K. Terada, T. Inoue, K. Katoh, T. Tsujii, R. Sanuki, D. Kurokawa, S. Aizawa, Y. Tamaki, and T. Furukawa. 2011. An essential role for RAX homeoprotein and NOTCH-HES signaling in Otx2 expression in embryonic retinal photoreceptor cell fate determination. *J Neurosci* 31:16792–16807.

Murdoch, B., and A. J. Roskams. 2007. Olfactory epithelium progenitors: insights from transgenic mice and in vitro biology. *J Mol Histol* 38:581–599.

Murko, C., and M. E. Bronner. 2017. Tissue specific regulation of the chick Sox10E1 enhancer by different Sox family members. *Dev Biol* 422:47–57.

Murphy, A. E., and S. Harvey. 2001. Extrapituitary beta TSH and GH in early chick embryos. *Mol Cell Endocrinol* 185:161–171.
Murray, R. C., D. Navi, J. Fesenko, A. D. Lander, and A. L. Calof. 2003. Widespread defects in the primary olfactory pathway caused by loss of Mash1 function. *J Neurosci* 23:1769–1780.
Muske, L. E., and F. L. Moore. 1988. The nervus terminalis in amphibians: anatomy, chemistry and relationship with the hypothalamic gonadotropin-releasing hormone system. *Brain Behav Evol* 32:141–150.
Musser, J. M., and D. Arendt. 2017. Loss and gain of cone types in vertebrate ciliary photoreceptor evolution. *Dev Biol* 431:26–35.
Nagiel, A., D. Andor-Ardo, and A. J. Hudspeth. 2008. Specificity of afferent synapses onto plane-polarized hair cells in the posterior lateral line of the zebrafish. *J Neurosci* 28:8442–8453.
Nagy, V., T. Cole, C. Van Campenhout, T. M. Khoung, C. Leung, S. Vermeiren, M. Novatchkova, D. Wenzel, D. Cikes, A. A. Polyansky, I. Kozieradzki, A. Meixner, E. J. Bellefroid, G. G. Neely, and J. M. Penninger. 2015. The evolutionarily conserved transcription factor PRDM12 controls sensory neuron development and pain perception. *Cell Cycle* 14:1799–1808.
Nakashima, A., H. Takeuchi, T. Imai, H. Saito, H. Kiyonari, T. Abe, M. Chen, L. S. Weinstein, C. R. Yu, D. R. Storm, H. Nishizumi, and H. Sakano. 2013. Agonist-independent GPCR activity regulates anterior-posterior targeting of olfactory sensory neurons. *Cell* 154:1314–1325.
Nakata, K., T. Nagai, J. Aruga, and K. Mikoshiba. 1998. *Xenopus* Zic family and its role in neural and neural crest development. *Mech Dev* 75:43–51.
Narayanan, C. H., and Y. Narayanan. 1980. Neural crest and placodal contributions in the development of the glossopharyngeal-vagal complex in the chick. *Anat Rec* 196:71–82.
Nasonkin, I. O., R. D. Ward, L. T. Raetzman, A. F. Seasholtz, T. L. Saunders, P. J. Gillespie, and S. A. Camper. 2004. Pituitary hypoplasia and respiratory distress syndrome in Prop1 knockout mice. *Hum Mol Genet* 13:2727–2735.
Nathans, J., D. Thomas, and D. S. Hogness. 1986. Molecular genetics of human color vision: the genes encoding blue, green, and red pigments. *Science* 232:193–202.
Nayagam, B. A., M. A. Muniak, and D. K. Ryugo. 2011. The spiral ganglion: connecting the peripheral and central auditory systems. *Hear Res* 278:2–20.
Nayak, G. D., H. S. Ratnayaka, R. J. Goodyear, and G. P. Richardson. 2007. Development of the hair bundle and mechanotransduction. *Int J Dev Biol* 51:597–608.
Nechiporuk, A., T. Linbo, K. D. Poss, and D. W. Raible. 2007. Specification of epibranchial placodes in zebrafish. *Development* 134:611–623.
Nechiporuk, A., T. Linbo, and D. W. Raible. 2005. Endoderm-derived Fgf3 is necessary and sufficient for inducing neurogenesis in the epibranchial placodes in zebrafish. *Development* 132:3717–3730.
Nechiporuk, A., and D. W. Raible. 2008. FGF-dependent mechanosensory organ patterning in zebrafish. *Science* 320:1774–1777.
Nei, M., Y. Niimura, and M. Nozawa. 2008. The evolution of animal chemosensory receptor gene repertoires: roles of chance and necessity. *Nat Rev Genet* 9:951–963.
Nelson, S. M., L. Park, and D. L. Stenkamp. 2009. Retinal homeobox 1 is required for retinal neurogenesis and photoreceptor differentiation in embryonic zebrafish. *Dev Biol* 328:24–39.
Neves, J., A. Kamaid, B. Alsina, and F. Giráldez. 2007. Differential expression of Sox2 and Sox3 in neuronal and sensory progenitors of the developing inner ear of the chick. *J Comp Neurol* 503:487–500.

Neves, J., C. Parada, M. Chamizo, and F. Giráldez. 2011. Jagged 1 regulates the restriction of Sox2 expression in the developing chicken inner ear: a mechanism for sensory organ specification. *Development* 138:735–744.

Neves, J., M. Uchikawa, A. Bigas, and F. Giráldez. 2012. The prosensory function of sox2 in the chicken inner ear relies on the direct regulation of atoh1. *PLoS One* 7:e30871.

Neves, J., I. Vachkov, and F. Giráldez. 2013. Sox2 regulation of hair cell development: incoherence makes sense. *Hear Res* 297:20–29.

New, J. G. 1997. The evolution of vertebrate electrosensory systems. *Brain Behav Evol* 50:244–252.

Ng, L., J. B. Hurley, B. Dierks, M. Srinivas, C. Salto, B. Vennstrom, T. A. Reh, and D. Forrest. 2001. A thyroid hormone receptor that is required for the development of green cone photoreceptors. *Nat Genet* 27:94–98.

Nguyen, M. Q., Y. Wu, L. S. Bonilla, L. J. von Buchholtz, and N. J. P. Ryba. 2017. Diversity amongst trigeminal neurons revealed by high throughput single cell sequencing. *PLoS One* 12:e0185543.

Nica, G., W. Herzog, C. Sonntag, and M. Hammerschmidt. 2004. Zebrafish pit1 mutants lack three pituitary cell types and develop severe dwarfism. *Mol Endocrinol* 18:1196–1209.

Nica, G., W. Herzog, C. Sonntag, M. Nowak, H. Schwarz, A. G. Zapata, and M. Hammerschmidt. 2006. Eya1 is required for lineage-specific differentiation, but not for cell survival in the zebrafish adenohypophysis. *Dev Biol* 292:189–204.

Nichane, M., N. de Croze, X. Ren, J. Souopgui, A. H. Monsoro-Burq, and E. J. Bellefroid. 2008. Hairy2-Id3 interactions play an essential role in *Xenopus* neural crest progenitor specification. *Dev Biol* 15:355–367.

Nichane, M., X. Ren, J. Souopgui, and E. J. Bellefroid. 2008. Hairy2 functions through both DNA-binding and non DNA-binding mechanisms at the neural plate border in *Xenopus*. *Dev Biol* 322:368–380.

Nichols, D. H., S. Pauley, I. Jahan, K. W. Beisel, K. J. Millen, and B. Fritzsch. 2008. Lmx1a is required for segregation of sensory epithelia and normal ear histogenesis and morphogenesis. *Cell Tissue Res* 334:339–358.

Nicolay, D. J., J. R. Doucette, and A. J. Nazarali. 2006. Transcriptional regulation of neurogenesis in the olfactory epithelium. *Cell Mol Neurobiol* 26:803–821.

Nicolson, T. 2017. The genetics of hair-cell function in zebrafish. *J Neurogenet* 31:102–112.

Nieber, F., T. Pieler, and K. A. Henningfeld. 2009. Comparative expression analysis of the neurogenins in *Xenopus tropicalis* and *Xenopus laevis*. *Dev Dyn* 238:451–458.

Niehrs, C. 2004. Regionally specific induction by the Spemann-Mangold organizer. *Nat Rev Genet* 5:425–434.

Niehrs, C. 2010. On growth and form: a Cartesian coordinate system of Wnt and BMP signaling specifies bilaterian body axes. *Development* 137:845–857.

Niehrs, C. 2012. The complex world of WNT receptor signalling. *Nat Rev Mol Cell Biol* 13:767–779.

Nieuwenhuys, R. 2011. The structural, functional, and molecular organization of the brainstem. *Front Neuroanat* 5:33.

Nieuwkoop, P. D. 1963. Pattern formation in artificially activated ectoderm (*Rana pipiens* and *Ambystoma punctatum*). *Dev Biol* 7:255–279.

Nieuwkoop, P. D., A. G. Johnen, and B. Albers. 1985. *The inductive nature of early chordate development*. Cambridge: Cambridge University Press.

Nieuwkoop, P. D., and G. V. Nigtevecht. 1954. Neural activation and transformation in explants of competent ectoderm under the influence of fragments of anterior notochord in urodeles. *J Embryol Exp Morph* 2:175–193.

Niimura, Y. 2009. Evolutionary dynamics of olfactory receptor genes in chordates: interaction between environments and genomic contents. *Hum Genomics* 4:107–118.

Niimura, Y., A. Matsui, and K. Touhara. 2014. Extreme expansion of the olfactory receptor gene repertoire in African elephants and evolutionary dynamics of orthologous gene groups in 13 placental mammals. *Genome Res* 24:1485–1496.

Nikaido, M., K. Doi, T. Shimizu, M. Hibi, Y. Kikuchi, and K. Yamasu. 2007. Initial specification of the epibranchial placode in zebrafish embryos depends on the fibroblast growth factor signal. *Dev Dyn* 236:564–571.

Nikaido, M., J. Navajas Acedo, K. Hatta, and T. Piotrowski. 2017. Retinoic acid is required and Fgf, Wnt, and Bmp signaling inhibit posterior lateral line placode induction in zebrafish. *Dev Biol* 431:215–225.

Nikitina, N., T. Sauka-Spengler, and M. Bronner-Fraser. 2008. Dissecting early regulatory relationships in the lamprey neural crest gene network. *Proc Natl Acad Sci USA* 105:20083–20088.

Nimmo, R. A., G. E. May, and T. Enver. 2015. Primed and ready: understanding lineage commitment through single cell analysis. *Trends Cell Biol* 25:459–467.

Nimpf, S., G. C. Nordmann, D. Kagerbauer, E. P. Malkemper, L. Landler, A. Papadaki-Anastasopoulou, L. Ushakova, A. Wenninger-Weinzierl, M. Novatchkova, P. Vincent, T. Lendl, M. Colombini, M. J. Mason, and D. A. Keays. 2019. A putative mechanism for magnetoreception by electromagnetic induction in the pigeon inner ear. *Curr Biol* 29:4052–4059 e4.

Nishida, A., A. Furukawa, C. Koike, Y. Tano, S. Aizawa, I. Matsuo, and T. Furukawa. 2003. Otx2 homeobox gene controls retinal photoreceptor cell fate and pineal gland development. *Nat Neurosci* 6:1255–1263.

Nishimura, K., T. Noda, and A. Dabdoub. 2017. Dynamic expression of Sox2, Gata3, and Prox1 during primary auditory neuron development in the mammalian cochlea. *PLoS One* 12:e0170568.

Nissen, R. M., J. Yan, A. Amsterdam, N. Hopkins, and S. M. Burgess. 2003. Zebrafish foxi one modulates cellular responses to Fgf signaling required for the integrity of ear and jaw patterning. *Development* 130:2543–2554.

Niwa, H., J. Miyazaki, and A. G. Smith. 2000. Quantitative expression of Oct-3/4 defines differentiation, dedifferentiation or self-renewal of ES cells. *Nat Genet* 24:372–376.

Noden, D. M. 1980a. Somatotopic and functional organization of the avian trigeminal ganglion: an HRP analysis in the hatchling chick. *J Comp Neurol* 190:405–428.

Noden, D. M. 1980b. Somatotopic organization of the embryonic chick trigeminal ganglion. *J Comp Neurol* 190:429–444.

Noden, D. M. 1983. The role of the neural crest in patterning of avian cranial skeletal, connective and muscle tissues. *Dev Biol* 96:144–165.

Noden, D. M. 1991. Vertebrate craniofacial development: the relation between ontogenetic process and morphological outcome. *Brain Behav Evol* 38:190–225.

Noden, D. M., and P. A. Trainor. 2005. Relations and interactions between cranial mesoderm and neural crest populations. *J Anat* 207:575–601.

Nomura, T., S. Takahashi, and T. Ushiki. 2004. Cytoarchitecture of the normal rat olfactory epithelium: light and scanning electron microscopic studies. *Arch Histol Cytol* 67:159–170.

Nordström, U., E. Maier, T. M. Jessell, and T. Edlund. 2006. An early role for WNT signaling in specifying neural patterns of Cdx and Hox gene expression and motor neuron subtype identity. *PLoS Biol* 4:e252.

Norgren, R. B., Jr., N. Ratner, and R. Brackenbury. 1992. Development of olfactory nerve glia defined by a monoclonal antibody specific for Schwann cells. *Dev Dyn* 194:231–238.

Norris, D. O., and J. A. Carr. 2020. *Vertebrate endocrinology*. 6th ed. London: Academic Press.

Norris, H. W. 1925. Observations upon the peripheral distribution of the cranial nerves of certain ganoid fishes (*Amia*, *Lepidosteus*, *Polyodon*, *Scapirhynchus* and *Acipenser*). *J Comp Neurol* 39:345–432.

Norris, H. W., and S. P. Hughes. 1920. The cranial, occipital, and anterior spinal nerves of the dogfish, *Squalus acanthias*. *J Comp Neurol* 31:293–402.

Northcutt, R. G. 1989. The phylogenetic distribution and innervation of craniate mechanoreceptive lateral lines. In *The mechanosensory lateral line*, edited by S. Coombs, P. Görner and H. Münz, 17–78. New York: Springer.

Northcutt, R. G. 1992. The phylogeny of octavolateralis ontogenies: a reaffirmation of Garstang's hypothesis. In *The evolutionary biology of hearing*, edited by D. B. Webster, R. R. Fay and A. N. Popper, 21–47. New York: Springer.

Northcutt, R. G. 1993a. The primitive pattern of development of lateral line organs. *J Comp Physiol A* 173:717–718.

Northcutt, R. G. 1993b. A reassessment of Goodrichs model of cranial nerve phylogeny. *Acta Anat* 148:71–80.

Northcutt, R. G. 1997. Evolution of gnathostome lateral line ontogenies. *Brain Behav Evol* 50:25–37.

Northcutt, R. G. 2003. Development of the lateral line system in the channel catfish. In *The Big Fish Bang. Proceedings of the 26th Annual Larval Fish Conference*, edited by H. I. Browman and A. B. Skiftesvik, 137–159. Bergen: Institute for Marine Research.

Northcutt, R. G. 2004. Taste buds: development and evolution. *Brain Behav Evol* 64:198–206.

Northcutt, R. G., and W. E. Bemis. 1993. Cranial nerves of the coelacanth, *Latimeria chalumnae* [Osteichthyes: Sarcopterygii: Actinistia], and comparisons with other Craniata. *Brain Behav Evol* 42 Suppl.:1–76.

Northcutt, R. G., and K. Brändle. 1995. Development of branchiomeric and lateral line nerves in the axolotl. *J Comp Neurol* 355:427–454.

Northcutt, R. G., K. Brändle, and B. Fritzsch. 1995. Electroreceptors and mechanosensory lateral line organs arise from single placodes in axolotls. *Dev Biol* 168:358–373.

Northcutt, R. G., K. C. Catania, and B. B. Criley. 1994. Development of lateral line organs in the axolotl. *J Comp Neurol* 340:480–514.

Northcutt, R. G., and C. Gans. 1983. The genesis of neural crest and epidermal placodes: a reinterpretation of vertebrate origins. *Q Rev Biol* 58:1–28.

Northcutt, R. G., and L. E. Muske. 1994. Multiple embryonic origins of gonadotropin-releasing hormone (GnRH) immunoreactive neurons. *Brain Res Dev Brain Res* 78:279–290.

Novitch, B. G., A. I. Chen, and T. M. Jessell. 2001. Coordinate regulation of motorneuron subtype identity and pan-neuronal properties by the bHLH repressor Olig2. *Neuron* 31:773–789.

Nowakowski, B. E., and R. A. Maurer. 1994. Multiple Pit-1-binding sites facilitate estrogen responsiveness of the prolactin gene. *Mol Endocrinol* 8:1742–1749.

Nozaki, M. 2008. The hagfish pituitary gland and its putative adenohypophysial hormones. *Zoolog Sci* 25:1028–1036.

Nozaki, M. 2013. Hypothalamic-pituitary-gonadal endocrine system in the hagfish. *Front Endocrinol* 4:200.

O'Leary, C. E., C. Schneider, and R. M. Locksley. 2019. Tuft cells – systemically dispersed sensory epithelia integrating immune and neural circuitry. *Annu Rev Immunol* 37:47–72.

O'Neill, P., S. S. Mak, B. Fritzsch, R. K. Ladher, and C. V. Baker. 2012. The amniote paratympanic organ develops from a previously undiscovered sensory placode. *Nat Commun* 3:1041.

Ocampo Daza, D., and D. Larhammar. 2018. Evolution of the growth hormone, prolactin, prolactin 2 and somatolactin family. *Gen Comp Endocrinol* 264:94–112.

Ochocinska, M. J., and P. F. Hitchcock. 2009. NeuroD regulates proliferation of photoreceptor progenitors in the retina of the zebrafish. *Mech Dev* 126:128–141.

Ogino, H., M. Fisher, and R. M. Grainger. 2008. Convergence of a head-field selector Otx2 and Notch signaling: a mechanism for lens specification. *Development* 135:249–258.

Ogino, H., H. Ochi, H. M. Reza, and K. Yasuda. 2012. Transcription factors involved in lens development from the preplacodal ectoderm. *Dev Biol* 363:333–347.

Ogino, H., and K. Yasuda. 1998. Induction of lens differentiation by activation of a bZIP transcription factor, L-Maf. *Science* 280:115–118.

Ogura, T., S. A. Szebenyi, K. Krosnowski, A. Sathyanesan, J. Jackson, and W. Lin. 2011. Cholinergic microvillous cells in the mouse main olfactory epithelium and effect of acetylcholine on olfactory sensory neurons and supporting cells. *J Neurophysiol* 106:1274–1287.

Oh, E. C., N. Khan, E. Novelli, H. Khanna, E. Strettoi, and A. Swaroop. 2007. Transformation of cone precursors to functional rod photoreceptors by bZIP transcription factor NRL. *Proc Natl Acad Sci USA* 104:1679–1684.

Ohto, H., S. Kamada, K. Tago, S. Tominaga, H. Ozaki, S. Sato, and K. Kawakami. 1999. Cooperation of Six and Eya in activation of their target genes through nuclear translocation of Eya. *Mol Cell Biol* 19:6815–6824.

Ohto, H., T. Takizawa, T. Saito, M. Kobayashi, K. Ikeda, and K. Kawakami. 1998. Tissue and developmental distribution of Six family gene products. *Int J Dev Biol* 42:141–148.

Ohtsuka, T., M. Ishibashi, G. Gradwohl, S. Nakanishi, F. Guillemot, and R. Kageyama. 1999. Hesl and Hes5 as notch effectors in mammalian neuronal differentiation. *EMBO J* 18:2196–2207.

Ohyama, T., and A. K. Groves. 2004. Expression of mouse Foxi class genes in early craniofacial development. *Dev Dyn* 231:640–646.

Ohyama, T., O. A. Mohamed, M. M. Taketo, D. Dufort, and A. K. Groves. 2006. Wnt signals mediate a fate decision between otic placode and epidermis. *Development* 133:865–875.

Oikawa, T., K. Suzuki, T. R. Saito, K. W. Takahashi, and K. Taniguchi. 1998. Fine structure of three types of olfactory organs in *Xenopus laevis*. *Anat Rec* 252:301–310.

Oisi, Y., K. G. Ota, S. Kuraku, S. Fujimoto, and S. Kuratani. 2013. Craniofacial development of hagfishes and the evolution of vertebrates. *Nature* 493:175–180.

Oka, Y., and S. I. Korsching. 2011. Shared and unique G alpha proteins in the zebrafish versus mammalian senses of taste and smell. *Chem Senses* 36:357–365.

Oka, Y., L. R. Saraiva, and S. I. Korsching. 2012. Crypt neurons express a single V1R-related ora gene. *Chem Senses* 37:219–227.

Okada, T. S., Y. Ito, K. Watanabe, and G. Eguchi. 1975. Differentiation of lens in cultures of neural retinal cells of chick embryos. *Dev Biol* 45:318–329.

Okamoto, Y., N. Nishimura, K. Matsuda, D. C. Ranawakage, Y. Kamachi, H. Kondoh, and M. Uchikawa. 2018. Cooperation of Sall4 and Sox8 transcription factors in the regulation of the chicken Sox3 gene during otic placode development. *Dev Growth Differ* 60:133–145.

Okano, T., D. Kojima, Y. Fukada, Y. Shichida, and T. Yoshizawa. 1992. Primary structures of chicken cone visual pigments: vertebrate rhodopsins have evolved out of cone visual pigments. *Proc Natl Acad Sci USA* 89:5932–5936.

Oken, L. 1807. *Über die Bedeutung der Schädelknochen*. Jena: Göpferdt.

Okubo, T., C. Clark, and B. L. Hogan. 2009. Cell lineage mapping of taste bud cells and keratinocytes in the mouse tongue and soft palate. *Stem Cells* 27:442–450.

Okuda, Y., E. Ogura, H. Kondoh, and Y. Kamachi. 2010. B1 SOX coordinate cell specification with patterning and morphogenesis in the early zebrafish embryo. *PLoS Genet* 6:e1000936.

Oliver, G., F. Loosli, R. Köster, J. Wittbrodt, and P. Gruss. 1996. Ectopic lens induction in fish in response to the murine homeobox gene Six3. *Mech Dev* 60:233–239.

Oliver, G., A. Mailhos, R. Wehr, N. G. Copeland, N. A. Jenkins, and P. Gruss. 1995. Six3, a murine homologue of the sine oculis gene, demarcates the most anterior border of the developing neural plate and is expressed during eye d evelopment. *Development* 121:4045–4055.

Oliver, G., R. Wehr, A. Jenkins, N. G. Copeland, B. N. R. Cheyette, V. Hartenstein, S. L. Zipursky, and P. Gruss. 1995. Homeobox genes and connective tissue patterning. *Development* 121:693–705.

Olson, H. M., and A. V. Nechiporuk. 2018. Using zebrafish to study collective cell migration in development and disease. *Front Cell Dev Biol* 6:83.

Olson, L. E., J. Tollkuhn, C. Scafoglio, A. Krones, J. Zhang, K. A. Ohgi, W. Wu, M. M. Taketo, R. Kemler, R. Grosschedl, D. Rose, X. Li, and M. G. Rosenfeld. 2006. Homeodomain-mediated beta-catenin-dependent switching events dictate cell-lineage determination. *Cell* 125:593–605.

Omori, Y., K. Katoh, S. Sato, Y. Muranishi, T. Chaya, A. Onishi, T. Minami, T. Fujikado, and T. Furukawa. 2011. Analysis of transcriptional regulatory pathways of photoreceptor genes by expression profiling of the Otx2-deficient retina. *PLoS One* 6:e19685.

Omura, M., and P. Mombaerts. 2014. Trpc2-expressing sensory neurons in the main olfactory epithelium of the mouse. *Cell Rep* 8:583–595.

Omura, M., and P. Mombaerts. 2015. Trpc2-expressing sensory neurons in the mouse main olfactory epithelium of type B express the soluble guanylate cyclase Gucy1b2. *Mol Cell Neurosci* 65:114–124.

Onai, T., N. Irie, and S. Kuratani. 2014. The evolutionary origin of the vertebrate body plan: the problem of head segmentation. *Annu Rev Genomics Hum Genet* 15:443–459.

Onichtchouk, D., and W. Driever. 2016. Zygotic genome activators, developmental timing, and pluripotency. *Curr Top Dev Biol* 116:273–297.

Ooi, G. T., N. Tawadros, and R. M. Escalona. 2004. Pituitary cell lines and their endocrine applications. *Mol Cell Endocrinol* 228:1–21.

Organisciak, D., R. Darrow, L. Barsalou, C. Rapp, B. McDonald, and P. Wong. 2011. Light induced and circadian effects on retinal photoreceptor cell crystallins. *Photochem Photobiol* 87:151–159.

Oron-Karni, V., C. Farhy, M. Elgart, T. Marquardt, L. Remizova, O. Yaron, Q. Xie, A. Cvekl, and R. Ashery-Padan. 2008. Dual requirement for Pax6 in retinal progenitor cells. *Development* 135:4037–4047.

Osada, S. I., and C. V. E. Wright. 1999. *Xenopus* nodal-related signaling is essential for mesendodermal patterning during early embryogenesis. *Development* 126:3229–3240.

Owen, R. 1848. *On the archetype and homologies of the vertebrate skeleton*. London: J. van Voorst.

Ozair, M. Z., C. Kintner, and A. H. Brivanlou. 2013. Neural induction and early patterning in vertebrates. *Wiley Interdiscip Rev Dev Biol* 2:479–498.

Ozaki, H., K. Nakamura, J. Funahashi, K. Ikeda, G. Yamada, H. Tokano, H. O. Okamura, K. Kitamura, S. Muto, H. Kotaki, K. Sudo, R. Horai, Y. Iwakura, and K. Kawakami. 2004. Six1 controls patterning of the mouse otic vesicle. *Development* 131:551–562.

Ozaki, H., Y. Watanabe, K. Takahashi, K. Kitamura, A. Tanaka, K. Urase, T. Momoi, K. Sudo, J. Sakagami, M. Asano, Y. Iwakura, and K. Kawakami. 2001. Six4, a putative myogenin gene regulator, is not essential for mouse embryonal development. *Mol Cell Biol* 21:3343–3350.

Packard, A. I., B. Lin, and J. E. Schwob. 2016. Sox2 and Pax6 play counteracting roles in regulating neurogenesis within the murine olfactory epithelium. *PLoS One* 11:e0155167.

Packard, A., N. Schnittke, R. A. Romano, S. Sinha, and J. E. Schwob. 2011. DeltaNp63 regulates stem cell dynamics in the mammalian olfactory epithelium. *J Neurosci* 31:8748–8759.

Padanad, M. S., N. Bhat, B. Guo, and B. B. Riley. 2012. Conditions that influence the response to Fgf during otic placode induction. *Dev Biol* 364 (1):1–10.

Padanad, M. S., and B. B. Riley. 2011. Pax2/8 proteins coordinate sequential induction of otic and epibranchial placodes through differential regulation of foxi1, sox3 and fgf24. *Dev Biol* 351:90–98.

Pan, B., G. S. Geleoc, Y. Asai, G. C. Horwitz, K. Kurima, K. Ishikawa, Y. Kawashima, A. J. Griffith, and J. R. Holt. 2013. TMC1 and TMC2 are components of the mechanotransduction channel in hair cells of the mammalian inner ear. *Neuron* 79:504–515.

Pan, L., M. Deng, X. Xie, and L. Gan. 2008. ISL1 and BRN3B co-regulate the differentiation of murine retinal ganglion cells. *Development* 135:1981–1990.

Pan, N., I. Jahan, J. Kersigo, B. Kopecky, P. Santi, S. Johnson, H. Schmitz, and B. Fritzsch. 2011. Conditional deletion of Atoh1 using Pax2-Cre results in viable mice without differentiated cochlear hair cells that have lost most of the organ of Corti. *Hear Res* 275:66–80.

Pan, Y., R. I. Martinez-De Luna, C. H. Lou, S. Nekkalapudi, L. E. Kelly, A. K. Sater, and H. M. El-Hodiri. 2010. Regulation of photoreceptor gene expression by the retinal homeobox (Rx) gene product. *Dev Biol* 339:494–506.

Pan, Y., S. Nekkalapudi, L. E. Kelly, and H. M. El-Hodiri. 2006. The Rx-like homeobox gene (Rx-L) is necessary for normal photoreceptor development. *Invest Ophthalmol Vis Sci* 47:4245–4253.

Panaliappan, T. K., W. Wittmann, V. K. Jidigam, S. Mercurio, J. A. Bertolini, S. Sghari, R. Bose, C. Patthey, S. K. Nicolis, and L. Gunhaga. 2018. Sox2 is required for olfactory pit formation and olfactory neurogenesis through BMP restriction and Hes5 upregulation. *Development* 145:dev153791.

Panda, S., S. K. Nayak, B. Campo, J. R. Walker, J. B. Hogenesch, and T. Jegla. 2005. Illumination of the melanopsin signaling pathway. *Science* 307:600–604.

Pandey, R. N., R. Rani, E. J. Yeo, M. Spencer, S. Hu, R. A. Lang, and R. S. Hegde. 2010. The Eyes Absent phosphatase-transactivator proteins promote proliferation, transformation, migration, and invasion of tumor cells. *Oncogene* 29:3715–3722.

Pandit, T., V. K. Jidigam, and L. Gunhaga. 2011. BMP-induced L-Maf regulates subsequent BMP-independent differentiation of primary lens fibre cells. *Dev Dyn* 240:1917–1928.

Pandur, P. D., and S. A. Moody. 2000. *Xenopus* Six1 gene is expressed in neurogenic cranial placodes and maintained in differentiating lateral lines. *Mech Dev* 96:253–257.

Pangršič, T., E. Reisinger, and T. Moser. 2012. Otoferlin: a multi-C2 domain protein essential for hearing. *Trends Neurosci* 35:671–680.

Paraiso, K. D., J. S. Cho, J. Yong, and K. W. Y. Cho. 2020. Early *Xenopus* gene regulatory programs, chromatin states, and the role of maternal transcription factors. *Curr Top Dev Biol* 139:35–60.

Park, B. Y., and J. P. Saint-Jeannet. 2010. *Induction and segregation of the vertebrate cranial placodes*. San Rafael: Morgan and Claypool.

Park, D., and H. L. Eisthen. 2003. Gonadotropin releasing hormone (GnRH) modulates odorant responses in the peripheral olfactory system of axolotls. *J Neurophysiol* 90:731–738.

Parker, D. E. 1980. The vestibular apparatus. *Sci Am* 243:118–137.

Parlier, D., V. Moers, C. Van Campenhout, J. Preillon, L. Leclere, A. Saulnier, M. Sirakov, H. Busengdal, S. Kricha, J. C. Marine, F. Rentzsch, and E. J. Bellefroid. 2013. The *Xenopus* doublesex-related gene Dmrt5 is required for olfactory placode neurogenesis. *Dev Biol* 373:39–52.

Parrilla, M., I. Chang, A. Degl'Innocenti, and M. Omura. 2016. Expression of homeobox genes in the mouse olfactory epithelium. *J Comp Neurol* 524:2713–2739.

Paschaki, M., L. Cammas, Y. Muta, Y. Matsuoka, S. S. Mak, M. Rataj-Baniowska, V. Fraulob, P. Dolle, and R. K. Ladher. 2013. Retinoic acid regulates olfactory progenitor cell fate and differentiation. *Neural Dev* 8:13.

Patrick, A. N., J. H. Cabrera, A. L. Smith, X. S. Chen, H. L. Ford, and R. Zhao. 2013. Structure-function analyses of the human SIX1-EYA2 complex reveal insights into metastasis and BOR syndrome. *Nat Struct Mol Biol* 20:447–453.

Patthey, C., H. Clifford, W. Haerty, C. P. Ponting, S. M. Shimeld, and J. Begbie. 2016. Identification of molecular signatures specific for distinct cranial sensory ganglia in the developing chick. *Neural Dev* 11:3.

Patthey, C., T. Edlund, and L. Gunhaga. 2009. Wnt-regulated temporal control of BMP exposure directs the choice between neural plate border and epidermal fate. *Development* 136:73–83.

Pattyn, A., C. Goridis, and J. F. Brunet. 2000. Specification of the central noradrenergic phenotype by the homeobox gene Phox2b. *Mol Cell Neurosci* 15:235–243.

Pattyn, A., M. R. Hirsch, C. Goridis, and J. F. Brunet. 2000. Control of hindbrain motor neuron differentiation by the homeobox gene Phox2b. *Development* 127 (7):1349–1358.

Pattyn, A., X. Morin, H. Cremer, C. Goridis, and J. F. Brunet. 1997. Expression and interactions of the two closely related homeobox genes Phox2a and Phox2b during neurogenesis. *Development* 124:4065–4075.

Pattyn, A., X. Morin, H. Cremer, C. Goridis, and J. F. Brunet. 1999. The homeobox gene Phox2b is essential for the development of autonomic neural crest derivatives. *Nature* 399:366–370.

Peng, A. W., F. T. Salles, B. Pan, and A. J. Ricci. 2011. Integrating the biophysical and molecular mechanisms of auditory hair cell mechanotransduction. *Nat Commun* 2:523.

Peng, G. H., and S. Chen. 2005. Chromatin immunoprecipitation identifies photoreceptor transcription factor targets in mouse models of retinal degeneration: new findings and challenges. *Vis Neurosci* 22:575–586.

Peng, G. H., and S. Chen. 2007. Crx activates opsin transcription by recruiting HAT-containing co-activators and promoting histone acetylation. *Hum Mol Genet* 16:2433–2452.

Peng, Y. R., K. Shekhar, W. Yan, D. Herrmann, A. Sappington, G. S. Bryman, T. van Zyl, M. T. H. Do, A. Regev, and J. R. Sanes. 2019. Molecular classification and comparative taxonomics of foveal and peripheral cells in primate retina. *Cell* 176:1222–1237e22.

Pera, E., S. Stein, and M. Kessel. 1999. Ectodermal patterning in the avian embryo: epidermis versus neural plate. *Development* 126:63–73.

Perez, S. E., S. Rebelo, and D. J. Anderson. 1999. Early specification of sensory neuron fate revealed by expression and function of neurogenins in the chick embryo. *Development* 126:1715–1728.

Perl, K., R. Shamir, and K. B. Avraham. 2018. Computational analysis of mRNA expression profiling in the inner ear reveals candidate transcription factors associated with proliferation, differentiation, and deafness. *Hum Genomics* 12:30.

Perron, M., S. Kanekar, M. L. Vetter, and W. A. Harris. 1998. The genetic sequence of retinal development in the ciliary margin of the *Xenopus* eye. *Dev Biol* 199:185–200.

Perron, M., K. Opdecamp, K. Butler, W. A. Harris, and E. J. Bellefroid. 1999. X-ngnr-1 and Xath3 promote ectopic expression of sensory neuron markers in the neurula ectoderm and have distinct inducing properties in the retina. *Proc Natl Acad Sci USA* 96:14996–15001.

Peter, I. S., and E. H. Davidson. 2015. *Genomic control process: development and evolution.* Amsterdam: Elsevier.

Petersen, C. P., and P. W. Reddien. 2009. Wnt signaling and the polarity of the primary body axis. *Cell* 139:1056–1068.

Peterson, E. H., J. R. Cotton, and J. W. Grant. 1996. Structural variation in ciliary bundles of the posterior semicircular canal. Quantitative anatomy and computational analysis. *Ann N Y Acad Sci* 781:85–102.

Petitpre, C., H. Wu, A. Sharma, A. Tokarska, P. Fontanet, Y. Wang, F. Helmbacher, K. Yackle, G. Silberberg, S. Hadjab, and F. Lallemend. 2018. Neuronal heterogeneity and stereotyped connectivity in the auditory afferent system. *Nat Commun* 9:3691.

Pevny, L. H., and S. K. Nicolis. 2010. Sox2 roles in neural stem cells. *Int J Biochem Cell Biol* 42:421–424.

Pevny, L. H., S. Sockanathan, M. Placzek, and R. Lovell-Badge. 1998. A role for SOX1 in neural determination. *Development* 125:1967–1978.

Pevny, L., and M. Placzek. 2005. SOX genes and neural progenitor identity. *Curr Opin Neurobiol* 15:7–13.

Phillips, B. T., H. J. Kwon, C. Melton, P. Houghtaling, A. Fritz, and B. B. Riley. 2006. Zebrafish msxB, msxC and msxE function together to refine the neural-nonneural border and regulate cranial placodes and neural crest development. *Dev Biol* 294:376–390.

Phillips, B. T., K. Bolding, and B. B. Riley. 2001. Zebrafish fgf3 and fgf8 encode redundant functions required for otic placode induction. *Dev Biol* 235:351–365.

Phillips, B. T., E. M. Storch, A. C. Lekven, and B. B. Riley. 2004. A direct role for Fgf but not Wnt in otic placode induction. *Development* 131:923–931.

Piatigorsky, J. 1998. Multifunctional lens crystallins and corneal enzymes. More than meets the eye. *Ann NY Acad Sci* 842:7–15.

Piatigorsky, J., W. E. O'Brien, B. L. Norman, K. Kalumuck, G. J. Wistow, T. Borras, J. M. Nickerson, and E. F. Wawrousek. 1988. Gene sharing by delta-crystallin and argininosuccinate lyase. *Proc Natl Acad Sci USA* 85:3479–3483.

Pieper, M., K. Ahrens, E. Rink, A. Peter, and G. Schlosser. 2012. Differential distribution of competence for panplacodal and neural crest induction to non-neural and neural ectoderm. *Development* 139:1175–1187.

Pieper, M., G. W. Eagleson, W. Wosniok, and G. Schlosser. 2011. Origin and segregation of cranial placodes in *Xenopus laevis*. *Dev Biol* 360:257–275.

Pignoni, F., B. Hu, K. H. Zavitz, J. Xiao, P. A. Garrity, and S. L. Zipursky. 1997. The eye-specification proteins So and Eya form a complex and regulate multiple steps in *Drosophila* eye development. *Cell* 91:881–891.

Pilon, N., K. Oh, J. R. Sylvestre, N. Bouchard, J. Savory, and D. Lohnes. 2006. Cdx4 is a direct target of the canonical Wnt pathway. *Dev Biol* 289:55–63.

Piotrowski, T., and C. V. Baker. 2014. The development of lateral line placodes: taking a broader view. *Dev Biol* 389:68–81.

Piotrowski, T., and R. G. Northcutt. 1996. The cranial nerves of the Senegal bichir, *Polypterus senegalus* [Osteichthyes: Actinopterygii: Cladistia]. *Brain Behav Evol* 47:55–102.

Piotrowski, T., and C. Nüsslein-Volhard. 2000. The endoderm plays an important role in patterning the segmented pharyngeal region in zebrafish (*Danio rerio*). *Dev Biol* 225:339–356.

Pisani, D., S. M. Mohun, S. R. Harris, J. O. McInerney, and M. Wilkinson. 2006. Molecular evidence for dim-light vision in the last common ancestor of the vertebrates. *Curr Biol* 16:R318–R319.

Pispa, J., and I. Thesleff. 2003. Mechanisms of ectodermal organogenesis. *Dev Biol* 262:195–205.

Pistocchi, A., C. G. Feijoo, P. Cabrera, E. J. Villablanca, M. L. Allende, and F. Cotelli. 2009. The zebrafish prospero homolog prox1 is required for mechanosensory hair cell differentiation and functionality in the lateral line. *BMC Dev Biol* 9:58.

Pla, P., M. R. Hirsch, S. Le Crom, S. Reiprich, V. R. Harley, and C. Goridis. 2008. Identification of Phox2b-regulated genes by expression profiling of cranial motoneuron precursors. *Neural Dev* 3:14.

Pla, P., and A. H. Monsoro-Burq. 2018. The neural border: induction, specification and maturation of the territory that generates neural crest cells. *Dev Biol* 444 Suppl 1:S36–S46.

Plachetzki, D. C., C. R. Fong, and T. H. Oakley. 2010. The evolution of phototransduction from an ancestral cyclic nucleotide gated pathway. *Proc Biol Sci* 277:1963–1969.

Platt, J. B. 1894. Ontogenetische Differenzierung des Ektoderms in *Necturus*. *Archiv Mikrosk Anat* 43:911–966.

Plessy, C., G. Pascarella, N. Bertin, A. Akalin, C. Carrieri, A. Vassalli, D. Lazarevic, J. Severin, C. Vlachouli, R. Simone, G. J. Faulkner, J. Kawai, C. O. Daub, S. Zucchelli, Y. Hayashizaki, P. Mombaerts, B. Lenhard, S. Gustincich, and P. Carninci. 2012. Promoter architecture of mouse olfactory receptor genes. *Genome Res* 22:486–497.

Plouhinec, J. L., S. Medina-Ruiz, C. Borday, E. Bernard, J. P. Vert, M. B. Eisen, R. M. Harland, and A. H. Monsoro-Burq. 2017. A molecular atlas of the developing ectoderm defines neural, neural crest, placode, and nonneural progenitor identity in vertebrates. *PLoS Biol* 15:e2004045.

Plouhinec, J. L., D. D. Roche, C. Pegoraro, A. L. Figueiredo, F. Maczkowiak, L. J. Brunet, C. Milet, J. P. Vert, N. Pollet, R. M. Harland, and A. H. Monsoro-Burq. 2014. Pax3 and Zic1 trigger the early neural crest gene regulatory network by the direct activation of multiple key neural crest specifiers. *Dev Biol* 386:461–472.

Plouhinec, J. L., T. Sauka-Spengler, A. Germot, C. Le Mentec, T. Cabana, G. Harrison, C. Pieau, J. Y. Sire, G. Veron, and S. Mazan. 2003. The mammalian Crx genes are highly divergent representatives of the Otx5 gene family, a gnathostome orthology class of orthodenticle-related homeogenes involved in the differentiation of retinal photoreceptors and circadian entrainment. *Mol Biol Evol* 20:513–521.

Poggi, L., M. Vitorino, I. Masai, and W. A. Harris. 2005. Influences on neural lineage and mode of division in the zebrafish retina in vivo. *J Cell Biol* 171:991–999.

Pogoda, H. M., and M. Hammerschmidt. 2009. How to make a teleost adenohypophysis: molecular pathways of pituitary development in zebrafish. *Mol Cell Endocrinol* 312:2–13.

Pogoda, H. M., Hardt S. von der, W. Herzog, C. Kramer, H. Schwarz, and M. Hammerschmidt. 2006. The proneural gene ascl1a is required for endocrine differentiation and cell survival in the zebrafish adenohypophysis. *Development* 133:1079–1089.

Pohl, B. S., and W. Knöchel. 2004. Isolation and developmental expression of *Xenopus* FoxJ1 and FoxK1. *Dev Genes Evol* 214:200–205.

Pohl, B. S., A. Rössner, and W. Knöchel. 2005. The Fox gene family in *Xenopus laevis*: FoxI2, FoxM1 and FoxP1 in early development. *Int J Dev Biol* 49:53–58.

Polevoy, H., Y. E. Gutkovich, A. Michaelov, Y. Volovik, Y. M. Elkouby, and D. Frank. 2019. New roles for Wnt and BMP signaling in neural anteroposterior patterning. *EMBO Rep* 20:e45842.

Politis, P. K., G. Makri, D. Thomaidou, M. Geissen, H. Rohrer, and R. Matsas. 2007. BM88/CEND1 coordinates cell cycle exit and differentiation of neuronal precursors. *Proc Natl Acad Sci USA* 104:17861–17866.

Pommereit, D., T. Pieler, and T. Hollemann. 2001. Xpitx3: a member of the Rieg/Pitx gene family expressed during pituitary and lens formation in *Xenopus laevis*. *Mech Dev* 102:255–257.

Popper, A. N., C. Platt, and P. L. Edds. 1992. Evolution of the vertebrate inner ear: an overview of ideas. In *The evolutionary biology of hearing*, edited by D. B. Webster, A. N. Popper and R. R. Fay, 49–57. New York: Springer.

Porras-Gallo, M. I., A. Pena-Meliaan, F. Viejo, T. Hernaandez, E. Puelles, D. Echevarria, and J. Ramon Sanudo. 2019. Overview of the history of the cranial nerves: from Galen to the 21st century. *Anat Rec* 302:381–393.

Porter, M. L. 2016. Beyond the eye: molecular evolution of extraocular photoreception. *Integr Comp Biol* 56:842–852.

Postigo, A. A., J. L. Depp, J. J. Taylor, and K. L. Kroll. 2003. Regulation of Smad signaling through a differential recruitment of coactivators and corepressors by ZEB proteins. *EMBO J* 22:2453–2462.

Poulin, G., B. Turgeon, and J. Drouin. 1997. NeuroD1/beta2 contributes to cell-specific transcription of the proopiomelanocortin gene. *Mol Cell Biol* 17:6673–6682.

Proske, U., and S. C. Gandevia. 2012. The proprioceptive senses: their roles in signaling body shape, body position and movement, and muscle force. *Physiol Rev.* 92:1651–1697.

Puelles, L. 2019. Survey of midbrain, diencephalon, and hypothalamus neuroanatomic terms Whose prosomeric definition conflicts with columnar tradition. *Front Neuroanat* 13:20.

Puelles, L., M. Harrison, G. Paxinos, and C. Watson. 2013. A developmental ontology for the mammalian brain based on the prosomeric model. *Trends Neurosci* 36:570–578.

Pujades, C., A. Kamaid, B. Alsina, and F. Giráldez. 2006. BMP-signaling regulates the generation of hair-cells. *Dev Biol* 292:55–67.

Pulichino, A. M., S. Vallette-Kasic, C. Couture, Y. Gauthier, T. Brue, M. David, G. Malpuech, C. Deal, G. Van Vliet, M. De Vroede, F. G. Riepe, C. J. Partsch, W. G. Sippell, M. Berberoglu, B. Atasay, and J. Drouin. 2003. Human and mouse TPIT gene mutations cause early onset pituitary ACTH deficiency. *Genes Dev* 17:711–716.

Pulichino, A. M., S. Vallette-Kasic, J. P. Tsai, C. Couture, Y. Gauthier, and J. Drouin. 2003. Tpit determines alternate fates during pituitary cell differentiation. *Genes Dev* 17:738–747.

Puligilla, C., A. Dabdoub, S. D. Brenowitz, and M. W. Kelley. 2010. Sox2 induces neuronal formation in the developing mammalian cochlea. *J Neurosci* 30:714–722.

Puligilla, C., and M. W. Kelley. 2017. Dual role for Sox2 in specification of sensory competence and regulation of Atoh1 function. *Dev Neurobiol* 77:3–13.

Purves, D., G. J. Augustine, D. Fitzpatrick, Hall. W. C., A. S. LaMantia, J. O. McNamara, and L.E. White. 2008. *Neuroscience*. 4th ed. Sunderland: Sinauer.

Putnam, N. H., T. Butts, D. E. Ferrier, R. F. Furlong, U. Hellsten, T. Kawashima, M. Robinson-Rechavi, E. Shoguchi, A. Terry, J. K. Yu, E. L. Benito-Gutierrez, I. Dubchak, J. Garcia-Fernandez, J. J. Gibson-Brown, I. V. Grigoriev, A. C. Horton, P. J. de Jong, J. Jurka, V. V. Kapitonov, Y. Kohara, Y. Kuroki, E. Lindquist, S. Lucas, K. Osoegawa, L. A. Pennacchio, A. A. Salamov, Y. Satou, T. Sauka-Spengler, J. Schmutz, I. Shin, A. Toyoda, M. Bronner-Fraser, A. Fujiyama, L. Z. Holland, P. W. Holland, N. Satoh, and D. S. Rokhsar. 2008. The amphioxus genome and the evolution of the chordate karyotype. *Nature* 453:1064–1071.

Qian, J., N. Esumi, Y. Chen, Q. Wang, I. Chowers, and D. J. Zack. 2005. Identification of regulatory targets of tissue-specific transcription factors: application to retina-specific gene regulation. *Nucleic Acids Res* 33:3479–3491.

Qian, Y., B. Fritzsch, S. Shirasawa, C. L. Chen, Y. Choi, and Q. Ma. 2001. Formation of brainstem (nor)adrenergic centers and first-order relay visceral sensory neurons is dependent on homeodomain protein Rnx/Tlx3. *Genes Dev* 15:2533–2545.

Qian, Y., S. Shirasawa, C. L. Chen, L. Cheng, and Q. Ma. 2002. Proper development of relay somatic sensory neurons and D2/D4 interneurons requires homeobox genes Rnx/Tlx-3 and Tlx-1. *Genes Dev* 16:1220–1233.

Quan, X. J., L. Yuan, L. Tiberi, A. Claeys, N. De Geest, J. Yan, R. van der Kant, W. R. Xie, T. J. Klisch, J. Shymkowitz, F. Rousseau, M. Bollen, M. Beullens, H. Y. Zoghbi, P. Vanderhaeghen, and B. A. Hassan. 2016. Post-translational control of the temporal dynamics of transcription factor activity regulates neurogenesis. *Cell* 164:460–475.

Quina, L. A., L. Tempest, Y. W. Hsu, T. C. Cox, and E. E. Turner. 2012. Hmx1 is required for the normal development of somatosensory neurons in the geniculate ganglion. *Dev Biol* 365:152–163.

Radde-Gallwitz, K., L. Pan, L. Gan, X. Lin, N. Segil, and P. Chen. 2004. Expression of Islet1 marks the sensory and neuronal lineages in the mammalian inner ear. *J Comp Neurol* 477:412–421.

Radosevic, M., A. Robert-Moreno, M. Coolen, L. Bally-Cuif, and B. Alsina. 2011. Her9 represses neurogenic fate downstream of Tbx1 and retinoic acid signaling in the inner ear. *Development* 138:397–408.

Raetzman, L. T., S. A. Ross, S. Cook, S. L. Dunwoodie, S. A. Camper, and P. Q. Thomas. 2004. Developmental regulation of Notch signaling genes in the embryonic pituitary: Prop1 deficiency affects Notch2 expression. *Dev Biol* 265:329–340.

Raetzman, L. T., R. Ward, and S. A. Camper. 2002. Lhx4 and Prop1 are required for cell survival and expansion of the pituitary primordia. *Development* 129:4229–4239.

Raft, S., and A. K. Groves. 2015. Segregating neural and mechanosensory fates in the developing ear: patterning, signaling, and transcriptional control. *Cell Tissue Res* 359:315–332.

Raft, S., E. J. Koundakjian, H. Quinones, C. S. Jayasena, L. V. Goodrich, J. E. Johnson, N. Segil, and A. K. Groves. 2007. Cross-regulation of Ngn1 and Math1 coordinates the production of neurons and sensory hair cells during inner ear development. *Development* 134:4405–4415.

Raft, S., S. Nowotschin, J. Liao, and B. E. Morrow. 2004. Suppression of neural fate and control of inner ear morphogenesis by Tbx1. *Development* 131:1801–1812.

Rajab, A., D. Kelberman, S. C. de Castro, H. Biebermann, H. Shaikh, K. Pearce, C. M. Hall, G. Shaikh, D. Gerrelli, A. Grueters, H. Krude, and M. T. Dattani. 2008. Novel mutations in LHX3 are associated with hypopituitarism and sensorineural hearing loss. *Hum Mol Genet* 17:2150–2159.

Ramirez, M. D., A. N. Pairett, M. S. Pankey, J. M. Serb, D. I. Speiser, A. J. Swafford, and T. H. Oakley. 2016. The last common ancestor of most bilaterian animals possessed at least nine opsins. *Genome Biol Evol* 8:3640–3652.

Ranade, S. S., D. Yang-Zhou, S. W. Kong, E. C. McDonald, T. A. Cook, and F. Pignoni. 2008. Analysis of the Otd-dependent transcriptome supports the evolutionary conservation of CRX/OTX/OTD functions in flies and vertebrates. *Dev Biol* 315:521–534.

Rao, N. A., S. Saraswathy, G. Pararajasegaram, and S. P. Bhat. 2012. Small heat shock protein alphaA-crystallin prevents photoreceptor degeneration in experimental autoimmune uveitis. *PLoS One* 7:e33582.

Rao, N. A., S. Saraswathy, G. S. Wu, G. S. Katselis, E. F. Wawrousek, and S. Bhat. 2008. Elevated retina-specific expression of the small heat shock protein, alphaA-crystallin, is associated with photoreceptor protection in experimental uveitis. *Invest Ophthalmol Vis Sci* 49:1161–1171.

Rao, P. V., and R. Maddala. 2006. The role of the lens actin cytoskeleton in fiber cell elongation and differentiation. *Semin Cell Dev Biol* 17:698–711.

Rawson, N. E., F. W. Lischka, K. K. Yee, A. Z. Peters, E. S. Tucker, D. W. Meechan, M. Zirlinger, T. M. Maynard, G. B. Burd, C. Dulac, L. Pevny, and A. S. Lamantia. 2010. Specific mesenchymal/epithelial induction of olfactory receptor, vomeronasal, and gonadotropin-releasing hormone (GnRH) neurons. *Dev Dyn* 239:1723–1738.

Rebay, I. 2016. Multiple functions of the Eya phosphotyrosine phosphatase. *Mol Cell Biol* 36:668–677.

Rebay, I., S. J. Silver, and T. L. Tootle. 2005. New vision from eyes absent: transcription factors as enzymes. *Trends Genet* 21:163–171.

Rebelo, S., Z. F. Chen, D. J. Anderson, and D. Lima. 2006. Involvement of DRG11 in the development of the primary afferent nociceptive system. *Mol Cell Neurosci* 33:236–246.

Reece, J. B., L. A. Urry, M. L. Cain, S. A. Wasserman, P. V. Minorsky, and R. B. Jackson. 2011. *Campbell Biology*. 9th ed. San Francisco: Pearson Benjamin Cummings.

Regadas, I., R. Soares-Dos-Reis, M. Falcao, M. R. Matos, F. A. Monteiro, D. Lima, and C. Reguenga. 2014. Dual role of Tlx3 as modulator of Prrxl1 transcription and phosphorylation. *Biochim Biophys Acta* 1839:1121–1131.

Reh, T. A. 2018. The development of the retina. In *Ryan's retina*, edited by A. P. Schachat, C. P. Wilkinson, D. R. Hinton, S.R. Sadda and P. Wiedemann, 375–386. Edinburgh: Elsevier.

Reichert, S., R. A. Randall, and C. S. Hill. 2013. A BMP regulatory network controls ectodermal cell fate decisions at the neural plate border. *Development* 140:4435–4444.

Reim, G., T. Mizoguchi, D. Y. Stainier, Y. Kikuchi, and M. Brand. 2004. The POU domain protein spg (pou2/Oct4) is essential for endoderm formation in cooperation with the HMG domain protein casanova. *Dev Cell* 6:91–101.

Reimer, K., P. Urbanek, M. Busslinger, and G. Ehret. 1996. Normal brainstem auditory evoked potentials in Pax5-deficient mice despite morphologic alterations in the auditory midbrain region. *Audiology* 35:55–61.

Ressler, K. J., S. L. Sullivan, and L. B. Buck. 1994. Information coding in the olfactory system: evidence for a stereotyped and highly organized epitope map in the olfactory bulb. *Cell* 79:1245–1255.

Rex, M., A. Orme, D. Uwanogo, K. Tointon, P. M. Wigmore, P. T. Sharpe, and P. J. Scotting. 1997. Dynamic expression of chicken Sox2 and Sox3 genes in ectoderm induced to form neural tissues. *Dev Dyn* 209:323–332.

Reyer, R. W. 1958. Studies on lens induction in *Amblystoma punctatum* and *Triturus viridescens viridescens*. I. Transplants of prospective belly ectoderm. *J Exp Zool* 138:505–555.

Reza, H. M., H. Nishi, K. Kataoka, Y. Takahashi, and K. Yasuda. 2007. L-Maf regulates p27kip1 expression during chick lens fiber differentiation. *Differentiation* 75:737–744.

Reza, H. M., H. Ogino, and K. Yasuda. 2002. L-Maf, a downstream target of Pax6, is essential for chick lens development. *Mech Dev* 116:61–73.

Reza, H. M., A. Urano, N. Shimada, and K. Yasuda. 2007. Sequential and combinatorial roles of maf family genes define proper lens development. *Mol Vis* 13:18–30.

Reza, H. M., and K. Yasuda. 2004. Roles of Maf family proteins in lens development. *Dev Dyn* 229:440–448.

Rheaume, B. A., A. Jereen, M. Bolisetty, M. S. Sajid, Y. Yang, K. Renna, L. Sun, P. Robson, and E. F. Trakhtenberg. 2018. Single cell transcriptome profiling of retinal ganglion cells identifies cellular subtypes. *Nat Commun* 9:2759.

Rhinn, M., K. Lun, R. Ahrendt, M. Geffarth, and M. Brand. 2009. Zebrafish gbx1 refines the midbrain-hindbrain boundary border and mediates the Wnt8 posteriorization signal. *Neural Dev* 4:12.

Riddiford, N., and G. Schlosser. 2016. Dissecting the pre-placodal transcriptome to reveal presumptive direct targets of Six1 and Eya1 in cranial placodes. *Elife* 5:e17666.

Riddiford, N., and G. Schlosser. 2017. Six1 and Eya1 both promote and arrest neuronal differentiation by activating multiple Notch pathway genes. *Dev Biol* 431:152–167.

Riley, B. B., M. Y. Chiang, L. Farmer, and R. Heck. 1999. The deltaA gene of zebrafish mediates lateral inhibition of hair cells in the inner ear and is regulated by pax2.1. *Development* 126:5669–5678.

Rinehart, J. F., and M. G. Farquhar. 1953. Electron microscopic studies of the anterior pituitary gland. *J Histochem Cytochem* 1:93–113.

Ring, B. Z., S. P. Cordes, P. A. Overbeek, and G. S. Barsh. 2000. Regulation of mouse lens fiber cell development and differentiation by the Maf gene. *Development* 127:307–317.

Ring, K. L., L. M. Tong, M. E. Balestra, R. Javier, Y. Andrews-Zwilling, G. Li, D. Walker, W. R. Zhang, A. C. Kreitzer, and Y. Huang. 2012. Direct reprogramming of mouse and human fibroblasts into multipotent neural stem cells with a single factor. *Cell Stem Cell* 11:100–109.

Riviere, S., L. Challet, D. Fluegge, M. Spehr, and I. Rodriguez. 2009. Formyl peptide receptor-like proteins are a novel family of vomeronasal chemosensors. *Nature* 459:574–577.

Rizzoti, K. 2015. Genetic regulation of murine pituitary development. *J Mol Endocrinol* 54:R55–R73.

Rizzoti, K., H. Akiyama, and R. Lovell-Badge. 2013. Mobilized adult pituitary stem cells contribute to endocrine regeneration in response to physiological demand. *Cell Stem Cell* 13:419–32.

Rizzoti, K., and R. Lovell-Badge. 2005. Early development of the pituitary gland: induction and shaping of Rathke's pouch. *Rev Endocr Metab Disord* 6:161–172.

Rizzoti, K., and R. Lovell-Badge. 2007. SOX3 activity during pharyngeal segmentation is required for craniofacial morphogenesis. *Development* 134:3437–3448.

Roach, G., R. Heath Wallace, A. Cameron, R. Emrah Ozel, C. F. Hongay, R. Baral, S. Andreescu, and K. N. Wallace. 2013. Loss of ascl1a prevents secretory cell differentiation within the zebrafish intestinal epithelium resulting in a loss of distal intestinal motility. *Dev Biol* 376:171–186.

Roberts, M. R., M. Srinivas, D. Forrest, G. Morreale de Escobar, and T. A. Reh. 2006. Making the gradient: thyroid hormone regulates cone opsin expression in the developing mouse retina. *Proc Natl Acad Sci USA* 103:6218–6223.

Robinson, D. R., and G. F. Gebhart. 2008. Inside information: the unique features of visceral sensation. *Mol Interv* 8:242–253.

Roch, G. J., E. R. Busby, and N. M. Sherwood. 2011. Evolution of GnRH: diving deeper. *Gen Comp Endocrinol* 171:1–16.

Roch, G. J., and N. M. Sherwood. 2014. Glycoprotein hormones and their receptors emerged at the origin of metazoans. *Genome Biol Evol* 6:1466–1479.

Rodriguez-Seguel, E., P. Alarcon, and J. L. Gómez-Skarmeta. 2009. The *Xenopus* Irx genes are essential for neural patterning and define the border between prethalamus and thalamus through mutual antagonism with the anterior repressors Fezf and Arx. *Dev Biol* 329 258–268.

Rodriguez, I., P. Feinstein, and P. Mombaerts. 1999. Variable patterns of axonal projections of sensory neurons in the mouse vomeronasal system. *Cell* 97:199–208.

Roellig, D., J. Tan-Cabugao, S. Esaian, and M. E. Bronner. 2017. Dynamic transcriptional signature and cell fate analysis reveals plasticity of individual neural plate border cells. *Elife* 6:e21620.

Rogers, C. D., T. C. Archer, D. D. Cunningham, T. C. Grammer, and E. M. Casey. 2008. Sox3 expression is maintained by FGF signaling and restricted to the neural plate by Vent proteins in the *Xenopus* embryo. *Dev Biol* 313:307–319.

Rogers, C. D., G. S. Ferzli, and E. S. Casey. 2011. The response of early neural genes to FGF signaling or inhibition of BMP indicate the absence of a conserved neural induction module. *BMC Dev Biol* 11:74.

Rogers, C. D., N. Harafuji, D. D. Cunningham, T. Archer, and E. S. Casey. 2009. *Xenopus* Sox3 activates sox2 and geminin and indirectly represses Xvent2 expression to induce neural progenitor formation at the expense of non-neural ectodermal derivatives. *Mech Dev* 126:42–55.

Rogers, C. D., S. A. Moody, and E. S. Casey. 2009. Neural induction and factors that stabilize a neural fate. *Birth Defects Res C Embryo Today* 87:249–262.

Rohde, K., T. Bering, T. Furukawa, and M. F. Rath. 2017. A modulatory role of the Rax homeobox gene in mature pineal gland function: investigating the photoneuroendocrine circadian system of a Rax conditional knockout mouse. *J Neurochem* 143:100–111.

Rohde, K., D. C. Klein, M. Moller, and M. F. Rath. 2011. Rax: developmental and daily expression patterns in the rat pineal gland and retina. *J Neurochem* 118:999–1007.

Röhlich, K. 1931. Gestaltungsbewegungen der präsumptiven Epidermis während der Neurulation und Kopfbildung bei *Triton taeniatus*. *Roux Arch Entw Mech* 124:66–81.

Romer, A. S. 1972. The vertebrate as a dual animal—somatic and visceral. In *Evol. Biol.*, edited by T. Dobzhansky, M. K. Hecht and W. C. Steere, 121–156. New York: Appleton-Century-Crofts.

Ronan, M., and D. Bodznick. 1991. Behavioral and neurophysiological demonstration of a lateralis skin photosensitivity in larval sea lampreys. *J Exp Biol* 161:97–117.

Ronan, M. C., and D. Bodznick. 1986. End buds: non-ampullary electroreceptors in adult lampreys. *J Comp Physiol A* 158:9–15.

Rothstein, M., D. Bhattacharya, and M. Simoes-Costa. 2018. The molecular basis of neural crest axial identity. *Dev Biol* 444 Suppl 1:S170–S180.

Rovsing, L., S. Clokie, D. M. Bustos, K. Rohde, S. L. Coon, T. Litman, M. F. Rath, M. Moller, and D. C. Klein. 2011. Crx broadly modulates the pineal transcriptome. *J Neurochem* 119:262–274.

Rowley, J. C., 3rd, D. T. Moran, and B. W. Jafek. 1989. Peroxidase backfills suggest the mammalian olfactory epithelium contains a second morphologically distinct class of bipolar sensory neuron: the microvillar cell. *Brain Res* 502:387–400.

Rubel, E. W., and B. Fritzsch. 2002. Auditory system development: primary auditory neurons and their targets. *Annu Rev Neurosci* 25:51–101.

Rubenstein, J. L., K. Shimamura, S. Martinez, and L. Puelles. 1998. Regionalization of the prosencephalic neural plate. *Annu Rev Neurosci* 21:445–477.

Rubinson, K., and H. Cain. 1989. Neural differentiation in the retina of the larval sea lamprey (*Petromyzon marinus*). *Vis Neurosci* 3:241–248.

Rugh, R. 1951. *The frog. Its reproduction and development.* Philadelphia: The Blakiston Company.

Russell, I. J. 1976. Amphibian lateral line receptors. In *Frog neurobiology*, edited by R. Llinas and W. Precht, 513–550. Berlin: Springer.

Ryba, N. J., and R. Tirindelli. 1997. A new multigene family of putative pheromone receptors. *Neuron* 19:371–379.

Rychlik, J. L., M. Hsieh, L. E. Eiden, and E. J. Lewis. 2005. Phox2 and dHAND transcription factors select shared and unique target genes in the noradrenergic cell type. *J Mol Neurosci* 27:281–292.

Sabado, V., P. Barraud, C. V. Baker, and A. Streit. 2012. Specification of GnRH-1 neurons by antagonistic FGF and retinoic acid signaling. *Dev Biol* 362:254–262.

Sadaghiani, B., and C. H. Thiebaud. 1987. Neural crest development in the *Xenopus laevis* embryo, studied by interspecific transplantation and scanning electron microscopy. *Dev Biol* 124:91–110.

Sahly, I., P. Andermann, and C. Petit. 1999. The zebrafish eya1 gene and its expression pattern during embryogenesis. *Dev Genes Evol* 209:399–410.

Sai, X., and R. K. Ladher. 2015. Early steps in inner ear development: induction and morphogenesis of the otic placode. *Front Pharmacol* 6:19.

Saigou, Y., Y. Kamimura, M. Inoue, H. Kondoh, and M. Uchikawa. 2010. Regulation of Sox2 in the pre-placodal cephalic ectoderm and central nervous system by enhancer N-4. *Dev Growth Differ* 52:397–408.

Saint-Germain, N., Y. H. Lee, Y. Zhang, T. D. Sargent, and J. P. Saint-Jeannet. 2004. Specification of the otic placode depends on Sox9 function in *Xenopus*. *Development* 131:1755–1763.

Saint-Jeannet, J. P., X. He, H. E. Varmus, and I. B. Dawid. 1997. Regulation of dorsal fate in the neuraxis by Wnt-1 and Wnt-3a. *Proc Natl Acad Sci USA* 94:13713–13718.

Saint-Jeannet, J. P., and S. A. Moody. 2014. Establishing the pre-placodal region and breaking it into placodes with distinct identities. *Dev Biol* 389:13–27.

Saito, T., A. Greenwood, Q. Sun, and D. J. Anderson. 1995. Identification by differential RT-PCR of a novel paired homeodomain protein specifically expressed in sensory neurons and a subset of their CNS targets. *Mol Cell Neurosci* 6:280–292.

Salinas, E., J. L. Quintanar, and J. A. Reig. 1999. Immunohistochemical study of Syntaxin-1 and SNAP-25 in the pituitaries of mouse, guinea pig and cat. *Acta Physiol Pharmacol Ther Latinoam* 49:61–64.

Sánchez-Arrones, L., J. L. Ferran, M. Hidalgo-Sánchez, and L. Puelles. 2015. Origin and early development of the chicken adenohypophysis. *Front Neuroanat* 9:7.

Sánchez-Arrones, L., A. Sandonis, M. J. Cardozo, and P. Bovolenta. 2017. Adenohypophysis placodal precursors exhibit distinctive features within the rostral preplacodal ectoderm. *Development* 144:3521–3532.

Sánchez-Cardenas, C., P. Fontanaud, Z. He, C. Lafont, A. C. Meunier, M. Schaeffer, D. Carmignac, F. Molino, N. Coutry, X. Bonnefont, L. A. Gouty-Colomer, E. Gavois, D. J. Hodson, P. Le Tissier, I. C. Robinson, and P. Mollard. 2010. Pituitary growth hormone network responses are sexually dimorphic and regulated by gonadal steroids in adulthood. *Proc Natl Acad Sci USA* 107:21878–21883.

Sánchez-Guardado, L. O., L. Puelles, and M. Hidalgo-Sánchez. 2014. Fate map of the chicken otic placode. *Development* 141:2302–2312.

Sandberg, M., M. Kallstrom, and J. Muhr. 2005. Sox21 promotes the progression of vertebrate neurogenesis. *Nat Neurosci* 8:995–1001.

Sandell, L. L., N. E. Butler Tjaden, A. J. Barlow, and P. A. Trainor. 2014. Cochleovestibular nerve development is integrated with migratory neural crest cells. *Dev Biol* 385:200–210.

Sanes, J. R., and R. H. Masland. 2015. The types of retinal ganglion cells: current status and implications for neuronal classification. *Annu Rev Neurosci* 38:221–246.

Sankar, S., D. Yellajoshyula, B. Zhang, B. Teets, N. Rockweiler, and K. L. Kroll. 2016. Gene regulatory networks in neural cell fate acquisition from genome-wide chromatin association of Geminin and Zic1. *Sci Rep* 6:37412.

Sansone, A., A. S. Syed, E. Tantalaki, S. I. Korsching, and I. Manzini. 2014. Trpc2 is expressed in two olfactory subsystems, the main and the vomeronasal system of larval *Xenopus laevis*. *J Exp Biol* 217:2235–2238.

Santagati, F., and F. M. Rijli. 2003. Cranial neural crest and the building of the vertebrate head. *Nat Rev Neurosci* 4:806–818.

Sapede, D., S. Dyballa, and C. Pujades. 2012. Cell lineage analysis reveals three different progenitor pools for neurosensory elements in the otic vesicle. *J Neurosci* 32:16424–16434.

Sapede, D., N. Gompel, C. Dambly-Chaudière, and A. Ghysen. 2002. Cell migration in the postembryonic development of the fish lateral line. *Development* 129:605–615.

Saraiva, L. R., X. Ibarra-Soria, M. Khan, M. Omura, A. Scialdone, P. Mombaerts, J. C. Marioni, and D. W. Logan. 2015. Hierarchical deconstruction of mouse olfactory sensory neurons: from whole mucosa to single-cell RNA-seq. *Sci Rep* 5:18178.

Saraiva, L. R., and S. I. Korsching. 2007. A novel olfactory receptor gene family in teleost fish. *Genome Res* 17:1448–1457.

Sarkar, A., and K. Hochedlinger. 2013. The sox family of transcription factors: versatile regulators of stem and progenitor cell fate. *Cell Stem Cell* 12:15–30.

Sarrazin, A. F., E. J. Villablanca, V. A. Nunez, P. C. Sandoval, A. Ghysen, and M. L. Allende. 2006. Proncural gene requirement for hair cell differentiation in the zebrafish lateral line. *Dev Biol* 295:534–545.

Sasai, N., K. Mizuseki, and Y. Sasai. 2001. Requirement of FoxD3-class signaling for neural crest determination in *Xenopus*. *Development* 128:2525–2536.

Sasai, Y., R. Kageyama, Y. Tagawa, R. Shigemoto, and S. Nakanishi. 1992. Two mammalian helix-loop-helix factors structurally related to *Drosophila* hairy and Enhancer of split. *Genes Dev* 6:2620–2634.

Sasaki, F., A. Doshita, Y. Matsumoto, S. Kuwahara, Y. Tsukamoto, and K. Ogawa. 2003. Embryonic development of the pituitary gland in the chick. *Cells Tissues Organs* 173:65–74.

Sato, A., S. Koshida, and H. Takeda. 2010. Single-cell analysis of somatotopic map formation in the zebrafish lateral line system. *Dev Dyn* 239:2058–2065.

Sato, A., and H. Takeda. 2013. Neuronal subtypes are specified by the level of neurod expression in the zebrafish lateral line. *J Neurosci* 33:556–562.

Sato, S., K. Ikeda, G. Shioi, H. Ochi, H. Ogino, H. Yajima, and K. Kawakami. 2010. Conserved expression of mouse Six1 in the pre-placodal region (PPR) and identification of an enhancer for the rostral PPR. *Dev Biol* 344:158–171.

Sato, S., T. Ishihara, and K. Kawakami. 2005. Evolutionarily conserved enhancers direct the expression of Six1 in cranial placodes. *Mech Dev* 122:S67.

Sato, T., N. Sasai, and Y. Sasai. 2005. Neural crest determination by co-activation of Pax3 and Zic1 genes in *Xenopus* ectoderm. *Development* 132:2355–2363.

Sato, Y., N. Miyasaka, and Y. Yoshihara. 2005. Mutually exclusive glomerular innervation by two distinct types of olfactory sensory neurons revealed in transgenic zebrafish. *J Neurosci* 25:4889–4897.

Satoh, T., and D. M. Fekete. 2005. Clonal analysis of the relationships between mechanosensory cells and the neurons that innervate them in the chicken ear. *Development* 132:1687–1697.

Sauka-Spengler, T., B. Baratte, L. Shi, and S. Mazan. 2001. Structure and expression of an Otx5-related gene in the dogfish *Scyliorhinus canicula*: evidence for a conserved role of Otx5 and Crx genes in the specification of photoreceptors. *Dev Genes Evol* 211:533–544.

Sauka-Spengler, T., and M. Bronner-Fraser. 2008. A gene regulatory network orchestrates neural crest formation. *Nat Rev Mol Cell Biol* 9:557–568.

Savage, J. J., B. C. Yaden, P. Kiratipranon, and S. J. Rhodes. 2003. Transcriptional control during mammalian anterior pituitary development. *Gene* 319:1–19.

Sax, C. M., and J. Piatigorsky. 1994. Expression of the alpha-crystallin/small heat-shock protein/molecular chaperone genes in the lens and other tissues. *Adv Enzymol Relat Areas Mol Biol* 69:155–201.

Saxena, A., B. N. Peng, and M. E. Bronner. 2013. Sox10-dependent neural crest origin of olfactory microvillous neurons in zebrafish. *Elife* 2:e00336.

Sbrogna, J. L., M. J. Barresi, and R. O. Karlstrom. 2003. Multiple roles for Hedgehog signaling in zebrafish pituitary development. *Dev Biol* 254:19–35.

Scardigli, R., C. Schuurmans, G. Gradwohl, and F. Guillemot. 2001. Crossregulation between Neurogenin2 and pathways specifying neuronal identity in the spinal cord. *Neuron* 31:203–217.

Scerbo, P., and A. H. Monsoro-Burq. 2020. The vertebrate-specific VENTX/NANOG gene empowers neural crest with ectomesenchyme potential. *Sci Adv* 6:eaaz1469.

Schaechinger, T. J., and D. Oliver. 2007. Nonmammalian orthologs of prestin (SLC26A5) are electrogenic divalent/chloride anion exchangers. *Proc Natl Acad Sci USA* 104:7693–7698.

Schang, A. L., A. Granger, B. Querat, C. Bleux, J. Cohen-Tannoudji, and J. N. Laverriere. 2013. GATA2-induced silencing and LIM-homeodomain protein-induced activation are mediated by a bi-functional response element in the rat GnRH receptor gene. *Mol Endocrinol* 27:74–91.

Scharrer, B. 1987. Neurosecretion: beginnings and new directions in neuropeptide research. *Annu Rev Neurosci* 10:1–17.

Scharrer, E., and B. Scharrer. 1945. Neurosecretion. *Physiol Rev* 25:171–181.

Scheffer, D. I., J. Shen, D. P. Corey, and Z. Y. Chen. 2015. Gene expression by mouse inner ear hair cells during development. *J Neurosci* 35:6366–6380.

Schier, A. F., and W. S. Talbot. 2005. Molecular genetics of axis formation in zebrafish. *Annu Rev Genet* 39:561–613.

Schietroma, C., K. Parain, A. Estivalet, A. Aghaie, J. Boutet de Monvel, S. Picaud, J. A. Sahel, M. Perron, A. El-Amraoui, and C. Petit. 2017. Usher syndrome type 1-associated cadherins shape the photoreceptor outer segment. *J Cell Biol* 216:1849–1864.

Schilling, T. F., and C. B. Kimmel. 1994. Segment and cell type lineage restrictions during pharyngeal arch development in the zebrafish embryo. *Development* 120:483–494.

Schilling, T. F., and R. D. Knight. 2001. Origins of anteroposterior patterning and Hox gene regulation during chordate evolution. *Philos Trans R Soc Lond B Biol Sci* 356:1599–1613.

Schimmang, T. 2007. Expression and functions of FGF ligands during early otic development. *Int J Dev Biol* 51:473–481.

Schlosser, G. 2002a. Development and evolution of lateral line placodes in amphibians. I. Development. *Zoology* 105:119–146.

Schlosser, G. 2002b. Development and evolution of lateral line placodes in amphibians. II. Evolutionary diversification. *Zoology* 105:177–193.

Schlosser, G. 2003. Hypobranchial placodes in *Xenopus laevis* give rise to hypobranchial ganglia, a novel type of cranial ganglia. *Cell Tissue Res* 312:21–29.

Schlosser, G. 2005. Evolutionary origins of vertebrate placodes: insights from developmental studies and from comparisons with other deuterostomes. *J Exp Zoolog B Mol Dev Evol* 304B:347–399.

Schlosser, G. 2006. Induction and specification of cranial placodes. *Dev Biol* 294:303–351.

Schlosser, G. 2008. Do vertebrate neural crest and cranial placodes have a common evolutionary origin? *Bioessays* 30:659–672.

Schlosser, G. 2010. Making senses: development of vertebrate cranial placodes. *Int Rev Cell Mol Biol* 283C:129–234.

Schlosser, G. 2021. *Evolutionary origin of sensory and neurosecretory cell types. Vertebrate cranial placodes*, Vol. 2. Boca Raton: CRC Press.

Schlosser, G.. 2014. Early embryonic specification of vertebrate cranial placodes. *WIREs Dev Biol.* 3:349-363.

Schlosser, G., and K. Ahrens. 2004. Molecular anatomy of placode development in *Xenopus laevis*. *Dev Biol* 271:439–466.

Schlosser, G., T. Awtry, S. A. Brugmann, E. D. Jensen, K. Neilson, G. Ruan, A. Stammler, D. Voelker, B. Yan, C. Zhang, M. W. Klymkowsky, and S. A. Moody. 2008. Eya1 and Six1 promote neurogenesis in the cranial placodes in a SoxB1-dependent fashion. *Dev Biol* 320:199–214.

Schlosser, G., and R. G. Northcutt. 2000. Development of neurogenic placodes in *Xenopus laevis*. *J Comp Neurol* 418:121–146.

Schlosser, G., and R. G. Northcutt. 2001. Lateral line placodes are induced during neurulation in the axolotl. *Dev Biol* 234:55–71.

Schlosser, G., and G. Roth. 1997. Evolution of nerve development in frogs. I. The development of the peripheral nervous system in *Discoglossus pictus* (Discoglossidae). *Brain Behav Evol* 50:61–93.

Schmitz, F., A. Königstorfer, and T. C. Südhof. 2000. RIBEYE, a component of synaptic ribbons: a protein's journey through evolution provides insight into synaptic ribbon function. *Neuron* 28:857–872.

Schneider-Maunoury, S., and C. Pujades. 2007. Hindbrain signals in otic regionalization: walk on the wild side. *Int J Dev Biol* 51:495–506.

Schohl, A., and F. Fagotto. 2002. Beta-catenin, MAPK and Smad signaling during early *Xenopus* development. *Development* 129:37–52.

Scholz, P., B. Kalbe, F. Jansen, J. Altmueller, C. Becker, J. Mohrhardt, B. Schreiner, G. Gisselmann, H. Hatt, and S. Osterloh. 2016. Transcriptome analysis of murine olfactory sensory neurons during development using single cell RNA-Seq. *Chem Senses* 41:313–323.

Schwanzel-Fukuda, M., and D. W. Pfaff. 1989. Origin of luteinizing hormone-releasing hormone neurons. *Nature* 338:161–164.

Schwarting, G. A., M. E. Wierman, and S. A. Tobet. 2007. Gonadotropin-releasing hormone neuronal migration. *Semin Reprod Med* 25:305–312.

Schweickert, A., H. Steinbeisser, and M. Blum. 2001. Differential gene expression of *Xenopus* Pitx1, Pitx2b and Pitx2c during cement gland, stomodeum and pituitary development. *Mech Dev* 107:191–194.

Schwob, J. E., J. M. Huard, M. B. Luskin, and S. L. Youngentob. 1994. Retroviral lineage studies of the rat olfactory epithelium. *Chem Senses* 19:671–682.

Schwob, J. E., W. Jang, E. H. Holbrook, B. Lin, D. B. Herrick, J. N. Peterson, and J. Hewitt Coleman. 2017. Stem and progenitor cells of the mammalian olfactory epithelium: taking poietic license. *J Comp Neurol* 525:1034–1054.

Scully, K. M., and M. G. Rosenfeld. 2002. Pituitary development: regulatory codes in mammalian organogenesis. *Science* 295:2231–2235.

Selleck, M. A. J., and M. Bronner-Fraser. 1995. Origins of the avian neural crest: the role of neural plate-epidermal interactions. *Development* 121:525–538.

Senzaki, K., S. Ozaki, M. Yoshikawa, Y. Ito, and T. Shiga. 2010. Runx3 is required for the specification of TrkC-expressing mechanoreceptive trigeminal ganglion neurons. *Mol Cell Neurosci* 43:296–307.

Seo, H. C., O. Drivenes, and A. Fjose. 1998. A zebrafish Six4 homologue with early expression in head mesoderm. *Biochim Biophys Acta* 1442:427–431.

Shaham, O., Y. Menuchin, C. Farhy, and R. Ashery-Padan. 2012. Pax6: a multi-level regulator of ocular development. *Prog Retin Eye Res* 31:351–376.

Shanmugalingam, S., C. Houart, A. Picker, F. Reifers, R. Macdonald, A. Barth, K. Griffin, M. Brand, and S. W. Wilson. 2000. Ace/Fgf8 is required for forebrain commissure formation and patterning of the telencephalon. *Development* 127:2549–2561.

Sharma, N., K. Flaherty, K. Lezgiyeva, D. E. Wagner, A. M. Klein, and D. D. Ginty. 2020. The emergence of transcriptional identity in somatosensory neurons. *Nature* 577:392-398.

Sharma, K., A. S. Syed, S. Ferrando, S. Mazan, and S. I. Korsching. 2019. The chemosensory receptor repertoire of a true shark Is dominated by a single olfactory receptor family. *Genome Biol Evol* 11:398–405.

Shekhar, K., S. W. Lapan, I. E. Whitney, N. M. Tran, E. Z. Macosko, M. Kowalczyk, X. Adiconis, J. Z. Levin, J. Nemesh, M. Goldman, S. A. McCarroll, C. L. Cepko, A. Regev, and J. R. Sanes. 2016. Comprehensive classification of retinal bipolar neurons by single-cell transcriptomics. *Cell* 166:1308–1323e30.

Shen, Y. C., and P. A. Raymond. 2004. Zebrafish cone-rod (crx) homeobox gene promotes retinogenesis. *Dev Biol* 269:237–251.

Sheng, G., M. dos Reis, and C. D. Stern. 2003. Churchill, a zinc finger transcriptional activator, regulates the transition between gastrulation and neurulation. *Cell* 115:603–613.

Sheng, H. Z., K. Moriyama, T. Yamashita, H. Li, S. S. Potter, K. A. Mahon, and H. Westphal. 1997. Multistep control of pituitary organogenesis. *Science* 278:1809–1812.

Shiau, C. E., and M. Bronner-Fraser. 2010. N-cadherin acts in concert with Slit1-Robo2 signaling in regulating aggregation of placode-derived cranial sensory neurons. *Development* 137:4155–4164.

Shiau, C. E., P. Y. Lwigale, R. M. Das, S. A. Wilson, and M. Bronner-Fraser. 2008. Robo2-Slit1 dependent cell-cell interactions mediate assembly of the trigeminal ganglion. *Nat Neurosci* 11:269–276.

Shichida, Y., and T. Matsuyama. 2009. Evolution of opsins and phototransduction. *Philos Trans R Soc Lond B Biol Sci* 364:2881–2895.

Shida, H., M. Mende, T. Takano-Yamamoto, N. Osumi, A. Streit, and Y. Wakamatsu. 2015. Otic placode cell specification and proliferation are regulated by Notch signaling in avian development. *Dev Dyn* 244:839–851.

Shimamura, K., and J. L. Rubenstein. 1997. Inductive interactions direct early regionalization of the mouse forebrain. *Development* 124:2709–2718.

Shimizu, T., and M. Hibi. 2009. Formation and patterning of the forebrain and olfactory system by zinc-finger genes Fezf1 and Fezf2. *Dev Growth Differ* 51:221–231.

Shimojo, H., A. Isomura, T. Ohtsuka, H. Kori, H. Miyachi, and R. Kageyama. 2016. Oscillatory control of Delta-like1 in cell interactions regulates dynamic gene expression and tissue morphogenesis. *Genes Dev* 30:102–116.

Shimojo, H., T. Ohtsuka, and R. Kageyama. 2008. Oscillations in notch signaling regulate maintenance of neural progenitors. *Neuron* 58:52–64.

Shimojo, H., T. Ohtsuka, and R. Kageyama. 2011. Dynamic expression of notch signaling genes in neural stem/progenitor cells. *Front Neurosci* 5:78.

Shimozaki, K. 2014. Sox2 transcription network acts as a molecular switch to regulate properties of neural stem cells. *World J Stem Cells* 6:485–490.

Shimozaki, K., K. Nakashima, H. Niwa, and T. Taga. 2003. Involvement of Oct3/4 in the enhancement of neuronal differentiation of ES cells in neurogenesis-inducing cultures. *Development* 130:2505–2512.

Shimozaki, K., C. L. Zhang, H. Suh, A. M. Denli, R. M. Evans, and F. H. Gage. 2012. SRY-box-containing gene 2 regulation of nuclear receptor tailless (Tlx) transcription in adult neural stem cells. *J Biol Chem* 287:5969–5978.

Shirasawa, N., Y. Mabuchi, E. Sakuma, O. Horiuchi, T. Yashiro, M. Kikuchi, Y. Hashimoto, Y. Tsuruo, D. C. Herbert, and T. Soji. 2004. Intercellular communication within the rat anterior pituitary gland: X. Immunohistocytochemistry of S-100 and connexin 43 of folliculo-stellate cells in the rat anterior pituitary gland. *Anat Rec A Discov Mol Cell Evol Biol* 278:462–473.

Shrestha, B. R., C. Chia, L. Wu, S. G. Kujawa, M. C. Liberman, and L. V. Goodrich. 2018. Sensory neuron diversity in the inner ear is shaped by activity. *Cell* 174:1229–1246e17.

Shykind, B. M., S. C. Rohani, S. O'Donnell, A. Nemes, M. Mendelsohn, Y. Sun, R. Axel, and G. Barnea. 2004. Gene switching and the stability of odorant receptor gene choice. *Cell* 117:801–815.

Siebel, C., and U. Lendahl. 2017. Notch signaling in development, tissue homeostasis, and disease. *Physiol Rev* 97:1235–1294.

Siegert, S., E. Cabuy, B. G. Scherf, H. Kohler, S. Panda, Y. Z. Le, H. J. Fehling, D. Gaidatzis, M. B. Stadler, and B. Roska. 2012. Transcriptional code and disease map for adult retinal cell types. *Nat Neurosci* 15:487–495.

Sienknecht, U. J. 2015. Current concepts of hair cell differentiation and planar cell polarity in inner ear sensory organs. *Cell Tissue Res* 361:25–32.

Silva, L., and A. Antunes. 2017. Vomeronasal receptors in vertebrates and the evolution of pheromone detection. *Annu Rev Anim Biosci* 5:353–370.

Silver, S. J., E. L. Davies, L. Doyon, and I. Rebay. 2003. Functional dissection of eyes absent reveals new modes of regulation within the retinal determination gene network. *Mol Cell Biol* 23:5989–5999.

Silver, S. J., and I. Rebay. 2005. Signaling circuitries in development: insights from the retinal determination gene network. *Development* 132:3–13.

Silvotti, L., A. Moiani, R. Gatti, and R. Tirindelli. 2007. Combinatorial co-expression of pheromone receptors, V2Rs. *J Neurochem* 103:1753–1763.

Simeone, A., D. Acampora, A. Mallamaci, A. Stornaiuolo, M. R. D'Apice, V. Nigro, and E. Boncinelli. 1993. A vertebrate gene related to orthodenticle contains a homeodomain of the bicoid class and demarcates anterior neuroectoderm in the gastrulating mouse embryo. *EMBO J* 12:2735–2747.

Simmons, D. M., J. W. Voss, H. A. Ingraham, J. M. Holloway, R. S. Broide, M. G. Rosenfeld, and L. W. Swanson. 1990. Pituitary cell phenotypes involve cell-specific Pit-1 mRNA translation and synergistic interactions with other classes of transcription factors. *Genes Dev* 4:695–711.

Simoes-Costa, M., and M. E. Bronner. 2015. Establishing neural crest identity: a gene regulatory recipe. *Development* 142:242–257.

Simoes-Costa, M. S., S. J. McKeown, J. Tan-Cabugao, T. Sauka-Spengler, and M. E. Bronner. 2012. Dynamic and differential regulation of stem cell factor FoxD3 in the neural crest is encrypted in the genome. *PLoS Genet* 8:e1003142.

Simoes-Costa, M., J. Tan-Cabugao, I. Antoshechkin, T. Sauka-Spengler, and M. E. Bronner. 2014. Transcriptome analysis reveals novel players in the cranial neural crest gene regulatory network. *Genome Res* 24:281–290.

Singh, S., and A. K. Groves. 2016. The molecular basis of craniofacial placode development. *Wiley Interdiscip Rev Dev Biol* 5:363–376.

Sive, H., and L. Bradley. 1996. A sticky problem: the *Xenopus* cement gland as a paradigm for anteroposterior patterning. *Dev Dyn* 205:265–280.

Sjödal, M., T. Edlund, and L. Gunhaga. 2007. Time of exposure to BMP signals plays a key role in the specification of the olfactory and lens placodes ex vivo. *Dev Cell* 13:141–149.

Slack, J. M. W. 1991. *From egg to embryo*, 2nd ed. Cambridge: Cambridge Univ. Press.

Sloop, K. W., B. C. Meier, J. L. Bridwell, G. E. Parker, A. M. Schiller, and S. J. Rhodes. 1999. Differential activation of pituitary hormone genes by human Lhx3 isoforms with distinct DNA binding properties. *Mol Endocrinol* 13:2212–2225.

Smith, S. C. 1996. Pattern formation in the urodele mechanoreceptive lateral line: what features can be exploited for the study of development and evolution. *Int J Dev B* 40:727–733.

Snir, M., R. Ofir, S. Elias, and D. Frank. 2006. *Xenopus laevis* POU91 protein, an Oct3/4 homologue, regulates competence transitions from mesoderm to neural cell fates. *EMBO J* 25:3664–3674.

So, W. K., H. F. Kwok, and W. Ge. 2005. Zebrafish gonadotropins and their receptors: II. Cloning and characterization of zebrafish follicle-stimulating hormone and luteinizing hormone subunits – their spatial-temporal expression patterns and receptor specificity. *Biol Reprod* 72:1382–1396.

Sokpor, G., E. Abbas, J. Rosenbusch, J. F. Staiger, and T. Tuoc. 2018. Transcriptional and epigenetic control of mammalian olfactory epithelium development. *Mol Neurobiol* 55:8306–8327.

Solbu, T. T., and T. Holen. 2012. Aquaporin pathways and mucin secretion of Bowman's glands might protect the olfactory mucosa. *Chem Senses* 37:35–46.

Solessio, E., and G. A. Engbretson. 1993. Antagonistic chromatic mechanisms in photoreceptors of the parietal eye of lizards. *Nature* 364:442–445.

Solomon, K. S., and A. Fritz. 2002. Concerted action of two dlx paralogs in sensory placode formation. *Development* 129:3127–3136.

Solomon, K. S., T. Kudoh, I. B. Dawid, and A. Fritz. 2003. Zebrafish foxi1 mediates otic placode formation and jaw development. *Development* 130:929–940.

Solomon, K. S., S. J. Kwak, and A. Fritz. 2004. Genetic interactions underlying otic placode induction and formation. *Dev Dyn* 230:419–433.

Solomon, K. S., J. M. Logsdon, Jr., and A. Fritz. 2003. Expression and phylogenetic analyses of three zebrafish FoxI class genes. *Dev Dyn* 228:301–307.

Somoza, G. M., L. A. Miranda, P. Strobl-Mazzulla, and L. G. Guilgur. 2002. Gonadotropin-releasing hormone (GnRH): from fish to mammalian brains. *Cell Mol Neurobiol* 22:589–609.

Song, J., and R. G. Northcutt. 1991. Morphology, distribution and innervation of the lateral-line receptors of the Florida gar, *Lepisosteus platirhynchus*. *Brain Behav Evol* 37:10–37.

Soufi, A., M. F. Garcia, A. Jaroszewicz, N. Osman, M. Pellegrini, and K. S. Zaret. 2015. Pioneer transcription factors target partial DNA motifs on nucleosomes to initiate reprogramming. *Cell* 161:555–568.

Sower, S. A., W. A. Decatur, K. N. Hausken, T. J. Marquis, S. L. Barton, J. Gargan, M. Freamat, M. Wilmot, L. Hollander, J. A. Hall, M. Nozaki, M. Shpilman, and B. Levavi-Sivan. 2015. Emergence of an ancestral glycoprotein hormone in the pituitary of the sea lamprey, a basal vertebrate. *Endocrinology* 156:3026–3037.

Sower, S. A., M. Freamat, and S. I. Kavanaugh. 2009. The origins of the vertebrate hypothalamic-pituitary-gonadal (HPG) and hypothalamic-pituitary-thyroid (HPT) endocrine systems: new insights from lampreys. *Gen Comp Endocrinol* 161:20–29.

Sower, S. A., S. Moriyama, M. Kasahara, A. Takahashi, M. Nozaki, K. Uchida, J. M. Dahlstrom, and H. Kawauchi. 2006. Identification of sea lamprey GTHbeta-like cDNA and its evolutionary implications. *Gen Comp Endocrinol* 148:22–32.

Spassky, N., and A. Meunier. 2017. The development and functions of multiciliated epithelia. *Nat Rev Mol Cell Biol* 18:423–436.

Speca, D. J., D. M. Lin, P. W. Sorensen, E. Y. Isacoff, J. Ngai, and A. H. Dittman. 1999. Functional identification of a goldfish odorant receptor. *Neuron* 23:487–498.

Spemann, H., and H. Mangold. 1924. Über induktion von embryonalanlagen durch implantation artfremder *Organisatoren Roux Arch Entwickl Mech* 100:599–638.

Spokony, R. F., Y. Aoki, N. Saint-Germain, E. Magner-Fink, and J. P. Saint-Jeannet. 2002. The transcription factor Sox9 is required for cranial neural crest development in *Xenopus*. *Development* 129:421–432.

Srinivas, M., L. Ng, H. Liu, L. Jia, and D. Forrest. 2006. Activation of the blue opsin gene in cone photoreceptor development by retinoid-related orphan receptor beta. *Mol Endocrinol* 20:1728–1741.

Stark, M. R., J. Sechrist, M. Bronner-Fraser, and C. Marcelle. 1997. Neural tube-ectoderm interactions are required for trigeminal placode formation. *Development* 124: 4287–4295.

Steel, K. P., and C. Barkway. 1989. Another role for melanocytes: their importance for normal stria vascularis development in the mammalian inner ear. *Development* 107:453–463.

Steevens, A. R., D. L. Sookiasian, J. C. Glatzer, and A. E. Kiernan. 2017. SOX2 is required for inner ear neurogenesis. *Sci Rep* 7:4086.

Steinberg, R. H., S. K. Fisher, and D. H. Anderson. 1980. Disc morphogenesis in vertebrate photoreceptors. *J Comp Neurol* 190:501–508.

Stern, C. D. 2001. Initial patterning of the central nervous system: how many organizers? *Nat Rev Neurosci* 2:92–98.

Stern, C. D. 2006. Neural induction: 10 years on since the 'default model'. *Curr Opin Cell Biol* 18:692–697.

Steventon, B., C. Araya, C. Linker, S. Kuriyama, and R. Mayor. 2009. Differential requirements of BMP and Wnt signalling during gastrulation and neurulation define two steps in neural crest induction. *Development* 136:771–779.

Steventon, B., and R. Mayor. 2012. Early neural crest induction requires an initial inhibition of Wnt signals. *Dev Biol* 365:196–207.

Steventon, B., R. Mayor, and A. Streit. 2012. Mutual repression between Gbx2 and Otx2 in sensory placodes reveals a general mechanism for ectodermal patterning. *Dev Biol* 367:55–65.

Steventon, B., R. Mayor, and A. Streit. 2014. Neural crest and placode interaction during the development of the cranial sensory system. *Dev Biol* 389:28–38.

Steventon, B., R. Mayor, and A. Streit. 2016. Directional cell movements downstream of Gbx2 and Otx2 control the assembly of sensory placodes. *Biol Open* 5:1620–1624.

Stojilkovic, S. S. 2006. Pituitary cell type-specific electrical activity, calcium signaling and secretion. *Biol Res* 39:403–423.

Stone, J. S., J. L. Shang, and S. Tomarev. 2003. Expression of Prox1 defines regions of the avian otocyst that give rise to sensory or neural cells. *J Comp Neurol* 460:487–502.

Stone, L. S. 1922. Experiments on the development of the cranial ganglia and the lateral line sense organs in Amblystoma punctatum *J Exp Zool* 35:421–496.

Storm, E. E., S. Garel, U. Borello, J. M. Hebert, S. Martinez, S. K. McConnell, G. R. Martin, and J. L. Rubenstein. 2006. Dose-dependent functions of Fgf8 in regulating telencephalic patterning centers. *Development* 133:1831–1844.

Streit, A. 2002. Extensive cell movements accompany formation of the otic placode. *Dev Biol* 249:237–254.

Streit, A. 2004. Early development of the cranial sensory nervous system: from a common field to individual placodes. *Dev Biol* 276:1–15.

Streit, A. 2007. The preplacodal region: an ectodermal domain with multipotential progenitors that contribute to sense organs and cranial sensory ganglia. *Int J Dev Biol* 51:447–461.

Streit, A. 2008. The cranial sensory nervous system: specification of sensory progenitors and placodes. In *StemBook*. Cambridge: Harvard Stem Cell Institute doi/10.3824/stembook.1.31.1.

Streit, A. 2018. Specification of sensory placode progenitors: signals and transcription factor networks. *Int J Dev Biol* 62:195–205.

Streit, A., A. J. Berliner, C. Papanayotou, A. Sirulnik, and C. D. Stern. 2000. Initiation of neural induction by FGF signalling before gastrulation. *Nature* 406:74–78.

Streit, A., K. J. Lee, I. Woo, C. Roberts, T. M. Jessell, and C. D. Stern. 1998. Chordin regulates primitive streak development and the stability of induced neural cells, but is not sufficient for neural induction in the chick embryo. *Development* 125:507–519.

Streit, A., and C. D. Stern. 1999. Establishment and maintenance of the border of the neural plate in the chick: involvement of FGF and BMP activity. *Mech Dev* 82:51–66.

Striedter, G. F., and R. G. Northcutt. 2020. *Brains through time. A natural history of vertebrates.* New York: Oxford University Press.

Strotmann, J., O. Levai, J. Fleischer, K. Schwarzenbacher, and H. Breer. 2004. Olfactory receptor proteins in axonal processes of chemosensory neurons. *J Neurosci* 24:7754–7761.

Stuhlmiller, T. J., and M. I. Garcia-Castro. 2012. FGF/MAPK signaling is required in the gastrula epiblast for avian neural crest induction. *Development* 139:289–300.

Suarez, R., D. Garcia-Gonzalez, and F. de Castro. 2012. Mutual influences between the main olfactory and vomeronasal systems in development and evolution. *Front Neuroanat* 6:50.

Sugahara, S., T. Fujimoto, H. Kondoh, and M. Uchikawa. 2018. Nasal and otic placode specific regulation of Sox2 involves both activation by Sox-Sall4 synergism and multiple repression mechanisms. *Dev Biol* 433:61–74.

Sullivan, C. H., H. D. Majumdar, K. M. Neilson, and S. A. Moody. 2019. Six1 and Irx1 have reciprocal interactions during cranial placode and otic vesicle formation. *Dev Biol* 446:68–79.

Sun, J., S. Rockowitz, Q. Xie, R. Ashery-Padan, D. Zheng, and A. Cvekl. 2015. Identification of in vivo DNA-binding mechanisms of Pax6 and reconstruction of Pax6-dependent gene regulatory networks during forebrain and lens development. *Nucleic Acids Res* 43:6827–6846.

Sun, S. K., C. T. Dee, V. B. Tripathi, A. Rengifo, C. S. Hirst, and P. J. Scotting. 2007. Epibranchial and otic placodes are induced by a common Fgf signal, but their subsequent development is independent. *Dev Biol* 303:675–686.

Sun, Y., I. M. Dykes, X. Liang, S. R. Eng, S. M. Evans, and E. E. Turner. 2008. A central role for Islet1 in sensory neuron development linking sensory and spinal gene regulatory programs. *Nat Neurosci* 11:1283–1293.

Surzenko, N., T. Crowl, A. Bachleda, L. Langer, and L. Pevny. 2013. SOX2 maintains the quiescent progenitor cell state of postnatal retinal Müller glia. *Development* 140:1445–1456.

Suzuki, A., N. Ueno, and A. Hemmati-Brivanlou. 1997. *Xenopus* msx1 mediates epidermal induction and neural inhibition by BMP4. *Development* 124:3037–3044.

Suzuki, D. G., and S. Grillner. 2018. The stepwise development of the lamprey visual system and its evolutionary implications. *Biol Rev Camb Philos Soc* 93:1461–1477.

Suzuki, J., and N. Osumi. 2015. Neural crest and placode contributions to olfactory development. *Curr Top Dev Biol* 111:351–374.

Suzuki, J., K. Yoshizaki, T. Kobayashi, and N. Osumi. 2013. Neural crest-derived horizontal basal cells as tissue stem cells in the adult olfactory epithelium. *Neurosci Res* 75:112–120.

Swaroop, A., D. Kim, and D. Forrest. 2010. Transcriptional regulation of photoreceptor development and homeostasis in the mammalian retina. *Nat Rev Neurosci* 11:563–576.

Sweet, E. M., S. Vemaraju, and B. B. Riley. 2011. Sox2 and Fgf interact with Atoh1 to promote sensory competence throughout the zebrafish inner ear. *Dev Biol* 358:113–121.

Syed, A. S., A. Sansone, W. Nadler, I. Manzini, and S. I. Korsching. 2013. Ancestral amphibian v2rs are expressed in the main olfactory epithelium. *Proc Natl Acad Sci USA* 110:7714–7719.

Szeto, D. P., C. Rodriguez-Esteban, A. K. Ryan, S. M. O'Connell, F. Liu, C. Kioussi, A. S. Gleiberman, J. C. Izpisua-Belmonte, and M. G. Rosenfeld. 1999. Role of the Bicoid-related homeodomain factor Pitx1 in specifying hindlimb morphogenesis and pituitary development. *Genes Dev* 13:484–494.

Szeto, D. P., A. K. Ryan, S. M. O'Connell, and M. G. Rosenfeld. 1996. P-OTX: a PIT-1-interacting homeodomain factor expressed during anterior pituitary gland development. *Proc Natl Acad Sci USA* 93:7706–7710.

Tadjuidje, E., and R. S. Hegde. 2013. The Eyes Absent proteins in development and disease. *Cell Mol Life Sci* 70:1897–1913.

Taira, M., W. P. Hayes, H. Otani, and I. B. Dawid. 1993. Expression of LIM class homeobox gene Xlim-3 in *Xenopus* development is limited to neural and neuroendocrine tissues. *Dev Biol* 159:245–256.

Takahashi, A., O. Nakata, M. Kasahara, S. A. Sower, and H. Kawauchi. 2005. Structures for the proopiomelanocortin family genes proopiocortin and proopiomelanotropin in the sea lamprey *Petromyzon marinus*. *Gen Comp Endocrinol* 144:174–181.

Takahashi, K., and S. Yamanaka. 2006. Induction of pluripotent stem cells from mouse embryonic and adult fibroblast cultures by defined factors. *Cell* 126:663–676.

Takahashi, K., and S. Yamanaka. 2016. A decade of transcription factor-mediated reprogramming to pluripotency. *Nat Rev Mol Cell Biol* 17:183–193.

Takahashi, S., C. Yokota, K. Takano, K. Tanegashima, Y. Onuma, J. Goto, and M. Asashima. 2000. Two novel nodal-related genes initiate early inductive events in *Xenopus* Nieuwkoop center. *Development* 127:5319–5329.

Takanaga, H., N. Tsuchida-Straeten, K. Nishide, A. Watanabe, H. Aburatani, and T. Kondo. 2009. Gli2 is a novel regulator of sox2 expression in telencephalic neuroepithelial cells. *Stem Cells* 27:165–174.

Takemoto, T., M. Uchikawa, M. Yoshida, D. M. Bell, R. Lovell-Badge, V. E. Papaioannou, and H. Kondoh. 2011. Tbx6-dependent Sox2 regulation determines neural or mesodermal fate in axial stem cells. *Nature* 470:394–398.

Takuma, N., H. Z. Sheng, Y. Furuta, J. M. Ward, K. Sharma, B. L. Hogan, S. L. Pfaff, H. Westphal, S. Kimura, and K. A. Mahon. 1998. Formation of Rathke's pouch requires dual induction from the diencephalon. *Development* 125:4835–4840.

Talikka, M., S. E. Perez, and K. Zimmerman. 2002. Distinct patterns of downstream target activation are specified by the helix-loop-helix domain of proneural basic helix-loop-helix transcription factors. *Dev Biol* 247:137–148.

Talikka, M., G. Stefani, A. H. Brivanlou, and K. Zimmerman. 2004. Characterization of *Xenopus* Phox2a and Phox2b defines expression domains within the embryonic nervous system and early heart field. *Gene Expr Patterns* 4:601–607.

Tan, L., Q. Li, and X. S. Xie. 2015. Olfactory sensory neurons transiently express multiple olfactory receptors during development. *Mol Syst Biol* 11:844.

Tan, L., C. Zong, and X. S. Xie. 2013. Rare event of histone demethylation can initiate singular gene expression of olfactory receptors. *Proc Natl Acad Sci USA* 110:21148–21152.

Tan, S. L., T. Ohtsuka, A. Gonzalez, and R. Kageyama. 2012. MicroRNA9 regulates neural stem cell differentiation by controlling Hes1 expression dynamics in the developing brain. *Genes Cells* 17:952–961.

Tan, X., J. L. Pecka, J. Tang, O. E. Okoruwa, Q. Zhang, K. W. Beisel, and D. Z. He. 2011. From zebrafish to mammal: functional evolution of prestin, the motor protein of cochlear outer hair cells. *J Neurophysiol* 105:36–44.

Tanaka, S., Y. Kamachi, A. Tanouchi, H. Hamada, N. Jing, and H. Kondoh. 2004. Interplay of SOX and POU factors in regulation of the Nestin gene in neural primordial cells. *Mol Cell Biol* 24:8834–8846.

Taniguchi, K., and K. Taniguchi. 2014. Phylogenic studies on the olfactory system in vertebrates. *J Vet Med Sci* 76:781–788.

Taranova, O. V., S. T. Magness, B. M. Fagan, Y. Wu, N. Surzenko, S. R. Hutton, and L. H. Pevny. 2006. SOX2 is a dose-dependent regulator of retinal neural progenitor competence. *Genes Dev* 20:1187–1202.

Teillet, M. A., C. Kalcheim, and N. M. Le Douarin. 1987. Formation of the dorsal root ganglia in the avian embryo: segmental origin and migratory behavior of neural crest progenitor cells. *Dev Biol* 120:329–347.

Theriault, F. M., H. N. Nuthall, Z. Dong, R. Lo, F. Barnabe-Heider, F. D. Miller, and S. Stifani. 2005. Role for Runx1 in the proliferation and neuronal differentiation of selected progenitor cells in the mammalian nervous system. *J Neurosci* 25:2050–2061.

Theveneau, E., B. Steventon, E. Scarpa, S. Garcia, X. Trepat, A. Streit, and R. Mayor. 2013. Chase-and-run between adjacent cell populations promotes directional collective migration. *Nat Cell Biol* 15:763–772.

Thirumangalathu, S., D. E. Harlow, A. L. Driskell, R. F. Krimm, and L. A. Barlow. 2009. Fate mapping of mammalian embryonic taste bud progenitors. *Development* 136:1519–1528.

Thomas, E. D., I. A. Cruz, D. W. Hailey, and D. W. Raible. 2015. There and back again: development and regeneration of the zebrafish lateral line system. *Wiley Interdiscip Rev Dev Biol* 4:1–16.

Thomas, P., and R. Beddington. 1996. Anterior primitive endoderm may be responsible for patterning the anterior neural plate in the mouse embryo. *Curr Biol* 6:1487–1496.

Thomson, M., S. J. Liu, L. N. Zou, Z. Smith, A. Meissner, and S. Ramanathan. 2011. Pluripotency factors in embryonic stem cells regulate differentiation into germ layers. *Cell* 145:875–889.

Tian, C., R. J. Ambroz, L. Sun, Y. Wang, K. Ma, Q. Chen, B. Zhu, and J. C. Zheng. 2012. Direct conversion of dermal fibroblasts into neural progenitor cells by a novel cocktail of defined factors. *Curr Mol Med* 12:126–137.

Tietjen, I., J. Rihel, and C. G. Dulac. 2005. Single-cell transcriptional profiles and spatial patterning of the mammalian olfactory epithelium. *Int J Dev Biol* 49:201–207.

Tietjen, I., J. M. Rihel, Y. Cao, G. Koentges, L. Zakhary, and C. Dulac. 2003. Single-cell transcriptional analysis of neuronal progenitors. *Neuron* 38:161–175.

Tiveron, M. C., M. R. Hirsch, and J. F. Brunet. 1996. The expression pattern of the transcription factor Phox2 delineates synaptic pathways of the autonomic nervous system. *J Neurosci* 16:7649–7660.

Tizzano, M., B. D. Gulbransen, A. Vandenbeuch, T. R. Clapp, J. P. Herman, H. M. Sibhatu, M. E. Churchill, W. L. Silver, S. C. Kinnamon, and T. E. Finger. 2010. Nasal chemosensory cells use bitter taste signaling to detect irritants and bacterial signals. *Proc Natl Acad Sci USA* 107:3210–3215.

Tomioka, M., M. Nishimoto, S. Miyagi, T. Katayanagi, N. Fukui, H. Niwa, M. Muramatsu, and A. Okuda. 2002. Identification of Sox-2 regulatory region which is under the control of Oct-3/4-Sox-2 complex. *Nucleic Acids Res* 30:3202–3213.

Toro, S., and Z. M. Varga. 2007. Equivalent progenitor cells in the zebrafish anterior preplacodal field give rise to adenohypophysis, lens, and olfactory placodes. *Semin Cell Dev Biol* 18:534–542.

Torres, M., E. G¢mez-Pardo, and P. Gruss. 1996. Pax2 contributes to inner ear patterning and optic nerve trajectory *Development* 122:3381–3391.

Torres, M., and F. Giráldez. 1998. The development of the vertebrate inner ear. *Mech Dev* 71:5–21.

Touhara, K., and L. B. Vosshall. 2009. Sensing odorants and pheromones with chemosensory receptors. *Annu Rev Physiol* 71:307–332.

Tour, E., G. Pillemer, Y. Gruenbaum, and A. Fainsod. 2002. Gbx2 interacts with Otx2 and patterns the anterior-posterior axis during gastrulation in *Xenopus*. *Mech Dev* 112:141–151.

Travaglini, K. J., A. N. Nabhan, L. Penland, R. Sinha, A. Gillich, R. V. Sit, S. Chang, S. D. Conle, Y. Mori, J. Seita, G. J. Berry, J. B. Shrager, R. J. Metzger, C. S. Kuo, N. Neff, I. L. Weissman, S. R. Quake, and M. A. Krasnow. 2020. A molecular cell atlas of the human lung from single cell RNA sequencing. Nature 587:619–625.

Treier, M., A. S. Gleiberman, S. M. O'Connell, D. P. Szeto, J. A. McMahon, A. P. McMahon, and M. G. Rosenfeld. 1998. Multistep signaling requirements for pituitary organogenesis in vivo. *Genes Dev* 12:1691–1704.

Treier, M., S. Oconnell, A. Gleiberman, J. Price, D. P. Szeto, R. Burgess, P. T. Chuang, A. P. McMahon, and M. G. Rosenfeld. 2001. Hedgehog signaling is required for pituitary gland development. *Development* 128 (3):377–386.

Tremblay, J. J., and J. Drouin. 1999. Egr-1 is a downstream effector of GnRH and synergizes by direct interaction with Ptx1 and SF-1 to enhance luteinizing hormone beta gene transcription. *Mol Cell Biol* 19:2567–2576.

Tremblay, J. J., C. Lanctot, and J. Drouin. 1998. The pan-pituitary activator of transcription, Ptx1 (pituitary homeobox 1), acts in synergy with SF-1 and Pit1 and is an upstream regulator of the Lim-homeodomain gene Lim3/Lhx3. *Mol Endocrinol* 12:428–441.

Tremblay, J. J., A. Marcil, Y. Gauthier, and J. Drouin. 1999. Ptx1 regulates SF-1 activity by an interaction that mimics the role of the ligand-binding domain. *EMBO J* 18:3431–3441.

Trevers, K. E., R. S. Prajapati, M. Hintze, M. J. Stower, A. C. Strobl, M. Tambalo, R. Ranganathan, N. Moncaut, M. A. F. Khan, C. D. Stern, and A. Streit. 2018. Neural induction by the node and placode induction by head mesoderm share an initial state resembling neural plate border and ES cells. *Proc Natl Acad Sci USA* 115:355–360.

Tribulo, C., M. J. Aybar, V. H. Nguyen, M. C. Mullins, and R. Mayor. 2003. Regulation of Msx genes by a Bmp gradient is essential for neural crest specification. *Development* 130:6441–6452.

Tripathi, V. B., Y. Ishii, M. M. Abu-Elmagd, and P. J. Scotting. 2009. The surface ectoderm of the chick embryo exhibits dynamic variation in its response to neurogenic signals. *Int J Dev Biol* 53:1023–1033.

Tsai, L., and G. Barnea. 2014. A critical period defined by axon-targeting mechanisms in the murine olfactory bulb. *Science* 344:197–200.

Tsunemoto, R., S. Lee, A. Szucs, P. Chubukov, I. Sokolova, J. W. Blanchard, K. T. Eade, J. Bruggemann, C. Wu, A. Torkamani, P. P. Sanna, and K. K. Baldwin. 2018. Diverse reprogramming codes for neuronal identity. *Nature* 557:375-380.

Tucker, E. S., M. K. Lehtinen, T. Maynard, M. Zirlinger, C. Dulac, N. Rawson, L. Pevny, and A. S. Lamantia. 2010. Proliferative and transcriptional identity of distinct classes of neural precursors in the mammalian olfactory epithelium. *Development* 137:2471–2481.

Uchida, K., S. Moriyama, H. Chiba, T. Shimotani, K. Honda, M. Miki, A. Takahashi, S. A. Sower, and M. Nozaki. 2010. Evolutionary origin of a functional gonadotropin in the pituitary of the most primitive vertebrate, hagfish. *Proc Natl Acad Sci USA* 107:15832–15837.

Uchida, K., S. Moriyama, S. A. Sower, and M. Nozaki. 2013. Glycoprotein hormone in the pituitary of hagfish and its evolutionary implications. *Fish Physiol Biochem* 39:75–83.

Uchida, K., Y. Murakami, S. Kuraku, S. Hirano, and S. Kuratani. 2003. Development of the adenohypophysis in the lamprey: evolution of epigenetic patterning programs in organogenesis. *J Exp Zoolog Part B Mol Dev Evol* 300:32–47.

Uchikawa, M., Y. Ishida, T. Takemoto, Y. Kamachi, and H. Kondoh. 2003. Functional analysis of chicken Sox2 enhancers highlights an array of diverse regulatory elements that are conserved in mammals *Dev Cell* 4:509–519.

Ueharu, H., S. Yoshida, T. Kikkawa, N. Kanno, M. Higuchi, T. Kato, N. Osumi, and Y. Kato. 2017. Gene tracing analysis reveals the contribution of neural crest-derived cells in pituitary development. *J Anat* 230:373–380.

Urban, N., D. L. van den Berg, A. Forget, J. Andersen, J. A. Demmers, C. Hunt, O. Ayrault, and F. Guillemot. 2016. Return to quiescence of mouse neural stem cells by degradation of a proactivation protein. *Science* 353:292–295.

Urbanek, P., Z. Q. Wang, I. Fetka, E. F. Wagner, and M. Busslinger. 1994. Complete block of early B cell differentiation and altered patterning of the posterior midbrain in mice lacking Pax5/BSAP. *Cell* 79:901–912.

Urness, L. D., C. N. Paxton, X. Wang, G. C. Schoenwolf, and S. L. Mansour. 2010. FGF signaling regulates otic placode induction and refinement by controlling both ectodermal target genes and hindbrain Wnt8a. *Dev Biol* 340:595–604.

Usoskin, D., A. Furlan, S. Islam, H. Abdo, P. Lonnerberg, D. Lou, J. Hjerling-Leffler, J. Haeggstrom, O. Kharchenko, P. V. Kharchenko, S. Linnarsson, and P. Ernfors. 2015. Unbiased classification of sensory neuron types by large-scale single-cell RNA sequencing. *Nat Neurosci* 18:145–153.

Van Raay, T. J., K. B. Moore, I. Iordanova, M. Steele, M. Jamrich, W. A. Harris, and M. L. Vetter. 2005. Frizzled 5 signaling governs the neural potential of progenitors in the developing *Xenopus* retina. *Neuron* 46:23–36.

van Wijhe, J. W. 1883. Über die Mesodermsegmente und die Entwicklung der Nerven des Selachierkopfes. *Verhandl Konink Akad Wetensch* 22:1–50.

Varga, Z. M., A. Amores, K. E. Lewis, Y. L. Yan, J. H. Postlethwait, J. S. Eisen, and M. Westerfield. 2001. Zebrafish smoothened functions in ventral neural tube specification and axon tract formation. *Development* 128 (18):3497–3509.

Vasconcelos, F. F., A. Sessa, C. Laranjeira, Aasf Raposo, V. Teixeira, D. W. Hagey, D. M. Tomaz, J. Muhr, V. Broccoli, and D. S. Castro. 2016. MyT1 counteracts the neural progenitor program to promote vertebrate neurogenesis. *Cell Rep* 17:469–483.

Vassalli, A., P. Feinstein, and P. Mombaerts. 2011. Homeodomain binding motifs modulate the probability of odorant receptor gene choice in transgenic mice. *Mol Cell Neurosci* 46:381–396.

Vassalli, A., A. Rothman, P. Feinstein, M. Zapotocky, and P. Mombaerts. 2002. Minigenes impart odorant receptor-specific axon guidance in the olfactory bulb. *Neuron* 35: 681–696.

Vassar, R., S. K. Chao, R. Sitcheran, J. M. Nunez, L. B. Vosshall, and R. Axel. 1994. Topographic organization of sensory projections to the olfactory bulb. *Cell* 79:981–991.

Veitch, E., J. Begbie, T. F. Schilling, M. M. Smith, and A. Graham. 1999. Pharyngeal arch patterning in the absence of neural crest. *Curr Biol* 9:1481–1484.

Venkatachalam, K., and C. Montell. 2007. TRP channels. *Annu Rev Biochem* 76:387–417.

Verschueren, K., J. E. Remacle, C. Collart, H. Kraft, B. S. Baker, P. Tylzanowski, L. Nelles, G. Wuytens, M. T. Su, R. Bodmer, J. C. Smith, and D. Huylebroeck. 1999. SIP1, a novel zinc finger/homeodomain repressor, interacts with Smad proteins and binds to 5'-CACCT sequences in candidate target genes. *J Biol Chem* 274:20489–20498.

Viczian, A. S., R. Vignali, M. E. Zuber, G. Barsacchi, and W. A. Harris. 2003. XOtx5b and XOtx2 regulate photoreceptor and bipolar fates in the *Xenopus* retina. *Development* 130:1281–1294.

Vierbuchen, T., A. Ostermeier, Z. P. Pang, Y. Kokubu, T. C. Südhof, and M. Wernig. 2010. Direct conversion of fibroblasts to functional neurons by defined factors. *Nature* 463:1035–1041.

Viets, K., K. Eldred, and R. J. Johnston, Jr. 2016. Mechanisms of photoreceptor patterning in vertebrates and invertebrates. *Trends Genet* 32:638–659.

Vigh, B., M. J. Manzano, A. Zadori, C. L. Frank, A. Lukats, P. Rohlich, A. Szel, and C. David. 2002. Nonvisual photoreceptors of the deep brain, pineal organs and retina. *Histol Histopathol* 17:555–590.

Vogt, W. 1929. Gestaltungsanalyse am Amphibienkeim mit örtlicher Vitalfärbung. II. Gastrulation und Mesodermbildung bei Urodelen und Anuren. *Wilhelm Roux Arch Dev Biol* 120:384–706.

Volland, S., L. C. Hughes, C. Kong, B. L. Burgess, K. A. Linberg, G. Luna, Z. H. Zhou, S. K. Fisher, and D. S. Williams. 2015. Three-dimensional organization of nascent rod outer segment disk membranes. *Proc Natl Acad Sci USA* 112:14870–14875.

Von Bartheld, C. S. 2004. The terminal nerve and its relation with extrabulbar "olfactory" projections: lessons from lampreys and lungfishes. *Microsc Res Tech* 65:13–24.

von Bubnoff, A., J. E. Schmidt, and D. Kimelman. 1996. The *Xenopus laevis* homeobox gene Xgbx-2 is an early marker of anteroposterior patterning in the ectoderm. *Mech Dev* 54:149–160.

von Kupffer, C. 1894. *Studien zur vergleichenden Entwicklungsgeschichte des Kopfes der Kranioten. II. Entwicklung des Kopfes von Ammocoetes planeri.* Munich: J. F. Lehmann.

Vosper, J. M., C. S. Fiore-Heriche, I. Horan, K. Wilson, H. Wise, and A. Philpott. 2007. Regulation of neurogenin stability by ubiquitin-mediated proteolysis. *Biochem J* 407: 277–284.

Wada, H., M. Iwasaki, and K. Kawakami. 2014. Development of the lateral line canal system through a bone remodeling process in zebrafish. *Dev Biol* 392:1–14.

Wagner, D. E., C. Weinreb, Z. M. Collins, J. A. Briggs, S. G. Megason, and A. M. Klein. 2018. Single-cell mapping of gene expression landscapes and lineage in the zebrafish embryo. *Science* 360:981–987.

Waise, T. M. Z., H. J. Dranse, and T. K. T. Lam. 2018. The metabolic role of vagal afferent innervation. *Nat Rev Gastroenterol Hepatol* 15:625–636.

Wakamatsu, Y. 2011. Mutual repression between Pax3 and Pax6 is involved in the positioning of ophthalmic trigeminal placode in avian embryo. *Dev Growth Differ* 53:994–1003.

Waldhaus, J., R. Durruthy-Durruthy, and S. Heller. 2015. Quantitative high-resolution cellular map of the organ of Corti. *Cell Rep* 11:1385–1399.

Wallis, D., M. Hamblen, Y. Zhou, K. J. Venken, A. Schumacher, H. L. Grimes, H. Y. Zoghbi, S. H. Orkin, and H. J. Bellen. 2003. The zinc finger transcription factor Gfi1, implicated in lymphomagenesis, is required for inner ear hair cell differentiation and survival. *Development* 130:221–232.

Walls, G. L. 1942. Duplicity and transmutation. In *The vertebrate eye and its adaptive radiation*, 163–168. Bloomfield Hills, Michigan: Cranbrook Institute of Science,.

Walters, B. J., E. Coak, J. Dearman, G. Bailey, T. Yamashita, B. Kuo, and J. Zuo. 2017. In vivo interplay between p27(Kip1), GATA3, ATOH1, and POU4F3 converts non-sensory cells to hair cells in adult mice. *Cell Rep* 19:307–320.

Wan, G., G. Corfas, and J. S. Stone. 2013. Inner ear supporting cells: rethinking the silent majority. *Semin Cell Dev Biol* 24:448–459.

Wang, L. H., and N. E. Baker. 2015. E proteins and ID proteins: helix-loop-helix partners in development and disease. *Dev Cell* 35:269–280.

Wang, Q. L., S. Chen, N. Esumi, P. K. Swain, H. S. Haines, G. Peng, B. M. Melia, I. McIntosh, J. R. Heckenlively, S. G. Jacobson, E. M. Stone, A. Swaroop, and D. J. Zack. 2004. QRX, a novel homeobox gene, modulates photoreceptor gene expression. *Hum Mol Genet* 13:1025–1040.

Wang, S. S., J. W. Lewcock, P. Feinstein, P. Mombaerts, and R. R. Reed. 2004. Genetic disruptions of O/E2 and O/E3 genes reveal involvement in olfactory receptor neuron projection. *Development* 131:1377–1388.

Wang, S. S., R. Y. Tsai, and R. R. Reed. 1997. The characterization of the Olf-1/EBF-like HLH transcription factor family: implications in olfactory gene regulation and neuronal development. *J Neurosci* 17:4149–4158.

Wang, S. W., L. Gan, S. E. Martin, and W. H. Klein. 2000. Abnormal polarization and axon outgrowth in retinal ganglion cells lacking the POU-domain transcription factor Brn-3b. *Mol Cell Neurosci* 16:141–156.

Wang, S. W., B. S. Kim, K. Ding, H. Wang, D. Sun, R. L. Johnson, W. H. Klein, and L. Gan. 2001. Requirement for math5 in the development of retinal ganglion cells. *Genes Dev* 15:24–29.

Wang, S. Z., J. Ou, L. J. Zhu, and M. R. Green. 2012. Transcription factor ATF5 is required for terminal differentiation and survival of olfactory sensory neurons. *Proc Natl Acad Sci USA* 109:18589–18594.

Wang, W., E. K. Chan, S. Baron, Water T. Van de, and T. Lufkin. 2001. Hmx2 homeobox gene control of murine vestibular morphogenesis. *Development* 128:5017–5029.

Wang, Y., P. M. Smallwood, M. Cowan, D. Blesh, A. Lawler, and J. Nathans. 1999. Mutually exclusive expression of human red and green visual pigment-reporter transgenes occurs at high frequency in murine cone photoreceptors. *Proc Natl Acad Sci USA* 96:5251–5256.

Washausen, S., and W. Knabe. 2013. Apoptosis contributes to placode morphogenesis in the posterior placodal area of mice. *Brain Struct Funct* 218:789–803.

Washausen, S., and W. Knabe. 2017. Pax2/Pax8-defined subdomains and the occurrence of apoptosis in the posterior placodal area of mice. *Brain Struct Funct* 222:2671–2695.

Washausen, S., and W. Knabe. 2018. Lateral line placodes of aquatic vertebrates are evolutionarily conserved in mammals. *Biol Open* 7:bio031815.

Washausen, S., B. Obermayer, G. Brunnett, H. J. Kuhn, and W. Knabe. 2005. Apoptosis and proliferation in developing, mature, and regressing epibranchial placodes. *Dev Biol* 278:86–102.

Wassarman, K. M., M. Lewandoski, K. Campbell, A. L. Joyner, J. L. Rubenstein, S. Martinez, and G. R. Martin. 1997. Specification of the anterior hindbrain and establishment of a normal mid/hindbrain organizer is dependent on Gbx2 gene function. *Development* 124:2923–2934.

Wawersik, S., P. Purcell, M. Rauchman, A. T. Dudley, E. J. Robertson, and R. Maas. 1999. BMP7 acts in murine lens placode development. *Dev Biol* 207:176–188.

Weasner, B., C. Salzer, and J. P. Kumar. 2007. Sine oculis, a member of the SIX family of transcription factors, directs eye formation. *Dev Biol* 303:756–771.

Webb, J. F., and D. M. Noden. 1993. Ectodermal placodes: contributions to the development of the vertebrate head. *Am Zool* 33:434–447.

Wegner, M. 2010. All purpose Sox: the many roles of Sox proteins in gene expression. *Int J Biochem Cell Biol* 42:381–390.

Wegner, M., and C. C. Stolt. 2005. From stem cells to neurons and glia: a Soxist's view of neural development. *Trends Neurosci* 28:583–588.

Weider, M., and M. Wegner. 2017. SoxE factors: transcriptional regulators of neural differentiation and nervous system development. *Semin Cell Dev Biol* 63:35–42.

Weng, P. L., M. Vinjamuri, and C. E. Ovitt. 2016. Ascl3 transcription factor marks a distinct progenitor lineage for non-neuronal support cells in the olfactory epithelium. *Sci Rep* 6:38199.

Wersäll, J. 1956. Studies on the structure and innervation of the sensory epithelium of the cristae ampulares in the guinea pig; a light and electron microscopic investigation. *Acta Otolaryngol Suppl* 126:1–85.

Wersäll, J., and D. Bagger-Sjöbäck. 1974. Morphology of the vestibular sense organs. In *Handbook of sensory physiology. Vol. 6. Vestibular system Pt L: basic mechanisms. l-676*, edited by H. H. Kornhauber, 123–170. Springer-Verlag, Berlin.

West, B. E., G. E. Parker, J. J. Savage, P. Kiratipranon, K. S. Toomey, L. R. Beach, S. C. Colvin, K. W. Sloop, and S. J. Rhodes. 2004. Regulation of the follicle-stimulating hormone beta gene by the LHX3 LIM-homeodomain transcription factor. *Endocrinology* 145:4866–4879.

Whitear, M., and E. B. Lane. 1983. Oligovillous cells of the epidermis: sensory elements of lamprey skin. *J Zool* 199:359–384.

Whitfield, T. T. 2015. Development of the inner ear. *Curr Opin Genet Dev* 32:112–118.

Whitfield, T. T., and K. L. Hammond. 2007. Axial patterning in the developing vertebrate inner ear. *Int J Dev Biol* 51:507–520.

Whitlock, K. E., N. Illing, N. J. Brideau, K. M. Smith, and S. Twomey. 2006. Development of GnRH cells: setting the stage for puberty. *Mol Cell Endocrinol* 254-255:39–50.

Whitlock, K. E., K. M. Smith, H. Kim, and M. V. Harden. 2005. A role for foxd3 and sox10 in the differentiation of gonadotropin-releasing hormone (GnRH) cells in the zebrafish *Danio rerio*. *Development* 132:5491–5502.

Whitlock, K. E., and M. Westerfield. 2000. The olfactory placodes of the zebrafish form by convergence of cellular fields at the edge of the neural plate. *Development* 127: 3645–3653.

Whitlock, K. E., C. D. Wolf, and M. L. Boyce. 2003. Gonadotropin-releasing hormone (GnRH) cells arise from cranial neural crest and adenohypophyseal regions of the neural plate in the zebrafish, *Danio rerio*. *Dev Biol* 257:140–152.

Whittington, N., D. Cunningham, T. K. Le, D. De Maria, and E. M. Silva. 2015. Sox21 regulates the progression of neuronal differentiation in a dose-dependent manner. *Dev Biol* 397:237–247.

Wichmann, C., and T. Moser. 2015. Relating structure and function of inner hair cell ribbon synapses. *Cell Tissue Res* 361:95–114.

Wigle, J. T., K. Chowdhury, P. Gruss, and G. Oliver. 1999. Prox1 function is crucial for mouse lens-fibre elongation. *Nat Genet* 21:318–322.

Wilson, S. I., E. Graziano, R. Harland, T. M. Jessell, and T. Edlund. 2000. An early requirement for FGF signalling in the acquisition of neural cell fate in the chick embryo. *Curr Biol* 10:421–429.

Wilson, S. I., A. Rydstrom, T. Trimborn, K. Willert, R. Nusse, T. M. Jessell, and T. Edlund. 2001. The status of Wnt signalling regulates neural and epidermal fates in the chick embryo. *Nature* 411:325–330.

Winklbauer, R. 1989. Development of the lateral line system in *Xenopus*. *Prog Neurobiol* 32:181–206.

Wistow, G. J., J. W. Mulders, and W. W. de Jong. 1987. The enzyme lactate dehydrogenase as a structural protein in avian and crocodilian lenses. *Nature* 326:622–624.

Wistow, G., and J. Piatigorsky. 1987. Recruitment of enzymes as lens structural proteins. *Science* 236:1554–1556.

Woda, J. M., J. Pastagia, M. Mercola, and K. B. Artinger. 2003. Dlx proteins position the neural plate border and determine adjacent cell fates. *Development* 130:331–342.

Wolff, G. 1895. Entwickelungsphysiologische Studien. I. Die Regeneration der Urodelenlinse. *Arch Entwicklungsmech* 1:380–390.

Wolpert, L., C. Tickle, and A. Martinez-Arias. 2019. *Principles of development*. 6th ed. Oxford: Oxford University Press.

Wong, E. Y., M. Ahmed, and P. X. Xu. 2013. EYA1-SIX1 complex in neurosensory cell fate induction in the mammalian inner ear. *Hear Res* 297:13–19.

Wood, H. B., and V. Episkopou. 1999. Comparative expression of the mouse Sox1, Sox2 and Sox3 genes from pre-gastrulation to early somite stages. *Mech Dev* 86:197–201.

Woo, S. H., E. A. Lumpkin, and A. Patapoutian. 2015. Merkel cells and neurons keep in touch. *Trends Cell Biol* 25:74–81.

Woods, C., M. Montcouquiol, and M. W. Kelley. 2004. Math1 regulates development of the sensory epithelium in the mammalian cochlea. *Nat Neurosci* 7:1310–1318.

Wray, S. 2010. From nose to brain: development of gonadotrophin-releasing hormone-1 neurones. *J Neuroendocrinol* 22:743–753.

Wray, S., P. Grant, and H. Gainer. 1989. Evidence that cells expressing luteinizing hormone-releasing hormone mRNA in the mouse are derived from progenitor cells in the olfactory placode. *Proc Natl Acad Sci USA* 86:8132–8136.

Wride, M. A. 2011. Lens fibre cell differentiation and organelle loss: many paths lead to clarity. *Philos Trans R Soc Lond B Biol Sci* 366:1219–1233.

Wright, K. D., A. A. Mahoney Rogers, J. Zhang, and K. Shim. 2015. Cooperative and independent functions of FGF and Wnt signaling during early inner ear development. *BMC Dev Biol* 15:33.

Wright, T. J., and S. L. Mansour. 2003. Fgf3 and Fgf10 are required for mouse otic placode induction. *Development* 130:3379–3390.

Wu, D. K., and M. W. Kelley. 2012. Molecular mechanisms of inner ear development. *Cold Spring Harb Perspect Biol* 4:a008409.

Wu, H. Y., M. Perron, and T. Hollemann. 2009. The role of *Xenopus* Rx-L in photoreceptor cell determination. *Dev Biol* 327:352–365.

Wu, M. Y., M. C. Ramel, M. Howell, and C. S. Hill. 2011. SNW1 is a critical regulator of spatial BMP activity, neural plate border formation, and neural crest specification in vertebrate embryos. *PLoS Biol* 9:e1000593.

Wu, X., A. A. Indzhykulian, P. D. Niksch, R. M. Webber, M. Garcia-Gonzalez, T. Watnick, J. Zhou, M. A. Vollrath, and D. P. Corey. 2016. Hair-cell mechanotransduction persists in TRP channel knockout mice. *PLoS One* 11:e0155577.

Wullimann, M., and B. Grothe. 2014. The central nervous organization of the lateral line system. In *The lateral line system*, edited by S. Coombs, H. Bleckmann, R. R. Fay and A. N. Popper, 195–251. New York: Springer.

Wurst, W., and L. Bally-Cuif. 2001. Neural plate patterning: upstream and downstream of the isthmic organizer. *Nat Rev Neurosci* 2:99–108.

Wysocki, C. J. 1979. Neurobehavioral evidence for the involvement of the vomeronasal system in mammalian reproduction. *Neurosci Biobehav Rev* 3:301–341.

Xi, J. H., F. Bai, and U. P. Andley. 2003. Reduced survival of lens epithelial cells in the alphaA-crystallin-knockout mouse. *J Cell Sci* 116:1073–1085.

Xiang, M., L. Gan, D. Li, Z. Y. Chen, L. Zhou, B. W. O'Malley, Jr., W. Klein, and J. Nathans. 1997. Essential role of POU-domain factor Brn-3c in auditory and vestibular hair cell development. *Proc Natl Acad Sci USA* 94:9445–9450.

Xiang, M., L. Gan, L. Zhou, W. H. Klein, and J. Nathans. 1996. Targeted deletion of the mouse POU domain gene Brn-3a causes selective loss of neurons in the brainstem and trigeminal ganglion, uncoordinated limb movement, and impaired suckling. *Proc Natl Acad Sci USA* 93:11950–11955.

Xiang, M., W. Q. Gao, T. Hasson, and J. J. Shin. 1998. Requirement for Brn-3c in maturation and survival, but not in fate determination of inner ear hair cells. *Development* 125:3935–3946.

Xiang, M., L. Zhou, J. P. Macke, T. Yoshioka, S. H. Hendry, R. L. Eddy, T. B. Shows, and J. Nathans. 1995. The Brn-3 family of POU-domain factors: primary structure, binding specificity, and expression in subsets of retinal ganglion cells and somatosensory neurons. *J Neurosci* 15:4762–4785.

Xie, Q., and A. Cvekl. 2011. The orchestration of mammalian tissue morphogenesis through a series of coherent feed-forward loops. *J Biol Chem* 286:43259–43271.

Xiong, W., N. M. Dabbouseh, and I. Rebay. 2009. Interactions with the abelson tyrosine kinase reveal compartmentalization of eyes absent function between nucleus and cytoplasm. *Dev Cell* 16:271–279.

Xu, H., C. M. Dude, and C. V. Baker. 2008. Fine-grained fate maps for the ophthalmic and maxillomandibular trigeminal placodes in the chick embryo. *Dev Biol* 317:174–186.

Xu, H., A. Viola, Z. Zhang, C. P. Gerken, E. A. Lindsay-Illingworth, and A. Baldini. 2007. Tbx1 regulates population, proliferation and cell fate determination of otic epithelial cells. *Dev Biol* 302:670–682.

Xu, J. C., D. L. Huang, Z. H. Hou, W. W. Guo, J. H. Sun, L. D. Zhao, N. Yu, W. Y. Young, D. Z. He, and S. M. Yang. 2012. Type I hair cell regeneration induced by Math1 gene transfer following neomycin ototoxicity in rat vestibular sensory epithelium. *Acta Otolaryngol* 132:819–828.

Xu, P. X. 2013. The EYA-SO/SIX complex in development and disease. *Pediatr Nephrol* 28:843–854.

Xu, P. X., J. Adams, H. Peters, M. C. Brown, S. Heaney, and R. Maas. 1999. Eya1-deficient mice lack ears and kidneys and show abnormal apoptosis of organ primordia. *Nature Genet* 23:113–117.

Xu, P. X., I. Woo, H. Her, D. R. Beier, and R. L. Maas. 1997. Mouse Eya homologues of the *Drosophila* eyes absent gene require Pax6 for expression in lens and nasal placode. *Development* 124:219–231.

Xu, P. X., W. Zheng, L. Huang, P. Maire, C. Laclef, and D. Silvius. 2003. Six1 is required for the early organogenesis of mammalian kidney. *Development* 130:3085–3094.

Xu, P. X., W. Zheng, C. Laclef, P. Maire, R. L. Maas, H. Peters, and X. Xu. 2002. Eya1 is required for the morphogenesis of mammalian thymus, parathyroid and thyroid. *Development* 129:3033–3044.

Xuan, S., C. A. Baptista, G. Balas, W. Tao, V. C. Soares, and E. Lai. 1995. Winged helix transcription factor BF-1 is essential for the development of the cerebral hemispheres. *Neuron* 14:1141–1152.

Yajima, H., M. Suzuki, H. Ochi, K. Ikeda, S. Sato, K. Yamamura, H. Ogino, N. Ueno, and K. Kawakami. 2014. Six1 is a key regulator of the developmental and evolutionary architecture of sensory neurons in craniates. *BMC Biol* 12:40.

Yamaguchi, T., J. Yamashita, M. Ohmoto, I. Aoude, T. Ogura, W. Luo, A. A. Bachmanov, W. Lin, I. Matsumoto, and J. Hirota. 2014. Skn-1a/Pou2f3 is required for the generation of Trpm5-expressing microvillous cells in the mouse main olfactory epithelium. *BMC Neurosci* 15:13.

Yamamoto, N., H. Uchiyama, H. Ohki-Hamazaki, H. Tanaka, and H. Ito. 1996. Migration of GnRH-immunoreactive neurons from the olfactory placode to the brain: a study using avian embryonic chimeras. *Brain Res Dev Brain Res* 95:234–244.

Yan, Y. L., J. Willoughby, D. Liu, J. G. Crump, C. Wilson, C. T. Miller, A. Singer, C. Kimmel, M. Westerfield, and J. H. Postlethwait. 2005. A pair of Sox: distinct and overlapping functions of zebrafish sox9 co-orthologs in craniofacial and pectoral fin development. *Development* 132:1069–1083.

Yanagi, Y., S. Takezawa, and S. Kato. 2002. Distinct functions of photoreceptor cell-specific nuclear receptor, thyroid hormone receptor beta2 and CRX in one photoreceptor development. *Invest Ophthalmol Vis Sci* 43:3489–3494.

Yang, C., H. S. Kim, H. Seo, C. H. Kim, J. F. Brunet, and K. S. Kim. 1998. Paired-like homeodomain proteins, Phox2a and Phox2b, are responsible for noradrenergic cell-specific transcription of the dopamine beta-hydroxylase gene. *J Neurochem* 71:1813–1826.

Yang, J., S. Bouvron, P. Lv, F. Chi, and E. N. Yamoah. 2012. Functional features of trans-differentiated hair cells mediated by Atoh1 reveals a primordial mechanism. *J Neurosci* 32:3712–3725.

Yang, L., P. O'Neill, K. Martin, J. C. Maass, V. Vassilev, R. Ladher, and A. K. Groves. 2013. Analysis of FGF-dependent and FGF-independent pathways in otic placode induction. *PLoS One* 8:e55011.

Yang, Q., N. A. Bermingham, M. J. Finegold, and H. Y. Zoghbi. 2001. Requirement of Math1 for secretory cell lineage commitment in the mouse intestine. *Science* 294:2155–2158.

Yang, Z., K. Ding, L. Pan, M. Deng, and L. Gan. 2003. Math5 determines the competence state of retinal ganglion cell progenitors. *Dev Biol* 264:240–254.

Ye, W., K. Shimamura, J. L. Rubenstein, M. A. Hynes, and A. Rosenthal. 1998. FGF and Shh signals control dopaminergic and serotonergic cell fate in the anterior neural plate. *Cell* 93:755–766.

Yi, S. H., A. Y. Jo, C. H. Park, H. C. Koh, R. H. Park, H. Suh-Kim, I. Shin, Y. S. Lee, J. Kim, and S. H. Lee. 2008. Mash1 and neurogenin 2 enhance survival and differentiation of neural precursor cells after transplantation to rat brains via distinct modes of action. *Mol Ther* 16:1873–1882.

Yntema, C. L. 1933. Experiments on the determination of the ear ectoderm in the embryo of *Amblystoma punctatum J Exp Zool* 65:317–357.

Yntema, C. L. 1937. An experimental study of the origin of the cells which constitute the VIIth and VIIIth cranial ganglia and nerves in the embryo of *Amblystoma punctatum J Exp Zool* 75:75–101.

Yntema, C. L. 1939. Self-differentiation of heterotopic ear ectoderm in the embryo of *Amblystoma punctatum. J Exp Zool* 80:1–17.

Yntema, C. L. 1943. An experimental study on the origin of the sensory neurones and sheath cells of the IXth and Xth cranial nerves in *Amblystoma punctatum. J Exp Zool* 92:93–119.

Yntema, C. L. 1944. Experiments on the origin of the sensory ganglia of the facial nerve in the chick. *J Comp Neurol* 81:147–167.

Yokoyama, S. 2000. Molecular evolution of vertebrate visual pigments. *Prog Retin Eye Res* 19:385–419.

York, J. R., and D. W. McCauley. 2020. The origin and evolution of vertebrate neural crest cells. *Open Biol* 10:190285.

Young, J. M., R. M. Luche, and B. J. Trask. 2011. Rigorous and thorough bioinformatic analyses of olfactory receptor promoters confirm enrichment of O/E and homeodomain binding sites but reveal no new common motifs. *BMC Genomics* 12:561.

Yu, T. T., J. C. McIntyre, S. C. Bose, D. Hardin, M. C. Owen, and T. S. McClintock. 2005. Differentially expressed transcripts from phenotypically identified olfactory sensory neurons. *J Comp Neurol* 483:251–262.

Yu, Y., E. Davicioni, T. J. Triche, and G. Merlino. 2006. The homeoprotein six1 transcriptionally activates multiple protumorigenic genes but requires ezrin to promote metastasis. *Cancer Res* 66:1982–1989.

Yu, Y., J. Khan, C. Khanna, L. Helman, P. S. Meltzer, and G. Merlino. 2004. Expression profiling identifies the cytoskeletal organizer ezrin and the developmental homeoprotein Six-1 as key metastatic regulators. *Nat Med* 10:175–181.

Zagami, C. J., M. Zusso, and S. Stifani. 2009. Runx transcription factors: lineage-specific regulators of neuronal precursor cell proliferation and post-mitotic neuron subtype development. *J Cell Biochem* 107:1063–1072.

Zaidi, A. U., K. W. Kafitz, C. A. Greer, and B. S. Zielinski. 1998. The expression of tenascin-C along the lamprey olfactory pathway during embryonic development and following axotomy-induced replacement of the olfactory receptor neurons. *Brain Res Dev Brain Res* 109:157–168.

Zaraisky, A. G., V. Ecochard, O. V. Kazanskaya, S. A. Lukyanov, I. V. Fesenko, and A. M. Duprat. 1995. The homeobox-containing gene XANF-1 may control development of the Spemann organizer. *Development* 121:3839–3847.

Zaret, K. S., and J. S. Carroll. 2011. Pioneer transcription factors: establishing competence for gene expression. *Genes Dev* 25:2227–2241.

Zeiske, E., A. Kasumyan, P. Bartsch, and A. Hansen. 2003. Early development of the olfactory organ in sturgeons of the genus *Acipenser*: a comparative and electron microscopic study. *Anat Embryol* 206:357–372.

Zelarayan, L. C., V. Vendrell, Y. Alvarez, E. Dominguez-Frutos, T. Theil, M. T. Alonso, M. Maconochie, and T. Schimmang. 2007. Differential requirements for FGF3, FGF8 and FGF10 during inner ear development. *Dev Biol* 308:379–391.

Zhang, C., T. Basta, L. Hernandez-Lagunas, P. Simpson, D. L. Stemple, K. B. Artinger, and M. W. Klymkowsky. 2004. Repression of nodal expression by maternal B1-type SOXs regulates germ layer formation in *Xenopus* and zebrafish. *Dev Biol* 273:23–37.

Zhang, C., and M. W. Klymkowsky. 2007. The Sox axis, Nodal signaling, and germ layer specification. *Differentiation* 75:536–545.

Zhang, G., W. B. Titlow, S. M. Biecker, A. J. Stromberg, and T. S. McClintock. 2016. Lhx2 determines odorant receptor expression frequency in mature olfactory sensory neurons. *eNeuro* 3:ENEURO.0230–16.2016.

Zhang, H., G. Hu, H. Wang, P. Sciavolino, N. Iler, M. M. Shen, and C. Abate-Shen. 1997. Heterodimerization of Msx and Dlx homeoproteins results in functional antagonism. *Mol Cell Biol* 17:2920–2932.

Zhang, J., R. Pacifico, D. Cawley, P. Feinstein, and T. Bozza. 2013. Ultrasensitive detection of amines by a trace amine-associated receptor. *J Neurosci* 33:3228–3239.

Zhang, K. D., and T. M. Coate. 2017. Recent advances in the development and function of type II spiral ganglion neurons in the mammalian inner ear. *Semin Cell Dev Biol* 65:80–87.

Zhang, L., N. Yang, J. Huang, R. J. Buckanovich, S. Liang, A. Barchetti, C. Vezzani, A. O'Brien-Jenkins, J. Wang, M. R. Ward, M. C. Courreges, S. Fracchioli, A. Medina, D. Katsaros, B. L. Weber, and G. Coukos. 2005. Transcriptional coactivator *Drosophila* eyes absent homologue 2 is up-regulated in epithelial ovarian cancer and promotes tumor growth. *Cancer Res* 65:925–932.

Zhang, S., and W. Cui. 2014. Sox2, a key factor in the regulation of pluripotency and neural differentiation. *World J Stem Cells* 6:305–311.

Zhang, Y., B. M. Knosp, M. Maconochie, R. A. Friedman, and R. J. Smith. 2004. A comparative study of Eya1 and Eya4 protein function and its implication in branchio-oto-renal syndrome and DFNA10. *J Assoc Res Otolaryngol* 5:295–304.

Zhang, Y., C. Pak, Y. Han, H. Ahlenius, Z. Zhang, S. Chanda, S. Marro, C. Patzke, C. Acuna, J. Covy, W. Xu, N. Yang, T. Danko, L. Chen, M. Wernig, and T. C. Südhof. 2013. Rapid single-step induction of functional neurons from human pluripotent stem cells. *Neuron* 78:785–798.

Zhang, Z., Y. Shi, S. Zhao, J. Li, C. Li, and B. Mao. 2014. *Xenopus* Nkx6.3 is a neural plate border specifier required for neural crest development. *PLoS One* 9:e115165.

Zhao, B., and U. Müller. 2015. The elusive mechanotransduction machinery of hair cells. *Curr Opin Neurobiol* 34:172–179.

Zhao, L., M. Bakke, Y. Krimkevich, L. J. Cushman, A. F. Parlow, S. A. Camper, and K. L. Parker. 2001. Steroidogenic factor 1 (SF1) is essential for pituitary gonadotrope function. *Development* 128:147–54.

Zhao, S., J. Nichols, A. G. Smith, and M. Li. 2004. SoxB transcription factors specify neuroectodermal lineage choice in ES cells. *Mol Cell Neurosci* 27:332–342.

Zhao, Y., D. C. Morales, E. Hermesz, W. K. Lee, S. L. Pfaff, and H. Westphal. 2006. Reduced expression of the LIM-homeobox gene Lhx3 impairs growth and differentiation of Rathke's pouch and increases cell apoptosis during mouse pituitary development. *Mech Dev* 123:605–613.

Zheng, J. L., and W. Q. Gao. 2000. Overexpression of Math1 induces robust production of extra hair cells in postnatal rat inner ears. *Nat Neurosci* 3:580–586.

Zheng, J., W. Shen, D. Z. He, K. B. Long, L. D. Madison, and P. Dallos. 2000. Prestin is the motor protein of cochlear outer hair cells. *Nature* 405:149–155.

Zheng, W., L. Huang, Z. B. Wei, D. Silvius, B. Tang, and P. X. Xu. 2003. The role of Six1 in mammalian auditory system development. *Development* 130:3989–4000.

Zhou, C., X. Yang, Y. Sun, H. Yu, Y. Zhang, and Y. Jin. 2016. Comprehensive profiling reveals mechanisms of SOX2-mediated cell fate specification in human ESCs and NPCs. *Cell Res* 26:171–189.

Zhou, H., L. Zhang, R. L. Vartuli, H. L. Ford, and R. Zhao. 2018. The Eya phosphatase: its unique role in cancer. *Int J Biochem Cell Biol* 96:165–170.

Zhu, C. C., M. A. Dyer, M. Uchikawa, H. Kondoh, O. V. Lagutin, and G. Oliver. 2002. Six3-mediated auto repression and eye development requires its interaction with members of the Groucho-related family of co-repressors. *Development* 129:2835–2849.

Zhu, M., and P. E. Ahlberg. 2004. The origin of the internal nostril of tetrapods. *Nature* 432:94–97.

Zhu, X., A. S. Gleiberman, and M. G. Rosenfeld. 2007. Molecular physiology of pituitary development: signaling and transcriptional networks. *Physiol Rev* 87:933–963.

Zhu, X., J. Wang, B. G. Ju, and M. G. Rosenfeld. 2007. Signaling and epigenetic regulation of pituitary development. *Curr Opin Cell Biol* 19:605–611.

Zhu, X., J. Zhang, J. Tollkuhn, R. Ohsawa, E. H. Bresnick, F. Guillemot, R. Kageyama, and M. G. Rosenfeld. 2006. Sustained Notch signaling in progenitors is required for sequential emergence of distinct cell lineages during organogenesis. *Genes Dev* 20:2739–2753.

Zilinski, C. A., R. Shah, M. E. Lane, and M. Jamrich. 2005. Modulation of zebrafish pitx3 expression in the primordia of the pituitary, lens, olfactory epithelium and cranial ganglia by hedgehog and nodal signaling. *Genesis* 41:33–40.

Zimmerman, A., L. Bai, and D. D. Ginty. 2014. The gentle touch receptors of mammalian skin. *Science* 346:950–954.

Zou, D., C. Erickson, E. H. Kim, D. Jin, B. Fritzsch, and P. X. Xu. 2008. Eya1 gene dosage critically affects the development of sensory epithelia in the mammalian inner ear. *Hum Mol Genet* 17:3340–3356.

Zou, D., D. Silvius, J. Davenport, R. Grifone, P. Maire, and P. X. Xu. 2006. Patterning of the third pharyngeal pouch into thymus/parathyroid by Six and Eya1. *Dev Biol* 293:499–512.

Zou, D., D. Silvius, B. Fritzsch, and P. X. Xu. 2004. Eya1 and Six1 are essential for early steps of sensory neurogenesis in mammalian cranial placodes. *Development* 131:5561–5572.

Zou, D., D. Silvius, S. Rodrigo-Blomqvist, S. Enerback, and P. X. Xu. 2006. Eya1 regulates the growth of otic epithelium and interacts with Pax2 during the development of all sensory areas in the inner ear. *Dev Biol* 298:430–441.

Zuber, M. E., G. Gestri, A. S. Viczian, G. Barsacchi, and W. A. Harris. 2003. Specification of the vertebrate eye by a network of eye field transcription factors. *Development* 130:5155–5167.

Zuccolo, J., J. Bau, S. J. Childs, G. G. Goss, C. W. Sensen, and J. P. Deans. 2010. Phylogenetic analysis of the MS4A and TMEM176 gene families. *PLoS One* 5:e9369.

Zygar, C. A., T. L. Cook, and R. M. Grainger. 1998. Gene activation during early stages of lens induction in *Xenopus*. *Development* 125:3509–3519.

Index (Volume I)

Note: Locators in *italics* represent figures

A

D

I

R

T

U

V